Numerical Methods and Scientific Computing

Numerical Methods and Scientific Computing

Using Software Libraries for Problem Solving

Norbert Köckler

University of Paderborn, Germany

CLARENDON PRESS · OXFORD

1994

Oxford University Press, Walton Street, Oxford OX2 6DP
Oxford New York Toronto
Delhi Bombay Calcutta Madras Karachi
Kuala Lumpur Singapore Hong Kong Tokyo
Nairobi Dar es Salaam Cape Town
Melbourne Auckland Madrid
and associated companies in
Berlin Ibadan

Oxford is a trade mark of Oxford University Press

Published in the United States by
Oxford University Press Inc., New York

A catalogue record for this book is available from the British Library

Library of Congress Cataloging in Publication Data
(Data available upon request)

ISBN 0 19 859698 7

Typeset by the author
Printed in Great Britain by Bookcraft (Bath) Ltd
Midsomer Norton, Avon

Preface

The idea for this text came to me after being continually astonished that engineers, physicists or mathematicians usually prefer to write their own programs rather than using existing reliable software. What is the reason for this situation?

First thesis:

'Software libraries and program packages are either reliable or user-friendly, but not both.'

This thesis sounds quite harsh, but I am sure it contains an element of truth. A number of user-friendly menu-driven numerical software packages exists for PCs. These lull the user into a false sense of security, but give little indication of the reliability of the solution. Some interesting examples may be found in [37].

On the other hand there are libraries like NAG and IMSL, containing hundreds of routines, which are carefully tested and documented and which are regularly updated. But the documentation for each of these two libraries fills at least five binders and some preparation is needed even when using one of the simpler routines. That deters many potential users, who do not see that the effort is rewarded with error-free programs, and either reliable results or warnings.

Second thesis :

'Every program with more than 50 lines is wrong.'

Unfortunately this statement is true more often than not. With the easy availability of reliable software libraries it has become unnecessary for the scientist and engineer to program standard numerical applications by themselves.

This text aims to ease the use of software libraries by outlining the theory of the underlying numerical methods, and by describing some programs and the library routines involved. The methods and their use should become clearer from the examples and real world applications given. For our programs we use the NAG Fortran Library. But it should be easy to modify them for similar libraries like IMSL. By using library routines within our programs we will attempt to make the second thesis less and less correct.

An understanding of the numerical methods underlying library routines is usually very helpful and often essential.

The introductory chapter will describe the structure of the text. Furthermore it contains some notes on structured programming and programming languages, and finally it gives a brief review of the calculus of errors.

The first steps toward this English version of my German textbook [61] were made in Oxford in Summer 1991 while visiting NAG Ltd, together with two of my students. This visit was made possible by the EC with their COMETT II program.

Within this project a new problem-solving environment PAN has been developed which runs the programs using window management tools running under the X Window System. It contains a tutorial text based on this book using the HyTEX-system, which was developed especially for this task and allows LATEX-previewing within a node reference system (see [53] and Appendix A for further details). In this way this text is published in two different forms in parallel.

Acknowledgements

The basis of this text is the German textbook [61]. Therefore the names of people helpful for that text should be repeated here: J. R. Bunch, Carmen Buschmeyer, Brian Ford, H. D. Mittelmann, Thomas Schulze, H. R. Schwarz, P. Spuhler, Werner van den Boom, R. Walden and Matthias Simon, who read this text again with great mathematical care.

During my stay in Oxford many people from NAG were a great help for me, I mention Brian Ford again, Peter Mayes and Neil Swindells for their excellent copy-editing, and Astrid, Bob, Del, Frances, Gareth, Ian and Ian, Jacqui, John, Julie, Mike, Richard, Sarah and Steve for quite different kinds of help. My students Thomas Herchenröder and Dierk Wendt accompanied me to NAG, and they did much good work. Dierk invented the new PAN. Back home again, I very much appreciate my students' help in developing PAN. Thanks then to Michael Braitmeier for the PLTMG/PAN interface, Ingo Dahm for the demo and guided tour, for preparing Appendix A of this text, and as man Friday, Jutta Hopp for additional applications, Jochen Hillebrand, Burkhard Kuhhoff, Markus Neth and Uta Rüsenberg for adding industrial multimedia parts and perfecting our hypertext system HyTEX, and finally Andreas Wegener for refreshing the programs and designing a template library. Donald Degenhardt and Richard Preston from Oxford University Press gave continuous support, even in the hectic times of rapidly approaching deadlines.

Norbert Köckler Oxford, May 1991; Paderborn, December 1993.

Contents

Guide to Notation

Abbreviations

BC	Boundary condition.
BLAS	Basic linear algebra subroutines.
CG	Conjugate gradients.
DFT	Discrete Fourier transform.
DVI	Device independent.
EC	European Community.
FFT	Fast Fourier transform.
ftp	File transfer protocol.
GKS	Graphical kernel system.
IC	Initial condition.
ICC	Incomplete Cholesky decomposition.
I/O	Input/output.
lss	Least-squares solution.
NAG	Numerical Algorithms Group.
PAN	Problem-solving environment for Algorithms in Numerical mathematics.
PC	Personal computer.
PCG	Preconditioned conjugate gradients.
PLTMG	Piecewise linear triangular multigrid.
PDE	Partial differential equation.
PSE	Problem-solving environment.
SPD	Symmetric positive-definite.
SVD	Singular value decomposition.
XGKS	*see* GKS.

Symbols

$\forall$	For all.
$\in$	Belongs to.
$x := y$	Assign value(y) to x.
$x \cong y$	x and y agree in first order.
$O(h^k)$	Expression of kth order in h.
$\mathbb{C}$	Complex numbers.
$\mathbb{C}^n$	Space of n-dimensional complex vectors.
$\mathrm{Im}(z)$	Imaginary part of a complex number z.
$\mathrm{Re}(z)$	Real part of a complex number z.
$\mathbb{N}$	Set of natural numbers.
$\mathbb{R}$	Set of real numbers.
$\mathbb{R}^n$	Space of n-dimensional real vectors.
$\mathbb{R}^{n,m}$	Space of real n by m matrices.
$\mathbb{Z}$	Set of rational numbers.
Δx	Absolute error of real number x.
ε_x	Relative error of real number x.
$fl(x)$	Floating point representation of x.
E	Base of a floating point number.
τ	Machine precision.
$\|\|x\|\|_2$	Euclidean or 2-norm of a vector x.
$\|\|x\|\|_1$	1-norm of a vector x.
$\|\|x\|\|_\infty$	Maximum or ∞-norm of a vector x.
$\mathrm{cond}(A)$	Condition number of matrix A.
$\det(A)$	Determinant of matrix A.
$\|A\|$	Determinant of matrix A.
$\mathrm{diag}(d_1, \ldots, d_n)$	Diagonal matrix.
I	Identity matrix.
A^{-1}	Inverse of matrix A.
$A^\dagger$	Pseudoinverse of matrix A.
A^T	Transpose of matrix A.
$C(X)$	Set of all functions continuous on X.
$C^n(X)$	Set of all functions having n continuous derivatives on X.
$[y_{i_0}, y_{i_1}, \ldots, y_{i_j}]$	General divided difference.
$D\varphi$	Jacobian of vector function φ.
Δ	Laplacian operator.
Ω	Open region (domain) in $\mathbb{R}^n$.
$\partial\Omega$	Boundary of $\Omega \in \mathbb{R}^n$.

List of Examples and Applications

Introduction

> The aim of numerical mathematics is to find computational solutions for mathematically formulated problems.
> (H. Rutishauser)

Numerical mathematicians develop algorithms for these computational solutions. These algorithms are normally implemented on computers. The most reliable implementations are mainly to be found in the large software libraries mentioned in the preface. That is why students of mathematics, science, engineering and computer science should learn something about software libraries as part of their mathematical education.

The Structure of this Text

We will deal with the type of mathematical problems which are normally the subject of a course in numerical mathematics. We will describe the mathematical problems, the numerical algorithms for their solution, and programs using NAG library routines to solve the problems.

The chapters – or the sections within the larger chapters – are structured as follows:

- We start with the problem and some related mathematics.
- Then the numerical methods follow. We mention most of the classical algorithms as well as some special methods used in software libraries, which are not often to be found in textbooks.
- The section describing our Fortran program will follow, consisting mostly of a description of the input file together with an example. The construction of the input file is simplified using the input templates developed for the problem-solving environment (PSE) PAN, see Appendix A.
- We have tried to find applications for most of the methods. We describe them in some detail to show typical real world problems and their numerical solution. Examples and applications are completed by the exercises at the end of each chapter.

In Appendix B you will find an overview over the Fortran programs written for this text with a short description of the subject of the program and the NAG routines called. This should give sufficient information about the programs and the NAG routines used. You will find at least one example for each input file within the text. Every user of the program has access to the program source with comments at the heads including a detailed input file description. If the PAN system is available even more information is available within the HyTEX-version of this text and all the input files can be more comfortably generated by templates.

All programs use the NAG Fortran Library, [74]. You will find an introduction to it in Appendix C.

Some of the programs show the results graphically using the NAG Graphics Library, [75]. Some examples of the use of that library may be found in Appendix D.

Appendix A describes the problem-solving environment PAN for numerical analysis, see [53], developed in cooperation with the NAG Ltd., Oxford, United Kingdom.

An enlarged electronic version of this book organized as a hypertext tutorial is a central part of PAN. Within the PAN system the tutorial is presented by our LaTEX previewer HyTEX. HyTEX uses a node system: The text is divided into several inter-connected nodes, which are referenced by keywords within the page header or framed keywords within the text. Thus, the user is able to "jump" to the topic he is interested in to get basic information and from there to other related nodes with algorithms or examples or program applications.

Furthermore PAN consists of an integrated graphical user interface for the development and management of the programs described here and the data necessary for running them. The programs are presented through a graphical user interface. Interactions are handled by different window- or mouse-events. The programs are hierarchically established according to the sectioning of this text. Users can run them, modify them, and add their individual programs. In addition the system offers option management like compiler commands and different utilities like a file manager.

Both abilities can be used in parallel. So you can use numerical solvers while reading the theory that rules them. The system is designed to run on UNIX™- and LINUX-machines under the X Window System in order to reach a wide range of users.

Libraries, Packages, Tools

First of all we mention the NAG Fortran Library which we have used for this text. It contains more than 1600 routines, about 1000 of which are callable by the user and described in the manuals. We will also give hints for the use of other numerical software, mainly for the IMSL Library. The IMSL Library [58] covers a similar area of numerical mathematics to the NAG Library, and we hope that an IMSL user will also find this text useful.

These two large software libraries are particularly useful where the user is faced with a choice of different algorithms for the same problem. In addition, we will give examples later to demonstrate the excellent modular construction of these libraries. To most users the routines of these libraries are unsurpassable as far as accuracy, reliability and efficiency are concerned.

Some software libraries are available in several different language versions. NAG produces libraries in several other languages: ALGOL 68, ADA, C and PASCAL. The development of a library with similar functionality to the NAG Fortran Library in a new language, including testing and documentation, may take upwards of 100 man-years.

An introduction to the NAG Fortran Library is given in Appendix C. Additional NAG products include a statistics package GENSTAT, a set of software tools for Fortran 77 program development called NAGWare f77 Tools and the NAG Graphics Library (see Appendix D).

IMSL produce three libraries, called the 'MATH/Library' (for mathematics), the 'STAT/Library' (for statistics), and the 'SPECFUN/Library' (for special functions). In addition, IMSL produce products based on a high-level problem-oriented language called PROTRAN (see [87]), derived from Fortran. 'LP/PROTRAN' is designed for solving linear optimization problems, and 'PDE/PROTRAN' for partial differential equations.

Many library routines require workspace, and IMSL and NAG provide this in different ways. IMSL provides a version of each routine which requires workspace with a shortened calling sequence containing no workspace parameters. Any workspace required is then taken from an internal array of fixed size. If this internal array is too small, then the user may either call an auxiliary routine to increase the size of the internal workspace, or else call the lower level routine directly. NAG routines contain all necessary workspace arrays as parameters, equivalent to the second type of IMSL routine. For small problems, the IMSL routines are easier to use, but the NAG routines do not use unnecessary space. For larger problems, the difference is less significant.

For more restricted problem areas there are a number of other libraries and packages of similar quality. An early but very significant numerical library is described in the 'Handbook for Automatic Computation II. Linear Algebra' of Wilkinson and Reinsch, [99], which is exemplary for its systematic documentation and program testing. The software described (written in ALGOL 60) is written to very high standards, and paved the way for a number of later and larger packages and libraries. Indeed, some later libraries were based on Fortran translations of some of these routines.

Another significant early development was the package LINPACK, [26], which contains most commonly occurring algorithms for the solution of simultaneous linear equations. EISPACK, [89], is a similar package for matrix eigenvalue problems. This was created mainly by translating the routines of [99] from ALGOL 60 to Fortran. Recently a package called LAPACK has been published, [2], which is designed to supersede LINPACK and EISPACK. It is especially efficient on modern high-performance computers, and shared memory vector multiprocessors in particular.

For numerical integration there is the package QUADPACK, [78], and for curve and surface fitting FITPACK, [25]. For one of the most important scientific problems, namely the solution of partial differential equations, there are a number of packages and systems, too numerous to mention. Most of these systems use the finite element method (FEM). NAG distributes the 'NAG/SERC Finite Element Library', developed at the Rutherford Appleton Laboratories, [47]. In Chapter 9 we briefly describe PLTMG, which is integrated in our PSE PAN. PLTMG is freely available, see [6] for details.

There are a number of textbooks which are accompanied by programs. These are not normally as reliable as libraries, but several are worthy of mention. A useful introduction to both PASCAL and numerical mathematics is [23]. The programs from the book Numerical Recipes, [80], are obtainable in Fortran, PASCAL or C.

There are several packages, developed mainly for personal computers, which can solve mathematical problems, produce graphics and which allow text to be edited. Two examples are MathCad and DERIVE. A package written mainly for numerical linear algebra, which, being based predominantly on LINPACK, is very reliable, and which also has excellent graphics capabilities, is MATLAB (see [70] or the introductory text [55]).

Finally there are a number of computer algebra systems, which complement the numerical software packages described above, rather than compete with them, e.g. MACSYMA, REDUCE, MAPLE, MATHEMATICA and AXIOM, which is distributed by NAG. Some of these are able to call numerical routines, but this is usually not as efficient as using a numerical library directly. Recently a group in Paderborn has developed a new computer algebra systems called MuPAD, see [36]. MuPAD is a universal system, which has a lot of similarities to MAPLE, but it additionally aims at the management of large data sets and uses parallel constructs for better efficiency on multi-processor machines.

Structured Programming

When Seymour Cray was asked which programming language will mostly be used in the year 2000, he answered:

> 'I don't know what language it will be, but I'm sure it will be called Fortran.'

People often argue about what is the 'best' programming language, but no one language is best for all applications. It will be easy to offend some readers with the following remarks, but I often meet two extremes of programmers. Firstly, there is the 'modern linguist', who only uses the very latest structured language†. Then there is the 'ignoramus', who cares nothing for the structure of his spaghetti-like code. To be over-simplistic, the first type works in universities, creating chaos by writing new, but probably incorrect, programs to solve the same old problems, while the second type create chaos by writing huge, inscrutable programs with no structure, and with no comments. However, if they were to sit down together, they would agree about the following:

- In each programming language it is possible to program in a structured and modular fashion. For a discussion on this point it is worth reading the introduction of [80].
- There do exist programming languages which are better than others, in that they force the user to use a structured programming style, and which contain important language constructs not available in all languages, such as *recursion*, *pointers* or elegant *string processing* facilities.
- It is not useful or economical to rewrite well-organized and documented programs as long as they run well.
- There is a huge body of software written in Fortran. So it is important for the enthusiast of 'modern' languages to be able to write hybrid programs. Fortran routines can, with care, be called by C or PASCAL programs. However, these hybrid programs are usually not portable to other machines.

When using software libraries with only a small amount of additional user programming, many problems will be avoided if the calling program and library routine are written in the same language. We therefore decided to write the programs for this textbook in Fortran. In any case, we tried to write structured programs, and to follow some conventions which we will describe in the introduction to Appendix B.

Calculus of Errors

In order to apply numerical methods to mathematical problems it is important to know something about errors, the way they propagate during a computation and the possible effects on the results this can cause. We will

† By the way, this is Fortran 90 right now!

therefore give a short review of the calculus of errors here. More details may be found in a number of standard textbooks on numerical mathematics. For numerical linear algebra there is an extra section about norms, condition numbers and errors, see Section 1.6.

Definition 0.1. *An algorithm is a sequence of elementary operations (including conditional statements and standard functions) which is uniquely determined for each case considered.*

The algorithm for solving a mathematical problem with a numerical method is only one link in the chain leading from the application problem to the results. We will look at these links on the basis of a physical problem:

Links	Example
1. Physical problem	Bridge building
2. Physical model	Theory of elasticity
3. Mathematical model	Differential equation
4. Numerical method	Choice, e.g. Finite Element Method
5. Algorithm	Computational organization, including I/O
6. Program	Use of software (choice)
7. Computation	Data organization
8. Error estimation	Calculation of estimates with given data

A scientist will usually only be involved in a few of these steps, and will usually have little influence on the sources of error in the chain. In large applications a realistic estimate of the error in the result is very difficult, or even impossible. But in any case it is very important to know something about all of these steps and the relationships between them.

Data errors
are errors which are introduced by inaccurate input data, such as rounded data or measurements. That is the reason why they are sometimes called *measurement errors.*

Discretization errors
are those errors which appear because a continuous problem is approximated by a discrete problem. For example, derivatives are approximated by finite differences.

Rounding errors
are created when computing with real numbers on a computing machine with a limited number of digits. These errors are propagated from one elementary operation to the next and this propagation of errors is the main source of numerical instability.

For a real number x and an approximation $\tilde{x} \in \mathbb{R}$ there are two types of error:

$$\Delta x := \tilde{x} - x \qquad \text{absolute error, and} \tag{0.1}$$

$$\varepsilon_x := \frac{\tilde{x} - x}{x} \qquad \text{relative error, provided that } x \neq 0. \tag{0.2}$$

The relative error is often given in percentages as $100\,\varepsilon_x$.
On a computing machine a real number $x \in \mathbb{R}$ is represented by the approximation

$$x = \text{sign}(x)\, a\, E^b. \tag{0.3}$$

Here E is the *base* (usually 2), b is the *exponent*, and a is the *mantissa*:

$$a = a_1 E^{-1} + a_2 E^{-2} + \cdots + a_k E^{-k}. \tag{0.4}$$

k is the length of the mantissa and a_i are base E digits. For example, a_i is 0 or 1 for base 2, and in general $0 \leq a_i \leq E - 1$. One may imagine the point in front of the first digit. If the number x is not equal to zero, then the first digit should be nonzero:

$$\text{either} \quad E^{-1} \leq a < 1 \quad \text{or} \quad a = 0. \tag{0.5}$$

In this case the representation of x is called a *normalized floating point number with k base E digits*. Computing with those numbers is called *computation with k significant digits*. The rounded floating point representation of a real number x is written as $\text{fl}(x)$.

On most computers the exact result of an elementary arithmetic operation is rounded[†] to the nearest machine representable number. Then the maximum error for each elementary operation is bounded by

$$|\text{fl}(x) - x| \leq \frac{E}{2} E^{-(k+1)} E^b \qquad \text{(max. absolute rounding error)} \tag{0.6}$$

and for $x \neq 0$

$$\frac{|\text{fl}(x) - x|}{|x|} \leq \frac{E}{2} E^{-k} \qquad \text{(max. relative rounding error)} \tag{0.7}$$

The upper bound for the relative rounding error defines the accuracy the machine is able to deliver and it is therefore called the *machine precision* ε:

$$\varepsilon := \frac{E}{2} E^{-k}. \tag{0.8}$$

Numerical errors are, in many cases, caused by propagation of errors after subtraction. When subtracting two numbers with several equal leading digits, the result will have that many leading zeros. Hence the relative error of the result is much larger than the relative error of the terms before. This

---

[†] Some machines truncate the correct result, which will not be considered here.

loss of precision is called *cancellation*. We will now consider an example of this effect.

Example 0.2. Here we have three arithmetic expressions for the same real number:

$$99 - 70\sqrt{2} = \sqrt{9801} - \sqrt{9800} = \frac{1}{\sqrt{9801} + \sqrt{9800}} = 0.00505063\ldots.$$

The three formulae represent three algorithms of different numerical stability as the following calculations with different numbers of significant digits will clearly show:

Digits	$99 - 70\sqrt{2}$	$\sqrt{9801} - \sqrt{9800}$	$\frac{1}{\sqrt{9801} + \sqrt{9800}}$
2	1.0	0.0	0.0050
4	0.02000	0.01000	0.005051
6	0.00530000	0.00510000	0.00505063
10	0.005050660000	0.005050640000	0.005050633884

The reason for the numerical instability of the first two algorithms is the subtraction, which cancels 3 or 4 digits. The results are obtained with rounding after each elementary arithmetic operation. ■

To consider (and prevent) these effects we look at an algorithm mathematically as a function mapping the input data onto the results:

$$\begin{aligned} \text{Input data}: &\quad x \in \mathbb{R}^m \\ \text{Results}: &\quad y \in \mathbb{R}^n \\ \text{Algorithm}: &\quad y = \varphi(x) \end{aligned} \tag{0.9}$$

In this way the algorithm is represented by a multivariate vector function mapping $\mathbb{R}^m$ into $\mathbb{R}^n$. A detailed analysis of an algorithm which considers data errors, rounding errors in each elementary operation, and the rounding of the result is called *differential error analysis*. We will only consider the simpler situation of the propagation of data errors into the result when computing exactly. This unrealistic model gives a quite realistic insight into the stability of algorithms in many cases.

Lemma 0.3. *Let $\Delta x := \tilde{x} - x$ be the absolute data error and $D\varphi = \left\{\frac{\partial \varphi_i}{\partial x_k}\right\}$ the Jacobian matrix of φ. If the function φ is computed exactly but with data $\tilde{x}$ instead of x, then for the resulting error of the algorithm $\tilde{y} = \varphi(\tilde{x})$ we have the first order representations*

$$\begin{aligned} \Delta y &\cong D\varphi \Delta x \\ \varepsilon_{y_i} &\cong \sum_{k=1}^{m} C_{ik}\varepsilon_{x_k}, \quad i = 1,\ldots,n, \end{aligned} \tag{0.10}$$

with the condition numbers

$$C_{ik} = \frac{\partial \varphi_i(x)}{\partial x_k}\frac{x_k}{\varphi_i(x)}, \quad i = 1,\ldots,n, \quad k = 1,\ldots,m. \tag{0.11}$$

We see that the condition numbers so defined are the amplification factors of the relative error of one input data into one result.

Software Availability

You will find the addresses of the software suppliers FITPACK, NAG, IMSL, and MATLAB given at appropriate places in the Bibliography; see [25], [47], [58], [74], and [75]. Information about our PAN and HyTEX systems can be found at the end of Appendix A. There is a lot of public domain software available through Netlib, a system which can be accessed via electronic mail. This includes PLTMG, see Page 259. In addition, we will list here some useful electronic mail addresses.

IMSL	support@imsl.com
Mathematica	info@wri.com
MATLAB	info@mathworks.com
MuPAD	MuPAD-distribution@uni-paderborn.de
NAG	infodesk@nag.co.uk
	infodesk@nag.com
NETLIB/PLTMG	netlib@research.att.com
PAN/HyTEX	PAN-distribution@uni-paderborn.de
PLTMG	netlib@research.att.com

Disclaimer of Warranty

The programs listed in this book are provided 'as is' without warranty of any kind. We make no warranties, explicit or implicit, that the programs are free of errors (see the first thesis of the preface), or are consistent with any particular standard of merchantability, or that they will meet your requirements for any particular application. They should not be relied on for solving a problem whose incorrect solution could result in injury to a person or loss of property. If you do use the programs or procedures in such a manner, it is at your own risk. The authors and publisher disclaim all liability for direct, incidental, or consequential damages resulting from your use of the programs, functions, or procedures in this book.

1
Simultaneous Linear Equations

Most application problems in numerical mathematics result in linear systems. There are many methods for their numerical solution. The decision which one to choose depends mainly on the structure of the coefficient matrix:

- Nonsingular matrix with no special structure: Gaussian elimination.
- Symmetric and positive-definite matrix (SPD): Cholesky's method.
- Band matrix: different methods.
- Sparse matrix: special methods like conjugate gradients.
- Singular matrix, least-squares problem: singular value decomposition.

For each of these items we have implemented a Fortran program using the respective NAG routines. At the end of this chapter some of the IMSL routines are described and a MATLAB example is given. There are two main types of methods:

- *Direct methods*, which yield the exact solution in a finite number of steps if all calculations are performed exactly (that is, without roundoff errors).
- *Iterative methods*, which are most important for large sparse systems and arise mainly in connection with the solution of partial differential equations.

Most of the direct methods use *elimination*: The coefficient matrix and the right-hand side are transformed using elementary operations into an equivalent system, which can then easily be solved by back substitution.

In Section 1.6 norms, condition numbers and error estimates are considered. Readers not familiar with the basic norm definitions should read that section first.

1.1 Nonsingular Systems – Gaussian Elimination

In this section we will describe Gaussian elimination for nonsingular linear systems. We start with the theory, then give two examples showing the necessity of row, column or full interchanges. The algorithm consists of the following parts: the elimination steps, the pivot strategy, which can be combined with scaling to become an implicit scaling, and the back substitution. The first solution can be refined by iteration. Gaussian elimination may be used to find the inverse of a matrix. We clarify our program, which uses NAG routine F04AEF, with an example. Finally as an application we consider the fluid flow through a plumbing network.

1.1.1 Introduction

Let $A \in \mathbb{R}^{n,n}$ be a given matrix of rank n, and let $b \in \mathbb{R}^n$ be a given vector. We wish to determine a vector $x \in \mathbb{R}^n$ such that

$$Ax = b. \tag{1.1}$$

Most of the linear systems arising from practical applications have full rank, which is equivalent to a non-vanishing determinant

$$\det(A) = |A| \neq 0. \tag{1.2}$$

We call these systems *nonsingular*. In this case a unique solution of $Ax = b$ always exists. For the remainder of this section we will assume this property. Systems with $\text{rank}(A) < n$ (or equivalently $\det(A) = 0$) are called *singular*. We will consider singular linear systems in Section 1.5.

Gaussian elimination leads to a factorization of A into two triangular matrices

$$A = LU \tag{1.3}$$

with a lower triangular matrix L and an upper triangular matrix U. Unfortunately such a factorization does not always exist, as Example 1.2 shows.

What can we do in such a case? The system remains unchanged if one rearranges the equations or equivalently interchanges the rows of A and b. For larger or ill-conditioned systems this is necessary for stability reasons anyway and is called *pivoting*. But not all rearrangements are useful. That is why there are different pivot strategies to be examined. The interchange can be done formally by multiplying A by a *permutation matrix* P:

$$PA = LU. \tag{1.4}$$

A permutation matrix is a matrix whose entries are all zero, except for a single '1' in each row and in each column, so that it is a 'permutation' of the identity matrix. Multiplication by a permutation matrix from left (right) leads to a rearrangement of the rows (columns) (see Example 1.3). A factorization (1.4) always exists:

Theorem 1.1. *Let $A \in \mathbb{R}^{n,n}$ be a nonsingular matrix. Then there exists a permutation matrix P, a lower triangular matrix L and an upper triangular* *matrix U such that*

$$PA = LU.$$

Example 1.2.

$$A = \begin{pmatrix} 0 & 1 \\ 1 & 1 \end{pmatrix}.$$

A is nonsingular, $\det(A) = -1$, but there are no triangular matrices L and U, such that $A = LU$. Otherwise we would have

$$l_{11}u_{11} = a_{11} = 0,$$

which means $l_{11} = 0$ or $u_{11} = 0$. But then either L or U must be singular, which contradicts the fact that A is nonsingular. ■

Example 1.3.

$$P = \begin{pmatrix} 0 & 1 \\ 1 & 0 \end{pmatrix}.$$

Then for the matrix of Example 1.1 we get

$$PA = \begin{pmatrix} 1 & 1 \\ 0 & 1 \end{pmatrix}.$$

For this matrix we easily find a factorization: let $L = I$ (the identity) and $U = PA$. ■

1.1.2 Elimination and Backward Substitution

For the sake of simplicity we firstly describe *Gaussian elimination* without pivoting. It consists of $n-1$ elimination steps.

The kth step reduces the kth column of A by subtracting a multiple of the kth row from the subsequent rows to give zeros below the kth diagonal element. The new elements of A and b may overwrite the old ones to give:

Elimination: Step No. k

for $i = k+1, \ldots, n$:

$$a_{ik} := \frac{a_{ik}}{a_{kk}}$$

for $j = k+1, \ldots, n$:

$$a_{ij} := a_{ij} - a_{ik}a_{kj}$$

$$b_i := b_i - a_{ik}b_k$$

After having finished the $n-1$ elimination steps ($k = 1, \ldots, n-1$) the matrices L (without the diagonal elements, which all are 1) and U may be found in A:

$$A = \begin{pmatrix} u_{11} & u_{12} & \cdots & u_{1n} \\ l_{21} & u_{22} & \cdots & u_{2n} \\ \vdots & \ddots & \ddots & \vdots \\ l_{n1} & \cdots & l_{n,n-1} & u_{nn} \end{pmatrix}. \tag{1.5}$$

This gives the new linear system

$$Ux = b \qquad \text{with} \qquad U = \begin{pmatrix} a_{11} & a_{12} & \cdots & a_{1n} \\ 0 & a_{22} & \cdots & a_{2n} \\ \vdots & \ddots & \ddots & \vdots \\ 0 & \cdots & 0 & a_{nn} \end{pmatrix}, \tag{1.6}$$

which is equivalent to the original system, and can easily be solved by backward substitution:

Backward Substitution

$x_n := b_n / a_{nn}$

for $i = n-1, n-2, \ldots, 1$:

$$x_i := \left(b_i - \sum_{k=i+1}^{n} a_{ik} x_k \right) \Big/ a_{ii} \tag{1.7}$$

1.1.3 Pivoting Strategies

The element a_{kk}, which all subsequent column elements have to be divided by within the kth step, is called the *pivot*. The size of the pivots determines the stability of the elimination algorithm as far as the development of roundoff errors is concerned. Small pivots have to be avoided whenever possible. Therefore when interchanging rows the element with the largest absolute value on or below the kth diagonal is taken. This is called *row pivoting*. If this element is zero, then the matrix is singular and the elimination is stopped. If it is smaller than an accuracy tolerance, then the matrix is called *numerically singular* and again the elimination procedure is stopped.

Row Pivoting: kth Step

Determine the number $l \geq k$ for which
$|a_{lk}| = \max_{i=k,\ldots,n} |a_{ik}|$ (a_{lk} is the pivot)

if $|a_{lk}| < \varepsilon$ then stop

Row Rearrangement: kth Step

$p_k := l$
if $k \neq l$ then
    for $j = 1, \ldots, n$:
        $s := a_{kj}$; $a_{kj} := a_{lj}$; $a_{lj} := s$
    $s := b_k$; $b_k := b_l$; $b_l := s$
endif

The permutations are represented by the vector p defined in that algorithm. It is used instead of the permutation matrix and holds all the information about the permutations.

Another pivot strategy is called *column pivoting*, which is equivalent to a renumbering of the unknowns and yields a factorization $AQ = LU$, where Q is a permutation matrix. In this case the renumbering of the unknowns must be reversed. The stability of column pivoting is not generally superior to that of row pivoting.

Row or column pivoting are also called *partial pivoting*. If more numerical stability is required, you may choose *full pivoting*:

Full Pivoting: kth Step

Determine the numbers $l \geq k$ and $m \geq k$ for which
$|a_{lm}| = \max_{i,j=k,\ldots,n} |a_{ij}|$ (a_{lm} is the pivot)

if $|a_{lm}| < \varepsilon$ then stop

In this case row l and row k have to be interchanged as well as column m and column k. The factorization can be written as $PAQ = LU$ where P and Q are permutation matrices. Two vectors are used to hold the necessary information about the permutations. Although this is the most stable form of pivoting, currently available software libraries usually use another strategy with similar stability properties called implicit scaling. In any case, for ill-conditioned systems one should always use iterative refinement.

1.1.4 Scaling

Rows or columns of strongly varying size may spoil the numerical stability of the elimination algorithm. This can be avoided by scaling the matrix:

each row and/or column is multiplied by a factor to yield rows and columns of similar magnitude. The scaling corresponds to a multiplication of A by two diagonal matrices:

$$A' = D_1 A D_2 \qquad \text{with} \tag{1.8}$$

$$D_i = \text{diag}(d_{i1}, \ldots, d_{in}), \quad i = 1, 2.$$

The explicit scaling should only be carried out with factors which are integer powers of the base of the computer's arithmetic to avoid additional roundoff errors. Because of the additional effort, implicit scaling – also called relative pivoting – is usually preferred.

1.1.5 Implicit Scaling = Relative Pivoting

It is possible to combine scaling and pivoting. This yields good numerical stability. It consists of row pivoting, but chooses the pivot elements relative to the norm of the row, which yields in addition the scaling effect required:

Relative Pivoting

Calculate the inverse row norms:
for $k = 1, \ldots, n$:

$$p_k := \left(\sum_{i=1}^{n} a_{ki}^2 \right)^{-1/2}$$

Row Pivoting: kth Step

Determine the number $l \geq k$ for which
$|a_{lk}|\, p_l = \max_{i=k,\ldots,n} |a_{ik}|\, p_i$ (a_{lk} is the pivot)
if $|a_{lk}| < \varepsilon$ then stop

The method of implicit scaling is to be found in most of the important software libraries, for example with NAG. As a programming trick, the inverses of the row norms are assigned to the vector p representing the permutation matrix at the end (see the complete Gaussian algorithm, Section 1.1.8).

1.1.6 Iterative Refinement

Any solution $\tilde{x}$ computed using Gaussian elimination with pivoting followed by backward substitution is normally not an exact solution of $Ax = b$. But it is possible to improve its accuracy if you are able to form inner products $x^T y = \sum_{i=1}^{n} x_i y_i$ in extended precision.

First the residual vector is calculated:

$$r := b - A\tilde{x}. \tag{1.9}$$

Its components $r_i := b_i - \sum_{j=1}^{n} a_{ij}\, x_j$ have to be accumulated in extended precision, but the result may be stored in standard precision. We then solve the linear system

$$Az = r. \tag{1.10}$$

$$\tilde{\tilde{x}} = \tilde{x} + z \tag{1.11}$$

is an improved solution, because without the influence of roundoff errors we would have

$$A(\tilde{x} + z) = b - r + r = b. \tag{1.12}$$

This process is called *iterative refinement.*

The linear system (1.10) can be solved easily because it has the same coefficient matrix A as the original system, so we may use the factorization $PA = LU$. Therefore one has to rearrange the right-hand sides using P and to solve

$$Ly = Pr \qquad \text{(Forward substitution).} \tag{1.13}$$

Using the y obtained we solve for z in

$$Uz = y \qquad \text{(Backward substitution).} \tag{1.14}$$

Together this gives the following algorithm (L without the ones on its diagonal and U are stored in A):

Rearranging the Right-Hand Side

for $j = 1, \ldots, n$:
    $s := r_j;\ r_j := r_{p_j};\ r_{p_j} := s$

Forward Substitution

$y_1 := r_1$
for $i = 2, \ldots, n$:

$$y_i := r_i - \sum_{k=1}^{i-1} a_{ik} y_k$$

Backward Substitution

$z_n := y_n / a_{nn}$
for $i = n-1, n-2, \ldots, 1$:

$$z_i := \Big(y_i - \sum_{k=i+1}^{n} a_{ik} z_k\Big) \Big/ a_{ii}$$

The same algorithm can be applied to any system with multiple right-hand sides and to invert a matrix (see Section 1.1.7).

This method of refinement can be applied iteratively, but normally one or at most two steps will give the best improvement possible.

If the standard precision in Fortran is defined as REAL*8 or DOUBLE PRECISION, then extended precision means calculation of the inner products $\sum_{j=1}^{n} a_{ij} x_j$ using REAL*16. This is not standard Fortran 77, and it is not supported by all systems. The NAG Fortran Library contains the routine X03AAF for this purpose, and this routine uses the floating-point arithmetic of the computing environment to reach the extended precision demand. It is vital that this routine is implemented correctly, yielding full extended precision accuracy, otherwise iterative refinement will not be effective.

1.1.7 Multiple Right-Hand Sides. Matrix Inversion

Iterative refinement makes use of the fact that after having factorized a matrix additional systems with the same coefficient matrix can easily be solved by forward and backward substitution, (1.13), (1.14). This method can be used for all linear systems with multiple right-hand sides.

One special problem of this kind is the inversion of a matrix $A \in \mathbb{R}^{n,n}$, where one looks for a matrix $X \in \mathbb{R}^{n,n}$ with

$$AX = I. \tag{1.15}$$

Here again I is the identity with the unit vectors $e_i \in \mathbb{R}^n$ as columns:

$$e_i = (0, \ldots, 0, \underset{i\text{th position}}{1}, 0, \ldots, 0)^T.$$

Hence the columns $x_i \in \mathbb{R}^n$ of the inverse X are obtained by solving the linear systems

$$Ax_i = e_i, \quad i = 1, \ldots, n. \tag{1.16}$$

This means that one has to solve n linear systems with the same matrix of coefficients, which is the easiest and most stable method for the inversion of a matrix.

1.1.8 The Gaussian Algorithm with Iterative Refinement

Initialize the permutation vector with the inverse row norms:

for $k = 1, \ldots, n$:

$$p_k := \left(\sum_{i=1}^{n} a_{ki}^2 \right)^{-1/2}$$

for $k = 1, \dots, n$:

Relative row pivoting

Determine the number $l \geq k$ for which

$|a_{lk}|p_l = \max_{i=k,\dots,n} |a_{ik}|p_i$

if $|a_{lk}| < \varepsilon$ then stop

Rearrangement of rows

if $k \neq l$ then

for $j = 1, \dots, n$:

$s := a_{kj};\ a_{kj} := a_{lj};\ a_{lj} := s$

$s := b_k;\ b_k := b_l;\ b_l := s$

$p_l := p_k;$

endif

$p_k := l$

Elimination

for $i = k+1, \dots, n$:

$a_{ik} := \dfrac{a_{ik}}{a_{kk}}$

for $j = k+1, \dots, n$

$a_{ij} := a_{ij} - a_{ik}a_{kj}$

$b_i := b_i - a_{ik}b_k$

Backward substitution

$x_n := b_n/a_{nn}$

for $i = n-1(-1)1$:

$$x_i := \left(b_i - \sum_{k=i+1}^{n} a_{ik}x_k\right) \Big/ a_{ii}$$

Iterative refinement

Calculate $r := b - Ax$ in extended precision

Rearrange r:

for $i = 1, \dots, n-1$:

$s = r(p(i));\ r(p(i)) = r(i);\ r(i) = s$

Solve $Ly = r$ (Forward Substitution)

Solve $Uz = y$ (Backward Substitution)

Let $x := x + z$ (Refinement)

Repeat iterative refinement until $||r|| < \varepsilon$.

1.1.9 Program and Example

The program GAUSS, see B.1.1, solves nonsingular linear systems calling NAG routine F04AEF. F04AEF calls F07ADG, a modified version of LAPACK routine SGETRF/F07ADF, which computes the LU decomposition, and F04AHF. F07ADG itself calls several other modified LAPACK routines and the BLAS routines DTRSM (=F06YJF) and DGEMM (=F06YAF) for the elimination. F07ADG computes the Crout factorization, which is equivalent to Gaussian elimination. Here it is the upper triangular matrix U which holds the unit diagonal instead of L. For the solution of the triangular systems with forward and backward substitution and for the iterative refinement stepsF04AHF is called. F04AHF organizes the storage and controls the refinement steps. For the solution of the triangular systems F04AHF calls the triangular solver F04AJF. F04AJF itself uses the BLAS routine DTRSV (=F06PJF). This modular structure is an example of the design of a good software library like NAG.

As an example we will solve a simple 3 by 3 system with two right-hand sides:

$$\begin{pmatrix} 10 & 8 & 1 \\ 0.001 & -0.001 & 0 \\ 1234 & 2334 & 56987 \end{pmatrix} \begin{pmatrix} x_{11} & x_{12} \\ x_{21} & x_{22} \\ x_{31} & x_{32} \end{pmatrix} = \begin{pmatrix} 19 & 29 \\ 0 & -0.001 \\ 60555 & 176863 \end{pmatrix}.$$

It has the solutions $x_1 = (1, 1, 1)^T$ and $x_2 = (1, 2, 3)^T$.

We use the input template of PAN to create an input file:

Input File for GAUSS

```
Input file for the PAN program GAUSS:
Matrix type (0 = Input, 1 = Random, 2 = Hilbert):
0
Row and column dimensions of Matrix A:
3 3
Matrix A:
10 8 1
0.001 -0.001 0
1234 2334 56987
Inverting matrix A (0 = no, 1 = yes):
0
Number of right-hand sides:
2
Dimension of  1.right-hand side:
3
 1.right-hand side:
19 0 60555
```

```
Dimension of  2.right-hand side:
3
 2.right-hand side:
29 -0.001 176863
```

After having called the program GAUSS we get the following results:

```
1. Solution:

1              1.00000000
2              1.00000000
3              1.00000000

2. Solution:

1              1.00000000
2              2.00000000
3              3.00000000
```

1.1.10 Application: Flow through a Plumbing Network

As an application problems we will solve a fluid flow problem (see [55]) with the program GAUSS, see B.1.1. The fluid through a plumbing networks obeys the following rules for laminar flow:

1. At a junction the fluid flowing in equals the fluid flowing out.
2. The pressure drop Δp across a length of pipe l is given by

$$\Delta p = \frac{8\eta l}{\pi r^4}\, q$$

   where r is the radius of the pipe, η is the viscosity, and q is the flow rate which we are searching for.
3. The pressure drop around any closed loop is zero.
4. Traversing a length of pipe opposite to the direction of flow increases the pressure, hence the pressure drop is negative.

We look at a simple network with two junctions A and B and six pipes $\boxed{1}$ to $\boxed{6}$ (see Figure 1.1). Let r_i, l_i, q_i and k_i, $i = 1, 2, \ldots, 6$, represent the radius, length, flow rate and effective resistance in the ith pipe, where

$$k_i := \frac{8\eta l_i}{\pi r_i^4}.$$

Applying the rules above and noting that $q_1 = q_6$ we can construct a system of five linear equations for the unknown q_i:

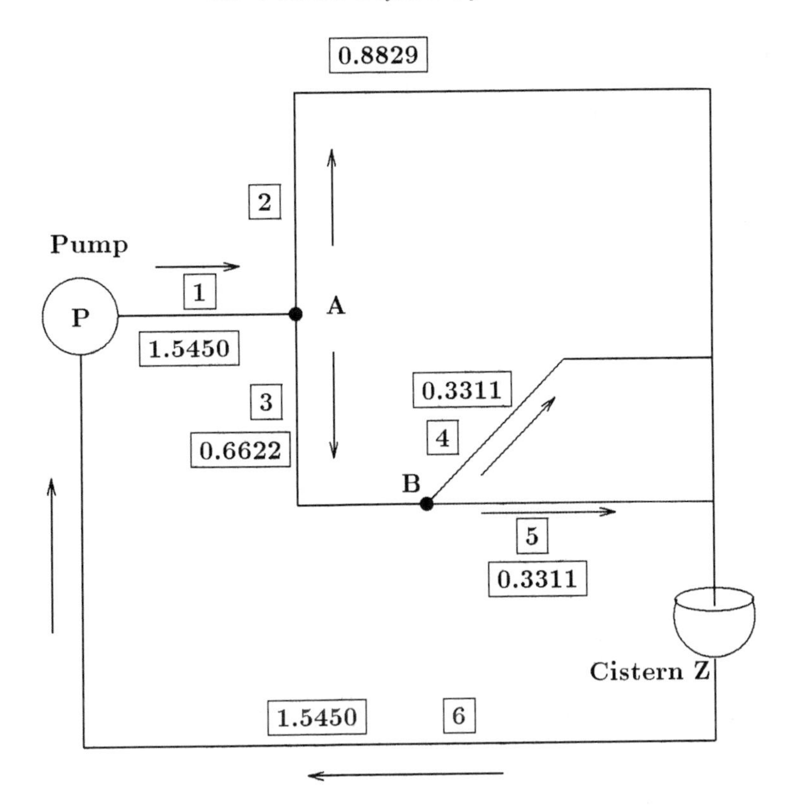

Figure 1.1. Plumbing network with resulting flow rates

$$\begin{array}{rcrcrcrcrcl}
q_1 & - & q_2 & - & q_3 & & & & & = & 0 \\
& & & & q_3 & - & q_4 & - & q_5 & = & 0 \\
(k_1+k_6)q_1 & + & k_2q_2 & & & & & & & = & P \\
& & k_2q_2 & - & k_3q_3 & & & - & k_5q_5 & = & 0 \\
& & & & & & k_4q_4 & - & k_5q_5 & = & 0
\end{array}$$

which for $P = 350\,000$, $\eta = 0.015$, $r_i \equiv r = 0.06$ and $l_i = (10, 12, 12, 8, 8, 60)$ can be solved with GAUSS. P is the pressure rise produced by the pump.

The results are also shown in Figure 1.1.

1.2 Symmetric Positive-Definite Systems Cholesky Factorization

The linear system (1.1) will now be solved with the additional assumption of being symmetric and positive-definite (SPD). Cholesky's method for factorization is then appropriate. SPD systems can be solved with our program CHOLESKY, which uses NAG routines F03AEF and F04AFF. We also show with an example, that positive-definite matrices can be very ill-conditioned. Additionally a network application is worked out.

1.2.1 Introduction

Definition 1.4. *A matrix $A \in \mathbb{R}^{n,n}$ is called symmetric if*

$$a_{ij} = a_{ji} \quad \text{for } i,j = 1,\dots n, \quad \text{or more briefly:}$$

$$A^T = A. \tag{1.17}$$

Definition 1.5. *A matrix $A \in \mathbb{R}^{n,n}$ is called positive-definite if*

$$x^T Ax > 0 \ \forall x \in \mathbb{R}^n \quad \text{with } x \neq 0. \tag{1.18}$$

This definition becomes clearer with the following theorem, which gives other characteristics of a positive-definite matrix:

Theorem 1.6.

1. *Let $A \in \mathbb{R}^{n,n}$ be symmetric. Then the following conditions are equivalent:*
   (a) *The matrix $A \in \mathbb{R}^{n,n}$ is positive-definite.*
   (b) *All principal minors are positive. These are the determinants*

$$|a_{11}|, \begin{vmatrix} a_{11} & a_{12} \\ a_{21} & a_{22} \end{vmatrix}, \dots, |A|.$$

   (c) *All eigenvalues (see Chapter 6) of A are real and positive.*
   (d) $\sqrt{x^T Ax}$ *is a norm.*
2. *Let $A \in \mathbb{R}^{n,n}$ be symmetric positive-definite. Then:*
   (a) $a_{ik}^2 < a_{ii} \cdot a_{kk} \quad \forall i \neq k.$
   (b) *The matrix element with the largest absolute value is a diagonal element.*

SPD systems appear quite frequently in applications, so this is not an artificial characteristic. Therefore a stable algorithm for solving those systems is of some importance. Simply using the method of Gaussian elimination would not take advantage of this characteristic. For SPD systems pivoting is not necessary and a symmetric factorization is always possible.

Theorem 1.7. *Let $A \in \mathbb{R}^{n,n}$ be symmetric positive-definite. Then there exists a unique lower triangular matrix L with positive diagonal elements such that*

$$LL^T = A. \tag{1.19}$$

1.2.2 The Cholesky Factorization

This methods computes the factorization according to Theorem 1.7:

> If $a_{11} < \varepsilon$ then stop
>
> $l_{11} := \sqrt{a_{11}}$
> for $k = 2, \ldots, n$:
> $$l_{k1} := \frac{a_{k1}}{l_{11}}$$
> for $i = 2, \ldots, n$:
> $$s := a_{ii} - \sum_{p=1}^{i-1} l_{ip}^2$$
> If $s < \varepsilon$ then stop
> $l_{ii} := \sqrt{s}$
> for $k = i+1, \ldots, n$:
> $$l_{ki} := \frac{1}{l_{ii}} \left(a_{ki} - \sum_{p=1}^{i-1} l_{ip} l_{kp} \right)$$

A frequently occurring alternative to this algorithm is the factorization $A = LDL^T$ where D is a diagonal matrix with positive entries and where all diagonal elements of L are 1. This factorization does not use square roots and it can be generalized to symmetric, but non-positive-definite matrices.

The Cholesky factorization is of optimal numerical stability. One indication of this stability is the fact that the elements of L cannot become large relative to those of A:

$$l_{ip}^2 \leq a_{ii} \quad \forall i, p.$$

That does not mean that you always get a very accurate solution, because the matrix A can be very ill-conditioned anyway.

To test the positive-definiteness of a symmetric matrix, it would not be wise to try to calculate all principal minors. The easiest way is simply to start the Cholesky factorization If one of the radical expressions s is not positive, the matrix is not positive-definite. Because of roundoff errors the test has to be made with ε instead of 0 as in the algorithm above.

1.2.3 Program and Example

Program CHOLESKY, see Section B.1.2, solves symmetric positive-definite linear systems. It calls NAG routines F03AEF and F04AFF.

The following example shows how poorly conditioned equations may occur, even in the case of positive-definite matrices. An example of an electric network is given as an application.

Ill-Conditioned Linear Equations

If you need to approximate a function or a table of function values with polynomials, you should always choose orthogonal polynomials (see for example Section 3.4.3).

We will see what happens if you do not choose orthogonal polynomials, but rather use standard polynomials instead. For computing the coefficients α_i of the approximating function

$$p(x) = \sum_{i=0}^{n} \alpha_i x^i ,$$

for the interval $[0, 1]$ you have to solve an ill-conditioned system of linear simultaneous equations, whose coefficient matrix is called the *Hilbert matrix*:

$$\begin{pmatrix} 1 & 1/2 & 1/3 & \cdots \\ 1/2 & 1/3 & 1/4 & \\ 1/3 & 1/4 & 1/5 & \\ \vdots & & & \ddots \end{pmatrix}$$

Its condition number grows dramatically with n:

n	3	5	9	12
cond(A)	524	4.8×10^5	4.9×10^{11}	1.7×10^{16}

The Hilbert matrices are notable examples for ill-conditioning and they are therefore very suitable for testing programs or routines for the solution of linear equations. We will solve the system

$$Ax = b$$

with $A \in \mathbb{R}^{n,n}$ the Hilbert matrix and the right-hand side calculated such that the exact solution is

$$x = (1, 1, \ldots, 1)^T .$$

For the program CHOLESKY, examples with Hilbert matrices are easily prepared with the input template or file for CHOLESKY. You only have to choose the order of the matrix and matrix type 'Hilbert'. We give the 2-norm of the errors $\Delta x := \tilde{x} - x$ of the solutions with the program CHOLESKY, as well as with MATLAB:

n	3	5	9	12	
$\|\|\Delta x\|\|_2$	1.6×10^{-14}	2.4×10^{-12}	3.3×10^{-5}	0.17	(CHOLESKY)
$\|\|\Delta x\|\|_2$	1.4×10^{-14}	9.7×10^{-13}	6.4×10^{-6}	1.1	(MATLAB)

MATLAB shows a warning that the system is too ill-conditioned for $n \geq 12$. For NAG routine F04AFF, which is called in CHOLESKY, the resulting value of IFAIL depends on the value for EPS.

1.2.4 Application: Network Theory

We will look at an example of network theory with different connections, voltage sources and resistances (see Figure 1.2).

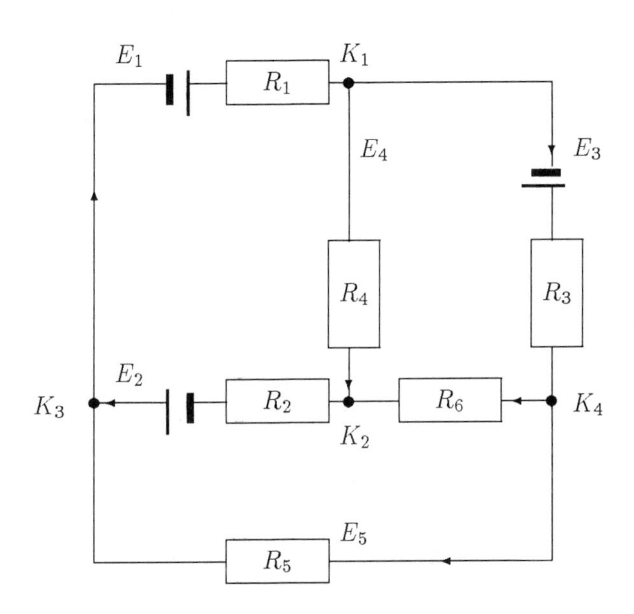

Figure 1.2. Network with four nodes and six branches

Using the laws of Kirchhoff and Ohm results in a symmetric positive-definite linear system for the potentials. From the potentials you can calculate the values for the voltage and for the current. More details may be found in [68]. We consider the example of Figure 1.2: As voltage values we take

$$E_1 = 10\,\text{V}, \quad E_2 = -15\,\text{V}, \quad E_3 = 20\,\text{V}, \quad E_4 = E_5 = 0\,\text{V}$$

and as resistances

$$R_1 = R_2 = \cdots = R_6 = 10\,\Omega,$$

where the voltage and resistance values E_j and R_j are assigned to branch j. Now we need the matrix of incidence $T \in \mathbb{R}^{4,6}$, which is built up by the connections between nodes and branches. Its element t_{ij} is zero, if node i is not connected to branch j. It has the value $+1$ if the direction of the current in branch j is towards node i, and -1 for the other way round. So for our example we get

$$T = \begin{pmatrix} 1 & 0 & -1 & -1 & 0 & 0 \\ 0 & -1 & 0 & 1 & 0 & 1 \\ -1 & 1 & 0 & 0 & 1 & 0 \\ 0 & 0 & 1 & 0 & -1 & -1 \end{pmatrix}.$$

Secondly we need the inverse of the diagonal matrix of the resistances:

$$R^{-1} := \mathrm{diag}(0.1, \ldots, 0.1), \quad R^{-1} \in \mathbb{R}^{6,6}.$$

Now we calculate the matrix

$$B = TR^{-1}T^T.$$

We then get the matrix A for the linear system to be solved by defining a reference node – we take K_4 here – and deleting the row and column belonging to that node. That gives

$$A = \begin{pmatrix} 0.3 & -0.1 & -0.1 \\ -0.1 & 0.3 & -0.1 \\ -0.1 & -0.1 & 0.3 \end{pmatrix}.$$

The right-hand side is given as $b = -TR^{-1}E$, where again the fourth element has to be deleted, which leads to

$$b = (1.0, -1.5, 2.5)^T.$$

With this procedure one always obtains a symmetric positive-definite matrix A. We will solve the resulting linear system with CHOLESKY. The relevant input file created by the template program of PAN is to be seen below. Running the program with it gives the solution:

```
1. Solution:


1              7.50000000
2              1.25000000
3             11.25000000
```

So the potentials are $P_1 = 7.5$, $P_2 = 1.25$, $P_3 = 11.25$, $P_4 = 0$. You have to solve the system for several right-hand sides if you want to calculate the net for different values of the voltages E_i.

Input File for CHOLESKY

```
Input file for the PAN program CHOLESKY:
Matrix type (0 = Input, 1 = Random, 2 = Hilbert):
0
Order of Matrix A:
3
Matrix A:
0.3 -0.1 -0.1
0.3 -0.1
0.3
Inverting matrix A (0 = no, 1 = yes):
0
Number of right-hand sides:
1
Dimension of  1.right-hand side:
3
 1.right-hand side:
1.0 -1.5 2.5
```

1.3 Systems with Band Matrices

In this section we consider different methods for solving linear systems with band matrices. We describe three kinds of band structure and give examples of them. For symmetric positive-definite band systems Cholesky's method should be chosen for variable as well as for fixed bandwidths. For the former case NAG routines F01MCF and F04MCF can be used, and F04ACF for the latter one.

For nonsymmetric band matrices Gauss' method will be applied using NAG routines F01LBF and F04LDF. The program BAND covers all three cases; a simple example of its use is given in Section 1.3.5.

1.3.1 Introduction

In many applications coefficient matrices with many zeros occur. Such matrices are called *sparse*. Often these sparse matrices are also banded, which means there are only zeros in the bottom left and top right 'corners' of A. We now consider (1.1) with banded matrices. Different types of band structure are possible:

Definition 1.8. *Let* $A \in \mathbb{R}^{n,n}$.

1. *A is a band matrix with (fixed) bandwidth* $2m+1$, *if* m *is the smallest number for which*

$$a_{ij} = 0, \quad \text{if } |i-j| > m. \tag{1.20}$$

2. *A has lower and upper bandwidths m_1 and m_2, if m_1 and m_2 are the smallest numbers for which*

$$a_{ij} = 0, \quad \text{if } i > j + m_1 \text{ or } j > i + m_2. \tag{1.21}$$

3. *A symmetric matrix A has variable bandwidth, if there are different numbers m_i, $i = 1, \ldots, n$, with*

$$a_{ij} = 0, \quad \text{if } j < i - m_i. \tag{1.22}$$

For the first definition the bandwidth is sometimes defined as m rather than $2m + 1$.

The difference between the total number of matrix elements and the number of non-zero elements only becomes significant for large orders. So the band algorithms should only be used if the number of elements to be stored is small compared to the order of A, because there are always some additional management calculations for the band algorithms.

If A is sparse – if the number of elements is small compared to n^2 – but not really a band matrix, you may use one of the methods for sparse systems (see Section 1.4). We will consider the special storage management needed for each of the three cases of Definition 1.8.

- Symmetric positive-definite matrix with variable bandwidth
- Symmetric positive-definite matrix with fixed bandwidth
- General nonsingular nonsymmetric band matrix

Thus the method of Cholesky will be applied to the first two cases and Gaussian elimination to the last one.

All three cases can be handled by our program BAND. The most important point when factorizing band matrices is not to fill non-zero elements in the zero-parts of the matrix. This is possible in the SPD case with Cholesky factorization, but not in the nonsymmetric case with Gaussian elimination due to row or column interchanges.

Example 1.9. The matrix

$$A = \begin{pmatrix} 4 & 0 & -1 & 0 & 0 & 0 \\ 0 & 4 & 0 & -1 & 0 & 0 \\ -1 & 0 & 4 & 0 & -1 & 0 \\ 0 & -1 & 0 & 4 & 0 & -1 \\ 0 & 0 & -1 & 0 & 4 & 0 \\ 0 & 0 & 0 & -1 & 0 & 4 \end{pmatrix}$$

has bandwidth 5, so it has $m = 2$. It is sufficient to store 24 instead of 36 matrix elements, but when making use of the symmetry only 15 elements have to be stored (see Section 1.3.3). ■

Example 1.10. The matrix

$$A = \begin{pmatrix} 4 & -1 & 0 & 0 & 0 & 0 \\ 0 & 4 & -1 & 0 & 0 & 0 \\ -1 & 0 & 4 & -1 & 0 & 0 \\ 0 & -1 & 0 & 4 & -1 & 0 \\ 0 & 0 & -1 & 0 & 4 & -1 \\ 0 & 0 & 0 & -1 & 0 & 4 \end{pmatrix}$$

also has a fixed bandwidth of 5, but when treated as a general band matrix with lower bandwidth $m_1 = 2$ and upper bandwidth $m_2 = 1$, only 20 elements have to be stored. ■

Example 1.11. Exploiting variable bandwidth is of use only if symmetric matrices have widely varying row bandwidths, for example for

$$A = \begin{pmatrix} 4 & -1 & 0 & 0 & 0 & -1 \\ -1 & 4 & 0 & -1 & 0 & 0 \\ 0 & 0 & 4 & 0 & -1 & 0 \\ 0 & -1 & 0 & 4 & 0 & 0 \\ 0 & 0 & -1 & 0 & 4 & -1 \\ -1 & 0 & 0 & 0 & -1 & 4 \end{pmatrix}$$

we have $(m_1, m_2, m_3, m_4, m_5) = (0, 1, 0, 2, 2, 5)$, and only 16 elements have to be stored. ■

1.3.2 Cholesky's Method for Variable Bandwidth

For symmetric positive definite matrices with variable bandwidth only the lower part of the matrix has to be stored, i.e. $m_i + 1$ values for the ith row. These row 'half'-lengths are stored in an integer vector NROW. A is then sequentially stored in a one-dimensional array with $m_i + 1$ elements per row. Therefore this array is of length

$$l = \sum_{i=1}^{n} (m_i + 1). \tag{1.23}$$

For Example 1.11 we get

$$\text{NROW} = (1, 2, 1, 3, 3, 6)$$

and

$$A = (4, \quad -1, 4, \quad 4, \quad -1, 0, 4, \quad -1, 0, 4, \quad -1, 0, 0, 0, -1, 4)$$

The method of Cholesky has to be reformulated. There is no fill-in outside the bandwidth of each row. Therefore L may be stored in a one-dimensional array of the same length as A.

1.3.3 Cholesky's Method for Fixed Bandwidth

A and L are stored in an n by $(m+1)$ array like the following one (the explicit one is for Example 1.9):

$$\begin{pmatrix} * & \cdots & * & a_{11} \\ \vdots & & a_{21} & a_{22} \\ \vdots & & \vdots & \vdots \\ * & & \vdots & \vdots \\ a_{m+1,1} & \cdots & a_{m+1,m} & a_{m+1,m+1} \\ \vdots & & \vdots & \vdots \\ a_{n,n-m} & \cdots & a_{n,n-1} & a_{nn} \end{pmatrix} = \begin{pmatrix} * & * & 4 \\ * & 0 & 4 \\ -1 & 0 & 4 \\ -1 & 0 & 4 \\ -1 & 0 & 4 \\ -1 & 0 & 4 \end{pmatrix}.$$

Elements marked $*$ are not used.

1.3.4 Gaussian Elimination for General Band Matrices

Here an array of dimension $(m_1 + m_2 + 1)$ by n is used for storing the matrix A. For efficiency reasons the matrix is stored with columns and rows interchanged, which means:
if for row i we define $m_l := \max(i - m_1, 1)$ and $m_u := \min(i + m_2, n)$, then

$$\text{for } \left.\begin{array}{l} i = 1, \dots, n \\ j = m_l, \dots, m_u \end{array}\right\} \left\{\begin{array}{ll} a_{ij} \to A(j, i), & \text{for } i \le m_1 \\ a_{ij} \to A(m_1 + j - i + 1, i) & \text{for } i > m_1. \end{array}\right.$$

As an example for $m_1 = 2$ and $m_2 = 1$ we have:

$$A \text{ is stored generally as } \begin{pmatrix} a_{11} & a_{21} & a_{31} & a_{42} & a_{53} & \cdots & a_{n,n-2} \\ a_{12} & a_{22} & a_{32} & a_{43} & a_{54} & \cdots & a_{n,n-1} \\ * & a_{23} & a_{33} & a_{44} & a_{55} & \cdots & a_{n,n} \\ * & * & a_{34} & a_{45} & a_{56} & \cdots & * \end{pmatrix},$$

$$\text{which means for Example 1.10: } \begin{pmatrix} 4 & 0 & -1 & -1 & -1 & -1 \\ -1 & 4 & 0 & 0 & 0 & 0 \\ * & -1 & 4 & 4 & 4 & 4 \\ * & * & -1 & -1 & -1 & * \end{pmatrix}.$$

Again elements marked $*$ are not used. The Gaussian elimination method can be carried out without fill-in, if no interchanges are necessary. Then the triangle matrices L and U can be stored in the corresponding elements of the array. But, as we have seen, interchanges are usually necessary, so we will use implicit scaling or relative pivoting as before.

For the kth elimination step the row interchanges have to be applied to the $(n-k+1)$ by n trailing submatrix. U may therefore have up to m_1 additional superdiagonals, and can therefore be stored in an $(m_1 + m_2 + 1)$

by n array. But this is just the size of the array for A, so U can overwrite A.

After interchanges L is normally no longer a band matrix. But because there are at most m_1 factors per row, L can be stored in an m_1 by n array (without the diagonal elements, which all have the value one). The vector representing the permutation matrix contains the necessary information about the interchanges.

1.3.5 Program and Example

The program BAND, see B.1.3, uses NAG routines F01MCF, F04MCF, F04ACF, F01LBF and F04LDF for solving one or more linear systems with a banded coefficient matrix.

The NAG library distinguishes between *accurate* and *approximate* routines. For Gaussian elimination for full nonsingular systems (Section 1.1) and Cholesky's method for full SPD systems (Section 1.2) we decided to use the accurate routines, i.e. those with iterative refinement and using the basic linear algebra subroutines (BLAS) with additional precision arithmetic (for example for the inner product routine X03AAF).

Most applications which lead to linear systems with band matrices deliver well-conditioned linear systems. Therefore for the examples of this section we use the approximate routines. Banded systems arise in many applications such as spline interpolation or approximation and the solution of partial differential equations.

The program is divided into three blocks which correspond to the three methods of Sections 1.3.2, 1.3.3 and 1.3.4. One of the methods is chosen according to the input file.

As an example we will solve the linear system (from Example 1.11):

$$\begin{pmatrix} 4 & -1 & 0 & 0 & 0 & -1 \\ -1 & 4 & 0 & -1 & 0 & 0 \\ 0 & 0 & 4 & 0 & -1 & 0 \\ 0 & -1 & 0 & 4 & 0 & 0 \\ 0 & 0 & -1 & 0 & 4 & -1 \\ -1 & 0 & 0 & 0 & -1 & 4 \end{pmatrix} \begin{pmatrix} x_1 \\ x_2 \\ x_3 \\ x_4 \\ x_5 \\ x_6 \end{pmatrix} = \begin{pmatrix} 2 \\ 2 \\ 3 \\ 3 \\ 2 \\ 2 \end{pmatrix}.$$

Its solution is $x = (1, 1, 1, 1, 1, 1)^T$. The necessary input file can be prepared using the template in the PAN system, which gives the following:

Input File for BAND

```
Input file for the PAN program BAND:
Order of Matrix A:
6
Band structure: (... , 1=spd, variable bw, ... )
```

```
1
Vector of row-widths of Matrix A:
1 2 1 3 3 6
Matrix A:
4
-1 4
4
-1 0 4
-1 0 4
-1 0 0 0 -1 4
Inverting matrix A (0 = no, 1 = yes):
0
Number of right-hand sides:
1
Dimension of  1.right-hand side:
6
1.Right-hand side:
2 2 3 3 2 2
```

1.3.6 Application: A Plane Truss

This application, taken from [55], leads to a linear system with a band matrix, and can therefore be solved with the program BAND.

Figure 1.3 represents a 13-member plane truss, in which the member forces are to be determined. The member forces are numbered 1 to 13 and the joints are indicated by the circled numbers. We assume that the truss members are fastened using frictionless pins. Since the number of joints j and the number of members m satisfy the relation $2j - 3 = m$, a

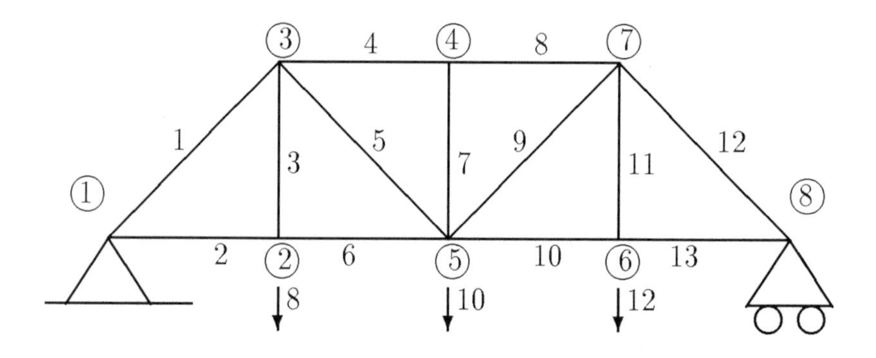

Figure 1.3. A 13-member plane-truss

theorem from mechanics tells us that the truss is statically determinant. This implies that the member forces are determined by the conditions of static equilibrium at the joints. Let F_x and F_y denote the horizontal and the vertical components respectively. Suppose that the angle between any pair of adjoining members is $\pi/4$ radians and let $t = \sin\pi/4 = \cos\pi/4$. We construct equilibrium equations at nonsupported joints 2 to 8 by resolving member forces into their vertical and horizontal components, and requiring that the sum of the horizontal and vertical forces equals zero. With member forces f_i the equilibrium conditions are given by the following equations:

$$\begin{array}{ll}
\text{Joint 2:} & \begin{cases} f_2 - f_6 = 0 \\ f_3 - 8 = 0 \end{cases} \\
\text{Joint 3:} & \begin{cases} -tf_1 + f_4 + tf_5 = 0 \\ -tf_1 - f_3 - tf_5 = 0 \end{cases} \\
\text{Joint 4:} & \begin{cases} -f_4 + f_8 = 0 \\ -f_7 = 0 \end{cases} \\
\text{Joint 5:} & \begin{cases} -tf_5 - f_6 + tf_9 + f_{10} = 0 \\ tf_5 + f_7 + tf_9 - 10 = 0 \end{cases} \\
\text{Joint 6:} & \begin{cases} -f_{10} + f_{13} = 0 \\ f_{11} - 12 = 0 \end{cases} \\
\text{Joint 7:} & \begin{cases} -f_8 - tf_9 + tf_{12} = 0 \\ -tf_9 - f_{11} - tf_{12} = 0 \end{cases} \\
\text{Joint 8:} & \{ \; -f_{13} - tf_{12} = 0
\end{array}$$

If we describe these equations in form of a matrix by leaving out the forces which can be determined directly, then we get the equation $Af = b$, where

$$A = \begin{pmatrix}
0 & 1 & 0 & 0 & -1 & 0 & 0 & 0 & 0 & 0 \\
-t & 0 & 1 & t & 0 & 0 & 0 & 0 & 0 & 0 \\
-t & 0 & 0 & -t & 0 & 0 & 0 & 0 & 0 & 0 \\
0 & 0 & -1 & 0 & 0 & 1 & 0 & 0 & 0 & 0 \\
0 & 0 & 0 & -t & -1 & 0 & t & 1 & 0 & 0 \\
0 & 0 & 0 & t & 0 & 0 & t & 0 & 0 & 0 \\
0 & 0 & 0 & 0 & 0 & 0 & 0 & -1 & 0 & 1 \\
0 & 0 & 0 & 0 & 0 & -1 & -t & 0 & t & 0 \\
0 & 0 & 0 & 0 & 0 & 0 & -t & 0 & -t & 0 \\
0 & 0 & 0 & 0 & 0 & 0 & 0 & 0 & -t & -1
\end{pmatrix}$$

$$f = (f_1, f_2, f_4, f_5, f_6, f_8, f_9, f_{10}, f_{12}, f_{13})^T$$

$$b = (0, 0, 8, 0, 0, 10, 0, 0, 12, 0)^T.$$

The matrix A has lower bandwidth 2 and upper bandwidth 4. For solving the linear system with the program BAND we created the file 'band_in' with the band template. The program gives the following results:

```
x( 1)=   -19.798989873223
x( 2)=    14.000000000000
x( 3)=   -20.000000000000
x( 4)=     8.4852813742386
x( 5)=    14.000000000000
x( 6)=   -20.000000000000
x( 7)=     5.6568542494924
x( 8)=    16.000000000000
x( 9)=   -22.627416997970
x( 10)=   16.000000000000
```

The solution of our problem is then:
$f = (-19.799,\ 14, 8,\ -20,\ 8.485,\ 14,\ 0,\ -20,\ 5.657,\ 16,\ 12,\ -22.627,\ 16)^T$.
Please remember that we left out the forces, which could be determined directly.

1.4 Large Sparse Systems

Sparse linear systems mainly appear when solving partial differential equations (PDEs). Relaxation methods are too slow. The method of conjugate gradients (CG) is also too slow by itself, but when combined with a preconditioner, for example with an incomplete Cholesky decomposition (ICC), the method becomes very attractive. The nonsymmetric case is not as easy to handle, so we only consider a special Gauss method. An example shows how the program SPARSE for sparse linear systems works. The application in Section 1.3.6 can be considered as a sparse system as well. This has been done in the appendix of additional applications, which is to be found only in the electronic version of this text.

1.4.1 Introduction

Sophisticated methods for sparse linear systems are beyond the scope of this text. Here we simply give a brief outline. Sparse systems are often banded, but the bandwidth may be large and there are normally many zeros within the band. Therefore special methods for the solution of this type of system are usually preferred. Sparse systems are often symmetric and positive-definite, but they can be solved with Cholesky's method for banded systems only if they are not too large.

Many methods for the solution of partial differential equations firstly construct a grid covering the domain for the solution of the partial differential equation. The linear systems consist of equations relating the solution at neighbouring grid points. The grid points are numbered, and for each grid point there is one solution value and one equation. Thus the order of the system is equal to the number of grid points. Whenever the points numbered i and j are not neighbouring, the element a_{ij} of the matrix is zero.

Example 1.12. For the grid of Figure 1.4 we have 23 inner grid points and obtain a matrix of the structure shown ($\times$ are the only elements $\neq 0$):

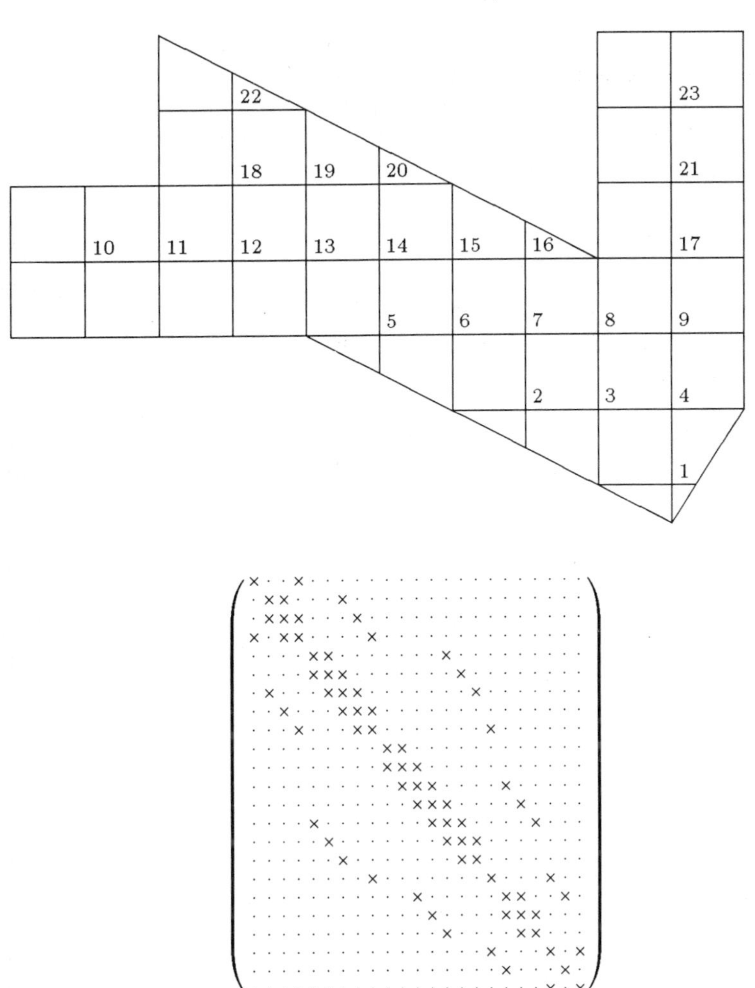

Figure 1.4. Grid and matrix for a two-dimensional domain

Using the symmetry of this system, we see that 51 elements are non-zero. For the Cholesky factorization with variable bandwidth 109 elements must be stored. This difference increases rapidly with the order of the matrix. For example, if we halve the grid size the system has order 115 (13225 elements in a full matrix). Using the symmetry again, there will be 308 non-zero elements, but 1282 elements must be stored for the variable bandwidth Cholesky factorization. ■

1.4.2 Relaxation Methods

For very large sparse systems direct methods usually are too inefficient, so iterative methods usually are used instead. We will describe relaxation methods and the method of conjugate gradients as simple standard methods. We will also briefly describe the effective method of preconditioning, which is often used in software libraries.

$$\begin{array}{lrcl} \text{The idea:} & Ax & = & b \\ \Longrightarrow & x & = & x + A^{-1}(b - Ax). \end{array} \qquad (1.24)$$

Construct an iterative method with (1.24). Replace A^{-1} by B^{-1}, where $B \approx A$, but where B^{-1} can be calculated easily.
The general relaxation method:

$$\begin{array}{lrcl} \text{Choose} & x^{(0)} & \in & \mathbb{R}^n \ \text{arbitrarily} \\ & x^{(i+1)} & := & x^{(i)} + B^{-1}(b - Ax^{(i)}) \\ & & = & (I - B^{-1}A)x^{(i)} + B^{-1}b, \quad i = 0, 1, 2, \ldots. \end{array} \qquad (1.25)$$

The simplest choice for B is

$$B := D := \operatorname{diag}(a_{11}, a_{22}, \ldots, a_{nn}). \qquad (1.26)$$

That leads to the well-known *Jacobi method*. In terms of components we have

$$x_k^{(i+1)} := \frac{1}{a_{kk}} \left(b_k - \sum_{\substack{j=1 \\ j \neq k}}^{n} a_{kj} x_j^{(i)} \right), \quad k = 1, 2, \ldots, n. \qquad (1.27)$$

To implement this algorithm we will take two vectors $x^{(\text{old})}$, $x^{(\text{new})} \in \mathbb{R}^n$:

$$x_k^{(\text{new})} := \frac{1}{a_{kk}} \left(b_k - \sum_{\substack{j=1 \\ j \neq k}}^{n} a_{kj} x_j^{(\text{old})} \right), \quad k = 1, 2, \ldots, n.$$

For the calculation of $x_k^{(\text{new})}$ we could have used $x_j^{(\text{new})}$ instead of $x_j^{(\text{old})}$ whenever possible. In this case we only need one vector $x \in \mathbb{R}^n$ in storage to obtain:

$$x_k := \frac{1}{a_{kk}} \left(b_k - \sum_{\substack{j=1 \\ j \neq k}}^{n} a_{kj} x_j \right), \quad k = 1, 2, \ldots, n.$$

This is the *Gauss–Seidel method*:

$$x_k^{(i+1)} := \frac{1}{a_{kk}} \left(b_k - \sum_{j=1}^{k-1} a_{kj} x_j^{(i+1)} - \sum_{j=k+1}^{n} a_{kj} x_j^{(i)} \right), \quad k = 1, 2, \ldots, n. \tag{1.28}$$

It corresponds to the choice $B := L$, where L is the lower triangular part of A.

The Jacobi and Gauss–Seidel methods converge linearly if A fulfills certain conditions (diagonal dominant, irreducible), but they converge unacceptably slowly.

1.4.3 Method of Conjugate Gradients

The idea for this method is of geometric structure. It minimizes the functions

$$F(x) := (Ax - b)^T A^{-1}(Ax - b) \quad \text{or} \quad \tilde{F}(x) := \frac{1}{2} x^T Ax - b^T x \tag{1.29}$$

with an iterative method $x^{(k+1)} := x^{(k)} + \alpha_k p^{(k)}$. The respective minimum problems are obviously equivalent and are minimized by the exact solution $\bar{x}$ of $Ax = b$. If the direction vectors $p^{(k)}$ satisfy

$$(p^{(k)}, Ap^{(j)}) = 0 \text{ for } k \neq j \tag{1.30}$$

they are called *conjugate* with respect to A. It can be shown that the method of conjugate gradients shown below converges for positive-definite matrices after at most n steps to $\bar{x}$. But this property is not useful in practice because n is very large for the problems considered. Therefore the method is modified for better efficiency by preconditioning (see next subsection), such that a good approximation to the exact solution may be obtained after far fewer than n steps.

The method mainly involves the calculation of scalar products. Therefore it is well suited for vector computers. The sparsity of the matrices should be exploited for the matrix multiplications.

CG-algorithm

Choose $x^{(0)} \in \mathbb{R}^n$ arbitrarily.

$p^{(0)} := r^{(0)} := b - Ax_0$
for $k = 0, 1, 2, \ldots$

$$\alpha_k := \frac{(r^{(k)}, r^{(k)})}{(p^{(k)}, Ap^{(k)})} \quad (\alpha_k \in \mathbb{R})$$

$$x^{(k+1)} := x^{(k)} + \alpha_k p^{(k)}$$

$$r^{(k+1)} := r^{(k)} - \alpha_k Ap^{(k)}$$

if $r^{(k+1)} < \varepsilon$ then stop

$$\beta_k := \frac{(r^{(k+1)}, r^{(k+1)})}{(r^{(k)}, r^{(k)})} \quad (\beta_k \in \mathbb{R})$$

$$p^{(k+1)} := r^{(k+1)} + \beta_k p^{(k)}$$

1.4.4 Preconditioned Conjugate Gradients

PCG-algorithm

Choose $x^{(0)} \in \mathbb{R}^n$ arbitrarily.
Calculate $r^{(0)} := b - Ax_0$
Solve $Bs^{(0)} = r^{(0)}, \quad p^{(0)} := s^{(0)}$
Iterate: for $k = 0, 1, 2, \ldots$

$$\alpha_k := \frac{(r^{(k)}, s^{(k)})}{(p^{(k)}, Ap^{(k)})} \quad (\alpha_k \in \mathbb{R})$$

$$x^{(k+1)} := x^{(k)} + \alpha_k p^{(k)}$$

$$r^{(k+1)} := r^{(k)} - \alpha_k Ap^{(k)}$$

if $r^{(k+1)} < \varepsilon$ then stop

Solve $Bs^{(k)} = r^{(k+1)}$

$$\beta_k := \frac{(r^{(k+1)}, s^{(k+1)})}{(r^{(k)}, s^{(k)})} \quad (\beta_k \in \mathbb{R})$$

$$p^{(k+1)} := s^{(k+1)} + \beta_k p^{(k)}$$

The convergence of the method of conjugate gradients depends on the condition number of A. Normally it converges too slowly. This behaviour can be improved by using preconditioning, which means that we replace $Ax = b$ by

$$B^{-1/2}AB^{-1/2}y = B^{-1/2}b, \quad \text{with } x = B^{-1/2}y \tag{1.31}$$

with a matrix B chosen such that $B \approx A$ and that a linear system with B as matrix of coefficients is easily to be solved. That leads to the PCG-algorithm shown here. In each step of this algorithm you have to solve a linear system of equations with B as matrix of coefficients. The computational effort for solving this system depends on the structure of B, which may be diagonal or tridiagonal or at least very sparse. A very useful preconditioning matrix is $B = LL^T$, where L is obtained by an incomplete Cholesky algorithm (see next section), which means no fill-in is permitted:

$$l_{ij} = 0 \text{ if } a_{ij} = 0,$$

or the fill-in is made minimal.

For $B = I$, i.e. for the standard conjugate gradient method, the convergence can be shown to be

$$\begin{aligned} \|x^{(k)} - \bar{x}\|_2 &\leq 2\sqrt{\kappa}\tau^{\kappa}\|x^{(0)} - \bar{x}\|_2 \\ \text{with} \quad \tau &:= (\sqrt{\kappa} - 1)/(\sqrt{\kappa} + 1), \end{aligned} \tag{1.32}$$

where $\bar{x}$ is the exact solution of $Ax = b$ and κ is the condition number of A (see Section 1.6.2).The larger the condition number, the closer τ is to 1 and the slower is the convergence. For the preconditioned algorithm κ is reduced to the condition number of $B^{-1}A \approx I$.

1.4.5 Incomplete Cholesky Factorization

This algorithm may be applied to symmetric positive-definite systems only. It is one of many preconditioned conjugate gradient methods, and will only be described briefly here. For a review of some more methods of this kind we refer to [5].

The method is implemented in NAG routines F01MAF and F04MAF, and is described in more detail in [73].

The many parameters of these routines will not be described in this text, but only the program SPARSE. Some of the more technical parameters of the NAG routines are assigned values in SPARSE, which are valid for many application examples. The sophisticated user will need to be familiar with the original routine descriptions and usually with the literature.

1. Construct a diagonal matrix $W \in \mathbb{R}^{n,n}$ such that all the diagonal elements of WAW are one.
2. Let $\quad WAW = C + E.$

   E contains those elements of A which are 'far' from the diagonal, and perhaps some diagonal elements such that C is positive-definite.
3. Calculate a Cholesky decomposition

$$C = PLDL^T P^T.$$

   L is a lower diagonal matrix, D a diagonal matrix and P a permutation matrix. P and P^T rearrange rows and columns symmetrically such that the fill-in is minimal.
4. Apply the method of conjugate gradients with the preconditioning matrix $B := W^{-1}CW^{-1}$ to the linear system

$$(WAW)(W^{-1}x) = Wb \iff Ax = b.$$

   Stop the iteration process when the residual vector $r = b - Ax$ satisfies:

$$||Wr||_2 \leq \varepsilon.$$

1.4.6 Nonsymmetric Sparse Systems

Most large sparse linear systems are symmetric and positive-definite. If the matrix is only slightly nonsymmetric the preconditioned CG method may still be applied, if the nonsymmetric parts of A are contained in E such that C is symmetric positive-definite.

If that is not possible or results in convergence which is too slow, you may choose a special Gauss algorithm with a pivoting strategy which stabilizes as well as minimizes the fill-in. Full pivoting is needed in this case:

$$PAQ = LU. \tag{1.33}$$

The method is described in detail in [29] and [28] and implemented in NAG routine F01BRF. The factorization is used by F04AXF for solving the system with forward and backward substitution. We restrict our description of this method to the use of the program SPARSE.

1.4.7 Program and Examples

The program SPARSE, see B.1.4, solves sparse linear systems. For symmetric positive-definite or only slightly non-positive-definite systems it uses the preconditioned conjugate gradients method. For non-positive-definite matrices it uses Gaussian elimination with full pivoting.

For these two cases the program calls NAG routines F01MAF and F04MAF or F01BRF and F04AXF respectively.

In the specification part of the program the parameter definitions for the dimensions and the routine parameters may be found. The actual parameters allow systems with an order up to NMAX = 150 and with fewer than MMAX = 600 non-zero elements to be stored. However, these parameters can easily be changed in the program.

There is an input template for this program, but for a large system a special program for building the input file will normally be used. We will show a typical example, but for an order n, which is much smaller than usually within this class of problems.

Example 1.13. Deflection of a diaphragm

For an important standard problem in partial differential equation (Poisson's equation with zero boundary values, see Chapter 9) there are a lot of applications like deflection of a diaphragm. If we have a grid with inner and boundary points, we can easily explain the structure of the matrix of coefficients A. An arbitrary grid leads to the following non-zero coefficients.

Each inner point determines one row of A. If its number is k with inner neighbouring points numbered p, q, r and s, then we get:

$$\begin{aligned} a_{kk} &= 4 \\ a_{kp} &= a_{kq} = a_{kr} = a_{ks} = -1 \\ a_{kj} &= 0 \text{ , otherwise.} \end{aligned}$$

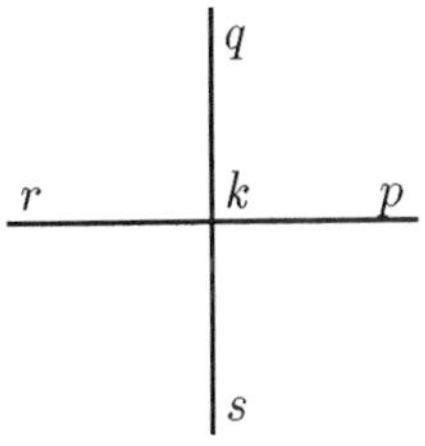

q

r k p

s

Neighbouring points at the boundary are not taken into account. If we apply this procedure to the domain and grid on the next page, we will get the matrix beside it. Again we define the right-hand side such that the solution is $x = (1,1,1,1,1,1)^T$. The necessary input file can be prepared with the template in the PAN system using the symmetry of A, which gives the following input file.

$$A = \begin{pmatrix} 4 & -1 & 0 & 0 & 0 & 0 \\ -1 & 4 & 0 & -1 & 0 & 0 \\ 0 & 0 & 4 & -1 & 0 & 0 \\ 0 & -1 & -1 & 4 & -1 & -1 \\ 0 & 0 & 0 & -1 & 4 & 0 \\ 0 & 0 & 0 & -1 & 0 & 4 \end{pmatrix}$$

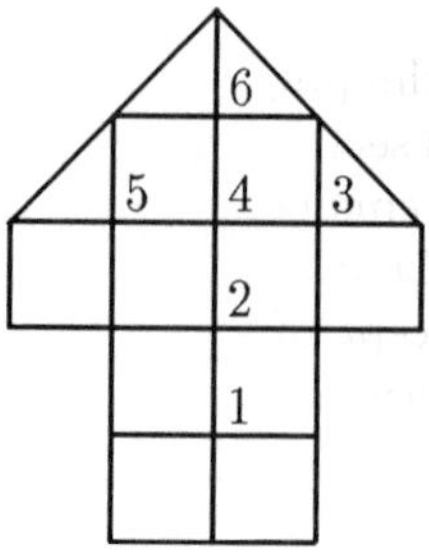

■

Input File for SPARSE

```
Input file for the PAN program SPARSE:
Order of matrix A:
6
Structure type (0=symm. pos.-def., 1=non-symm.):
0
Matrix A: (row no., col. no., element (end with '0 0 0'))
1 1 4
1 2 -1
2 2 4
2 4 -1
3 3 4
3 4 -1
4 4 4
4 5 -1
4 6 -1
5 5 4
6 6 4
0 0 0
Dimension of right-hand side b:
6
Right-hand side b:
3 2 3 0 3 3
```

We give some figures commenting on the incomplete Cholesky decomposition and the conjugate gradients method applied to the grid of Figure 1.4 in Example 1.12. The decomposition $A = \check{C} + E$ stores 30 of the 51 elements not equal to zero in C, the rest of them in E. Then there are 11 iteration steps necessary for a residual norm lower than 10^{-15}.

1.5 Singular System – The Method of Least Squares

In the previous sections we considered nonsingular linear systems only. In this section we will look at singular systems. Non-square and homogeneous systems belong to this class. After having defined a least-squares solution of $Ax = b$ we will consider the singular value decomposition and the related concept of a pseudoinverse of a matrix. The singular value decomposition is carried out by NAG routines F04JAF or F04JDF. These routines are used by our program LSS which is described in this section together with the classical curve fitting application of observing a celestial body.

1.5.1 Introduction

- *Homogeneous systems:*

$$Ax = 0 \qquad \text{with} \qquad A \in \mathbb{R}^{m,n}. \tag{1.34}$$

  For this system $x = 0$ always is a solution. It is the only one if $m = n$ and A is nonsingular. If A is singular or $m \neq n$, then either there are infinitely many solutions $x \neq 0$, or there is no such solution at all.

- *Non-square systems:*

$$Ax = b \qquad \text{with} \qquad A \in \mathbb{R}^{m,n}, \ b \in \mathbb{R}^m \ \text{ and } \ x \in \mathbb{R}^n. \tag{1.35}$$

  Such systems may be under- or overdetermined and there may be no solution or infinitely many solutions.

One or more solutions to all problems of these classes may be obtained if the problem is changed to one of least squares:

$$Ax = b \quad \longrightarrow \quad \min_x ||Ax - b||_2.$$

For this problem we have at least one solution. The solution is always unique if we look for the solution with the smallest norm.

Definition 1.14.

1. *A least-squares solution (LSS) of $Ax = b$ is a vector $\tilde{x} \in \mathbb{R}^n$ for which*

$$||A\tilde{x} - b||_2 = \min_{x \in \mathbb{R}^n} ||Ax - b||_2. \tag{1.36}$$

2. *The least-squares solution of minimal norm is that special solution $\hat{x}$ of (1.36), for which*

$$||\hat{x}||_2 = \min_{\tilde{x}} ||\tilde{x}||_2,$$

   or

$$\hat{x}: \quad ||\hat{x}|| = \min_{\tilde{x}} \left\{ ||\tilde{x}|| \;\middle|\; ||A\tilde{x} - b||_2 = \min_{x \in \mathbb{R}^n} ||Ax - b||_2 \right\}. \tag{1.37}$$

A least-squares solution of minimal norm also solves nonsingular linear systems, but the algorithm for a lss is usually too costly to be applied to such systems.

Theorem 1.15. *Let $\tilde{x} \in \mathbb{R}^n$ be a solution of* (1.36). *Then $\tilde{x}$ is a solution of*

$$A^T Ax = A^T b. \tag{1.38}$$

(1.38) is called the system of normal equations. The matrix $A^T A \in \mathbb{R}^{n,n}$ is square, symmetric and positive semidefinite. It is positive-definite if and only if $\text{rank}(A) = n$. That means that this is the only case where there is a unique solution. You should not solve the least-squares problem through the normal equations, because they are usually ill-conditioned. The reason is that the condition number of $A^T A$ is the square of the condition number of A (if A is nonsingular).

The most stable way to solve the least-squares problem is to compute a singular value decomposition (SVD) of A. Because the SVD is of some importance for numerical linear algebra it will be described in some detail.

If $m \geq n$ and $\text{rank}(A) = n$, then $A^T A$ is positive-definite and you can also solve the least-squares problem through a QR decomposition, with an orthogonal matrix Q and a nonsingular upper triangle matrix R. The QR decomposition with pivoting and iterative refinement is computed by NAG routine F04AMF. Sparse least-squares problems may be solved by NAG routine F04QAF. But we will concentrate on the singular value decomposition in what follows. A useful book about this class of problems with many algorithms and programs is [66].

1.5.2 Singular Value Decomposition and Pseudoinverse

Definition 1.16.

1. *A matrix $U \in \mathbb{R}^{n,n}$ is called orthogonal if*

$$U^T U = I, \tag{1.39}$$

i.e. $U^{-1} = U^T$. (For norms of orthogonal matrices see Section 1.6, Lemma 1.28.)

2. *Let $w \in \mathbb{R}^n$ and $||w||_2 = 1$. Then the symmetric orthogonal matrix*

$$U = I - 2ww^T \tag{1.40}$$

is called a Householder matrix.

Example 1.17. Let $w = \frac{1}{\sqrt{5}}\begin{pmatrix} 1 \\ 2 \end{pmatrix}$. Then we have

$$ww^T = \begin{pmatrix} 1/5 & 2/5 \\ 2/5 & 4/5 \end{pmatrix} \quad \text{and} \quad U = I - 2ww^T = \begin{pmatrix} 3/5 & -4/5 \\ -4/5 & -3/5 \end{pmatrix},$$

which obviously is symmetric and orthogonal: $U^2 = I$. ■

Theorem 1.18. *Let $A \in \mathbb{R}^{m,n}$ be a matrix of rank $r \le \min(m,n)$. Then there are two orthogonal matrices $U \in \mathbb{R}^{m,m}$ and $V \in \mathbb{R}^{n,n}$ and a diagonal matrix $S \in \mathbb{R}^{m,n}$, such that*

$$U^T AV = S \qquad \text{and} \qquad A = USV^T. \tag{1.41}$$

The diagonal elements of S may be ordered with non-increasing values, r of them are positive, the others zero:

$$S = \mathrm{diag}(\sigma_1, \sigma_2, \ldots, \sigma_r, 0, \ldots, 0) = \begin{bmatrix} \hat{S} & 0 \\ 0 & 0 \end{bmatrix} \qquad \text{with} \qquad \hat{S} \in \mathbb{R}^{r,r}. \tag{1.42}$$

The diagonal elements of S are called singular values *of A.*

The singular values of A are uniquely defined, but not the orthogonal matrices U and V.

Theorem 1.19. *Let the matrix $A \in \mathbb{R}^{m,n}$ have rank r, and let $A = USV^T$ be a singular value decomposition of A. Let the vector $d \in \mathbb{R}^m$ be defined as*

$$d = \begin{bmatrix} d_1 \\ d_2 \end{bmatrix} := U^T b, \quad d_1 \in \mathbb{R}^r, \; d_2 \in \mathbb{R}^{m-r}$$

and a new variable $y \in \mathbb{R}^n$ as

$$y = \begin{bmatrix} y_1 \\ y_2 \end{bmatrix} := V^T x, \quad y_1 \in \mathbb{R}^r, \; y_2 \in \mathbb{R}^{n-r}.$$

Let $\hat{y}_1 := \hat{S}^{-1} d_1$. Then:

1. *All least-squares solutions are of the form*

$$\tilde{x} = V \begin{bmatrix} \hat{y}_1 \\ y_2 \end{bmatrix} \quad \text{with } y_2 \text{ arbitrary.}$$

2. *Each such solution has the same residual vector*

$$\tilde{r} := b - A\tilde{x} = U \begin{bmatrix} 0 \\ d_2 \end{bmatrix} \qquad \text{with} \qquad \|\tilde{r}\|_2 = \|d_2\|_2 \,.$$

3. *The unique least-squares solution of minimal norm is*

$$\hat{x} = V \begin{bmatrix} \hat{y}_1 \\ 0 \end{bmatrix}.$$

Definition 1.20. *Let $A \in \mathbb{R}^{m,n}$. Then the pseudoinverse $A^\dagger \in \mathbb{R}^{n,m}$ of A is the matrix whose jth column z_j is the unique least-squares solution of minimal norm of $Az_j = e_j$. Here e_j is the jth column of the identity matrix I_m.*

Theorem 1.21.

1. *The least-squares solution of minimal norm of $Ax = b$ is given as*

$$\hat{x} = A^\dagger b. \tag{1.43}$$

2. *Let $B \in \mathbb{R}^{n,n}$ be nonsingular, i.e. rank $(B) = n$. Then*

$$B^\dagger = B^{-1}. \tag{1.44}$$

3. *Let $A \in \mathbb{R}^{m,n}$ have the singular value decomposition $A = USV^T$. Then the pseudoinverse of A is given as*

$$A^\dagger = VS^\dagger U^T, \tag{1.45}$$

with $S^\dagger = diag\ \{s_i^\dagger\}$ and

$$s_i^\dagger = \begin{cases} s_i^{-1} & \text{if } s_i > 0 \ (i = 1, \dots, r) \\ 0 & \text{if } s_i = 0. \end{cases}$$

We will describe the well known and widely used Golub–Reinsch algorithm for calculating the singular value decomposition, [43]. A useful review of algorithms and applications may be found in [41]. There are other methods which have recently been revived for parallel computers, but in [62] it is shown that the Golub–Reinsch algorithm is also competitive for parallel architectures.

For the matrix $A \in \mathbb{R}^{m,n}$ we will assume

$$m \geq n. \tag{1.46}$$

Then the singular value decomposition can be represented as

$$A = U \begin{bmatrix} \tilde{S} \\ 0 \end{bmatrix} V^T \quad \text{with} \quad A \in \mathbb{R}^{m,n}, U \in \mathbb{R}^{m,m}, \tilde{S} \in \mathbb{R}^{n,n}, V \in \mathbb{R}^{n,n}.$$

The case $m < n$ is easily obtained by transposition. These are the steps of the algorithm, which are given in more detail later:

1. Transformation of A into a bidiagonal matrix $\begin{bmatrix} B \\ 0 \end{bmatrix}$ with at most $2n-1$ Householder transformations.
2. Computation of a singular value decomposition of $B = \hat{U}\hat{S}\hat{V}^T$, but without ordering the singular values.
3. Ordering of $\hat{S}$ to a non-negative diagonal matrix S with non-increasing diagonal elements.
4. Composition of the computed matrices to become the singular value decomposition of $A = USV^T$.

Bidiagonalization of A

With $2n-1$ Householder transformations we obtain

$$\begin{bmatrix} B \\ 0 \end{bmatrix} = Q_n(\cdots((Q_1 A)H_2)\cdots H_n) = Q^T AH \tag{1.47}$$

with

$$B = \begin{bmatrix} q_1 & e_2 & & \\ & q_2 & \ddots & \\ & & \ddots & \ddots & \\ & & & & e_n \\ & & & & q_n \end{bmatrix} . \tag{1.48}$$

The matrices Q_i are chosen such that in the ith column the elements from row $i+1$ to row m become zero. The matrices H_i are chosen such that in the $(i-1)$th row the elements from column $i+1$ to column n become zero. Additionally $H_n := I$ (if $m \geq n$) and $Q_n := I$ (if $m \leq n$).

Singular Value Decomposition of B

We will compute the decomposition

$$B = \hat{U}\hat{S}\hat{V}^T . \tag{1.49}$$

Here B is the bidiagonal matrix (1.48). The singular value decomposition of B will be computed iteratively:

$$\begin{aligned} B_1 &= B . \\ B_{k+1} &= U_k^T B_k V_k, \quad k = 1, 2, \ldots . \end{aligned}$$

U_k and V_k are orthogonal, and all the B_k are bidiagonal. U_k and V_k are chosen so that the matrix

$$\hat{S} = \lim_{k\to\infty} B_k$$

exists and is diagonal. The elements of $\hat{S}$ may be negative and need not be ordered. In an additional step these diagonal elements will be sorted. The iteration is equivalent to the QR algorithm of Francis for the matrix $B^T B$, [33]. For the Golub–Reinsch method this algorithm was adapted to work with B instead of $B^T B$, [43].

Composition of the Singular Value Decomposition

The negative elements of $\hat{S}$ must be multiplied by -1. $\hat{S}$ is then sorted, giving S. These operations correspond to multiplications by a diagonal matrix and permutation matrices (from left and right). These matrices are also orthogonal. Finally all orthogonal matrices, which are applied to A from left and right within these different steps are multiplied together to give the orthogonal matrices U and V, which, together with S, form the singular value decomposition of A:

$$A = USV^T .$$

Solution of the Homogeneous System

The minimum-norm solution of a homogeneous problem is always $x = 0$, so one searches for a solution x with $||x||_2 = 1$ instead. But here it is first necessary to prove that there are any solutions $x \neq 0$. This is always the case if for a square matrix A its rank r is less than its order n. There are then infinitely many solutions and one normally looks for $n-r$ linear independent solution(s) with $||x||_2 = 1$. This is also possible with the method of least squares.

The solutions of the homogeneous problem are given immediately by Theorem 1.19. With $b = 0$ we have $\hat{y}_1 = 0$. In choosing the $n - r$ unit vectors one after the other for the free $(n-r)$-dimensional y_2 one sees that the last $n-r$ columns v_i $(i = n-r+1, \ldots, n)$ of V are linearly independent solutions of the homogeneous system. They are called a *basis* of the solution space, because each linear combination

$$\sum_{i=n-r+1}^{n} \alpha_i v_i$$

for any choice of α_i is also a solution. Conversely each solution can be represented as such a linear combination.

1.5.3 Program and Examples

The program LSS, see B.1.5, can solve non-square, singular and homogeneous linear systems with the method of least squares, in other words it determines least-squares solutions (lss). It calls NAG routine F04JAF, or for the case $m \leq n$ F04JDF. These routines call F02WEF to calculate a singular value decomposition of A. With this SVD the solution of the least-squares problem is computed.

Care is needed in choosing the tolerance TOL for determining the rank r of A. Setting this tolerance too small may lead to wrong solutions. Letting the routine choose the tolerance by itself (TOL = 0) is reasonable only if the singular values of A do not vary dramatically in size. That means if $\sigma_r/\sigma_1 \gg \varepsilon$ (machine precision).

There are many applications for the singular value decomposition and the method of least squares. Best known is the curve fitting of data in the calculus of observations. The aim is to determine a few functional parameters with a large number of observed values. We applied the program LSS to a practical problem from this area, [68].

Example 1.22. Orbit of a Celestial Body

A celestial body has been discovered. Its orbit around the sun has been observed at 10 positions. Their Cartesian coordinates (x_i, y_i), $i = 1, \ldots, 10$, represented in an adjusted coordinate system are given in the following table:

Observation No.	x_i	y_i
1	-1.024940	-0.389269
2	-0.949898	-0.322894
3	-0.866114	-0.265256
4	-0.773392	-0.216557
5	-0.671372	-0.177152
6	-0.559524	-0.147582
7	-0.437067	-0.128618
8	-0.302909	-0.121353
9	-0.155493	-0.127348
10	-0.007464	-0.148885

The orbit of the celestial body is an ellipse with the sun at one of its foci. Therefore we have to fit the given data with a curve of the form:

$$x^2 = ay^2 + bxy + cx + dy + e. \tag{1.50}$$

That leads to 10 equations with 5 unknowns:

$$\begin{pmatrix} y_1^2 & x_1y_1 & x_1 & y_1 & 1 \\ \vdots & \vdots & \vdots & \vdots & \vdots \\ y_{10}^2 & x_{10}y_{10} & x_{10} & y_{10} & 1 \end{pmatrix} \begin{pmatrix} a \\ b \\ c \\ d \\ e \end{pmatrix} = \begin{pmatrix} x_1^2 \\ \vdots \\ x_{10}^2 \end{pmatrix} \tag{1.51}$$

We solved this system by running our program LSS. It needs the following input file which can be created by the template program of PAN:

Input File for LSS

```
Input file for the PAN program LSS:
Matrix type (0 = Input, 1 = Random, 2 = Hilbert):
0
Row and column dimension of Matrix A:
10 5
Matrix A:
0.151530354 0.398977369 -1.024940 -0.389269 1
0.104260535 0.306716365 -0.949898 -0.322894 1
0.070360746 0.229741935 -0.866414 -0.265256 1
0.046896934 0.167483451 -0.773392 -0.216557 1
0.031382831 0.118934893 -0.671372 -0.177152 1
0.021780447 0.082575671 -0.559524 -0.147582 1
0.016542590 0.056214683 -0.437067 -0.128618 1
0.014726551 0.036758916 -0.302909 -0.121353 1
0.016217513 0.019801723 -0.155493 -0.127348 1
0.022166743 0.001111278 -0.007464 -0.148885 1
Solving a homogeneous system Ax = b (0 = no, 1 = yes):
0
Tolerance:
1e-8
Dimension of right-hand side b:
10
Right-hand side b:
1.050502005 0.902306210 0.750153461 0.598135186 0.450740362
0.313067107 0.191027563 0.091753862 0.024178073 5.511296e-5
```

Results

The results are the solution vector, the root-mean-square error and the determined rank of the matrix:

```
Solution x:
  1         -1.37049263
  2         -0.66946664
  3         -0.67185148
  4         -3.36894974
  5         -0.47503185

Standard error =    8.1515598486436D-04
Rank of matrix A =  5
```

We get the same results for TOL $= 0$, but a very different one for TOL $= 10^{-2}$, because rank three is determined now instead of five, which means that now the solution is constructed of three independent vectors only:

```
Solution x:
  1          0.41577485
  2          0.87560311
  3         -0.52836861
  4         -0.75232965
  5         -0.16517253

Standard error =    2.9829335630187D-02
Rank of matrix A =  3
```

Using NAG routine J06CCF from NAG Graphics Library (see Appendix D) the ellipse can be drawn by running a graphical program. The crosses mark the observed points. ■

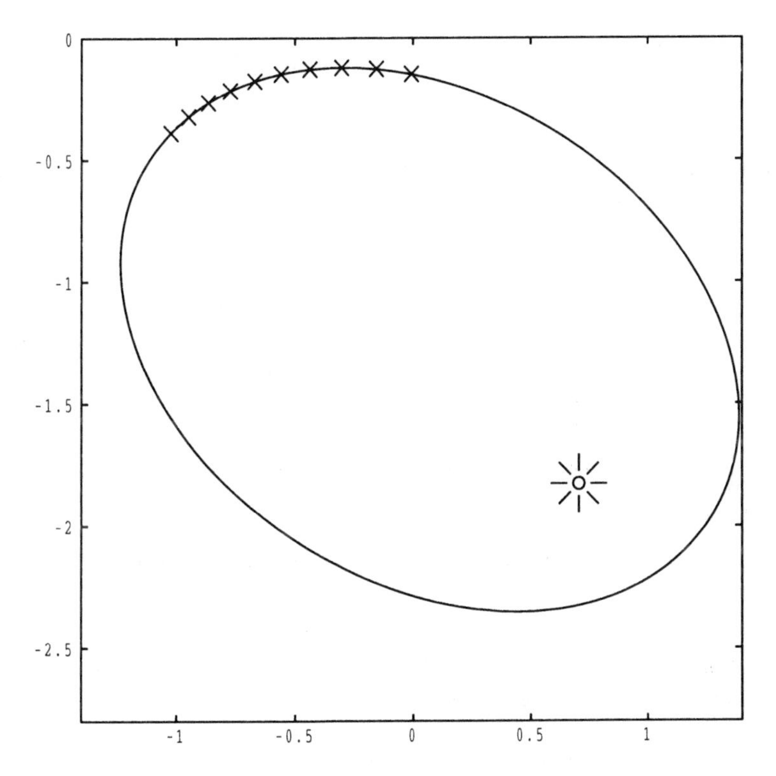

Figure 1.5. Orbit of a celestial body: an ellipse fitted to 10 observation points

1.6 Norms, Condition Number, Error Estimates

In this section we will define the most important vector and matrix norms, the condition number for linear systems and look at some related error estimates.

1.6.1 Vector and Matrix Norms

Definition 1.23.

1. *The Euclidean or 2-norm of a vector $x \in \mathbb{R}^n$ is defined as*

$$||x||_2 := \sqrt{x^T x} = \sqrt{x_1^2 + x_2^2 + \cdots + x_n^2}. \tag{1.52}$$

2. *The 1-norm of a vector $x \in \mathbb{R}^n$ is defined as*

$$||x||_1 := \sum_{i=1}^{n} |x_i|. \tag{1.53}$$

3. *The Maximum- or ∞-norm of a vector $x \in \mathbb{R}^n$ is defined as*

$$||x||_\infty := \max\{|x_1|, |x_2|, \ldots, |x_n|\}. \tag{1.54}$$

Definition 1.24. $||\cdot||$ *is called a matrix norm if for an arbitrary matrix $A \in \mathbb{R}^{m,n}$ it satisfies:*

$$\begin{aligned} ||A|| &\geq 0 \\ ||A|| &= 0 \iff A = 0 \\ ||\alpha A|| &= |\alpha|\ ||A|| \quad \forall \alpha \in \mathbb{R} \\ ||A + B|| &\leq ||A|| + ||B|| \quad \forall B \in \mathbb{R}^{m,n} \\ ||AB|| &\leq ||A||\ ||B|| \quad \forall B \in \mathbb{R}^{n,p}. \end{aligned} \tag{1.55}$$

The relationship between matrix norms and vector norms is interesting.

Definition 1.25.

1. *A matrix norm $||\cdot||_*$ is said to be consistent with a vector norm $||\cdot||_{**}$ if*

$$||Ax||_{**} \leq ||A||_*\ ||x||_{**} \quad \forall\ A \in \mathbb{R}^{m,n},\ x \in \mathbb{R}^n. \tag{1.56}$$

2. *A matrix norm $||\cdot||_*$ is said to be induced by a vector norm if it is defined for all $A \in \mathbb{R}^{m,n}$ by*

$$||A||_* := \max_{x \neq 0} \frac{||Ax||_*}{||x||_*} = \max_{||x||_*=1} ||Ax||_*. \tag{1.57}$$

The induced norm is also called lub-norm (lowest upper bound) because

$$\|Ax\|_* \le \|A\|_{**}\, \|x\|_* \tag{1.58}$$

for all matrix norms $\|\cdot\|_{**}$ *which are consistent with the vector norm* $\|\cdot\|_*$.

Example 1.26. We will give the most important norms as examples.

1. Absolute column sum norm:

$$\|A\|_1 = \max\left\{\sum_{i=1}^{m} |a_{ij}| \ :\ j = 1, 2, \ldots, n\right\}. \tag{1.59}$$

2. Absolute row sum norm:

$$\|A\|_\infty = \max\left\{\sum_{j=1}^{n} |a_{ij}| \ :\ i = 1, 2, \ldots, m\right\}. \tag{1.60}$$

3. Spectral norm:

$$\|A\|_2 = \sqrt{\max\{\lambda(A^T A)\}}\,, \tag{1.61}$$

with λ the eigenvalues of $A^T A$.

4. Euclidean or Schur norm:

$$\|A\|_E := \left(\sum_{i=1}^{m}\sum_{j=1}^{n} |a_{ij}|^2\right)^{1/2}. \tag{1.62}$$

The first three matrix norms are lub-norms induced by the respective vector norms. For the Euclidean norm there is no such vector norm. The most important norm is the 2-norm, which has the drawback that it is very hard to compute, because we have to solve an eigenvalue problem. The Schur norm is consistent with the 2-norm, but it normally delivers much bigger values than the 2-norm, as we may see from the following lemma. ■

Lemma 1.27. *Let* $A \in \mathbb{R}^{n,n}$. *Then the following relations hold.*

1. *If* $\|\cdot\|_*$ *is a lub-norm, then*

$$\|I\|_* = 1, \tag{1.63}$$

where $I \in \mathbb{R}^{n,n}$ *is the identity matrix.*

2. $\|I\|_E = \sqrt{n}$, *and in general*

$$\|A\|_2 \le \|A\|_E \le \sqrt{n}\, \|A\|_2. \tag{1.64}$$

3. *For all matrix norms and all eigenvalues* λ *of* A *we have*

$$\|A\| \ge |\lambda|\,. \tag{1.65}$$

4.

$$\|A\|_2^2 \le \|A\|_1 \, \|A\|_\infty. \tag{1.66}$$

Orthogonal matrices play a dominant role in numerical linear algebra. One reason is the stability of operations with them. This reason can be made clear by the following norm relations.

Lemma 1.28. *Let $U \in \mathbb{R}^{n,n}$ be an orthogonal matrix ($U^T U = I$), $x \in \mathbb{R}^n$ and $A \in \mathbb{R}^{n,m}$ arbitrarily chosen. Then:*

$$\begin{aligned} \|U\|_2 &= 1 \\ \|UA\|_2 &= \|A\|_2 \\ \|Ux\|_2 &= \|x\|_2 \quad \forall x \in \mathbb{R}^n \\ \|UA\|_E &= \|A\|_E. \end{aligned} \tag{1.67}$$

1.6.2 Condition Number

For an algorithm a condition number is defined as the fraction of the relative error of one of the results to the relative error of one of the input data.

For the problem of solving a linear system

$$Ax = b, \quad A \in \mathbb{R}^{n,n}$$

there are $n^2 + n$ data and n results, so there are $n^3 + n^2$ condition numbers. It is not possible in general, and in any case makes no sense, to compute or estimate all these numbers. Therefore we are looking for one number which gives a good measure of the stability of the system of linear equations considered. We will define such a number and see in the following subsection on error estimates that it is able to give a good insight into the stability of the solution process.

Definition 1.29. *The number*

$$\operatorname{cond}_*(A) := \|A\|_* \, \|A^{-1}\|_* \tag{1.68}$$

is called the condition number of the system of linear equations with A as its matrix of coefficients.

The condition number obviously depends on the norm chosen for its computation. And it is easily seen that the minimum possible value for the condition number is 1. With Lemma 1.28 we see that this minimum is taken on by orthogonal matrices when using the 2-norm, which once again demonstrates their good stability properties.

Standard examples of ill-conditioned systems are simultaneous equations with the Hilbert matrix, for which the condition number grows rapidly with the order.

1.6.3 Residual Vector and Error

The following theorem shows that the condition number is a good measure of the propagation of errors from the input data to the results when solving linear equations. It models the situation where we have perturbed data, but are able to perform calculations exactly. This is an unrealistic situation, but it shows the stability properties of the equations to be solved independently of the method and precision used.

Theorem 1.30. *Let $A \in \mathbb{R}^{n,n}$ be a non-singular matrix and $||\cdot||$ an lub-norm and let*

$$\begin{aligned} A\,x &= b \quad \textit{and} \\ (A + \Delta A)\,(x + \Delta x) &= b + \Delta b \\ \textit{with} \quad ||\Delta A||\,||A^{-1}|| &< 1. \end{aligned}$$

Then

$$\frac{||\Delta x||}{||x||} \leq \frac{\operatorname{cond}(A)}{1 - \operatorname{cond}(A)\frac{||\Delta A||}{||A||}} \left(\frac{||\Delta A||}{||A||} + \frac{||\Delta b||}{||b||}\right). \tag{1.69}$$

Condition numbers cannot normally be calculated before solving the linear system because one needs the inverse of A, the computation of which is equivalent to the solution of the linear system. But even after having solved the system it is much more effort to calculate the condition number and the estimate of Theorem 1.30 than the *residual vector*

$$r := b - Ax. \tag{1.70}$$

However, there is one important fact always to remember:

A small residual norm is not equivalent to a small error.

The following theorem shows that it is the condition number which governs whether a small residual norm also implies a small error in the solution.

Theorem 1.31. *Let $A \in \mathbb{R}^{n,n}$ be non-singular and $Ax = b$. Let $\tilde{x}$ be an approximate solution with*

$$r := b - A\tilde{x}.$$

$$\textit{Then} \quad ||\tilde{x} - x|| \leq ||A^{-1}||\,||r|| \tag{1.71}$$

$$\frac{||\tilde{x} - x||}{||x||} \leq \operatorname{cond}(A)\,\frac{||r||}{||b||}. \tag{1.72}$$

1.7 Solution of Linear Systems Using IMSL or MATLAB

The IMSL Library provides both single and double precision versions of each routine. In addition, two versions are provided for routines requiring workspace, one with automatic workspace and the other with workspace passed through the parameter list. Hence there are four versions of many IMSL routines. NAG, on the other hand, provides separate libraries for single and double precision, at least for most significant machine ranges. In addition, all workspace required is passed via the parameter list in the NAG Library.

1.7.1 General Real Linear Systems

The IMSL routine LSARG corresponds to NAG routine F04AEF, except that LSARG solves linear systems for one right-hand side only. To solve multiple systems with the same coefficient matrix, the matrix is first decomposed using LFTRG and then LFIRG is called for each right-hand side.

1.7.2 Symmetric Positive-Definite Systems

The IMSL routine LFCDS corresponds to NAG routine F03AEF. LFCDS distinguishes between the warnings 'The input matrix is algorithmically singular' and 'The input matrix is not positive-definite', while F03AEF gives the warning 'A is not positive-definite, possibly due to rounding errors' in both cases.

NAG routine F04AFF corresponds to the IMSL routine LFIDS, although the parameters EPS (machine precision) and IT (number of steps of iterative refinement) are absent from LFIDS. F04AFF can solve systems with multiple right-hand sides, whereas LFIDS has to be called repeatedly in this case. The error handling of these routines is similar.

1.7.3 Symmetric Positive-Definite Systems with a Band Matrix

The IMSL Library does not contain routines for systems with variable bandwidth. NAG routine F04ACF for systems with fixed bandwidth corresponds to the IMSL routine LSLQS. However, again the IMSL routine only solves systems with one right-hand side. For systems with multiple right-hand sides LFTQS is first called for the Cholesky decomposition and then LFSQS is called for each right-hand side. IMSL uses iterative refinement for systems with fixed bandwidth. LSLQS distinguishes between the warnings 'The input matrix is too ill-conditioned. The solution might not be accurate' and 'The input matrix is not positive-definite', while F04ACF only has the latter message.

1.7.4 Nonsingular Systems with Fixed Bandwidth

The IMSL routine LFTRB for the decomposition of a banded matrix corresponds to NAG routine F01LBF. LFTRB only has the error message 'The input matrix is singular'. The IMSL routine LFSRB for the solution of the decomposed system corresponds to NAG routine F04LDF. The IMSL routine LFIRB solves this system with iterative refinement. There is no corresponding routine in the NAG Library. Again in IMSL one has to call the solver once for each right-hand side.

1.7.5 Sparse Linear Systems

IMSL has no routines for the solution of nonsymmetric sparse systems. The IMSL routine PCGRC solves symmetric positive-definite sparse systems using a preconditioned conjugate gradient method. It corresponds to NAG routines F01MAF and F04MAF. PCGRC can also be used to solve negative-definite systems. PCGRC uses reverse communication, which means that the matrix A need not be stored. The user has to supply a routine for the computation of $z = Ax$ for any vector x.

1.7.6 Singular Linear Systems

IMSL routines LSQRR and LSBRR correspond to NAG routines F04JAF and F04JDF. However, the IMSL routines use a QR decomposition rather than the singular value decomposition. LSBRR uses iterative refinement, LSQRR does not. The NAG routine F04JAF (for $m \geq n$ if $A \in \mathbb{R}^{m,n}$) needs much less workspace ($32n$ Bytes) than LSQRR and LSBRR ($8mn+28n-8$ and $8mn + 28n + 2m - 8$ bytes respectively).

1.7.7 An Example using MATLAB

There is a big difference in working with libraries like NAG or IMSL compared with the interactive system MATLAB. We will demonstrate how MATLAB works with a small example (see also [70]). If we know an approximate solution $\tilde{x}$ for the system $Ax = b$, then the estimate (1.71) holds:

$$||x - \tilde{x}||_2 \leq ||A^{-1}||_2 ||b - A\tilde{x}||_2. \tag{1.73}$$

The system

$$\begin{pmatrix} 5.0 & 4.0 \\ 2.4 & 2.0 \end{pmatrix} \begin{pmatrix} x_1 \\ x_2 \end{pmatrix} = \begin{pmatrix} 14.0 \\ 6.8 \end{pmatrix}$$

has the solution $x = (2, 1)^T$.

Let $\tilde{x} = (1.99, 1.01)^T$ be an approximate solution. Then the following MATLAB dialogue gives the values of the estimate (1.73):

```
A=[5 4; 2.4 2]
                    A =
                           5.0000      4.0000
                           2.4000      2.0000
b=[14.6;6.8];
xs = [1.99;1.01];
r=b-A*xs
                    r =
                           0.0100
                           0.0040
A1=inv(A)
                    A1 =
                           5.0000    -10.0000
                          -6.0000     12.5000
norm(A1)*norm(r)
                    ans =
                           0.1918
```

Here the user input is printed on the left, and the system's response on the right. If a semicolon is used to terminate the command, then the system does not print the answer. The rest of the dialogue is self-explanatory.

1.8 Exercises

1.8.1 Electrical Networks

Write a program which creates the data for networks like that of the application in Section 1.2.4. As input your program should read the voltage values of the sources and the resistances as well as the number of nodes and their connections (i.e. the incidence matrix). Then it should create the matrix of coefficients and right-hand side for the linear system to be solved. The input data in detail are:

- The number m of nodes and n of branches.
- The starting points and endpoints, resistances and source voltage values for each branch.

All resistances R_i must be positive. The voltages E_i may be zero (i.e. no source within this branch). Take care with the directions of the current. Voltages E_i must be negative if their poles are directed against the direction of the current. Create the matrix of incidence $T \in \mathbb{R}^{m,n}$ and then $B = TR^{-1}T^T$, where R is the diagonal matrix of resistances. The elements of $B = TR^{-1}T^T$ can be computed in terms of components:

$$(B)_{ij} = \begin{cases} 0 & \text{if } i \neq j \text{ and no branch connects the nodes } i \text{ and } j \\ -R_k^{-1} & \text{if } i \neq j \text{ and branch } k \text{ connects the nodes } i \text{ and } j \\ \sum_l R_l^{-1} & \text{if } i = j, \\ & \text{sum over branches } l \text{ which start or end in node } i. \end{cases}$$

Make K_m the reference node. In this case A is obtained from B simply by deleting the last row and column of B. The right-hand side is obtained by computing $b = -TR^{-1}E$ and then deleting the last component. You may then solve the system $Ax = b$ with the program CHOLESKY. The solution gives the potentials $P = (x^T \quad 0)^T$. Using the potentials you obtain the voltage values as

$$E_{\text{branch}} := T^T P$$

and the currents S as

$$S := R^{-1}(E_{\text{branch}} + E).$$

Test your program with the values of Section 1.2.4 which should give the following results:

$$P = (7.5, 1.25, 11.25, 0)^T$$

$$E_{\text{branch}} = (-3.75, 10.0, -7.5, -6.25, 11.25, 1.25)^T$$

$$S = (0.625, -0.5, 1.25, -0.625, 1.125, 0.125).$$

1.8.2 Ill-Conditioning and Small Residual Norm

Let there be given

$$A = \begin{pmatrix} 1 & 0.5 \\ 2.001 & 0.999 \end{pmatrix} \quad \text{and} \quad b = \begin{pmatrix} 1.5 \\ 3 \end{pmatrix}.$$

A is non-singular. Therefore the linear system $Ax = b$ is uniquely solvable with solution vector

$$x = \begin{pmatrix} 1 \\ 1 \end{pmatrix}.$$

1. Show that A is ill-conditioned by inverting the matrix and calculating the condition number for some norm. How could one 'see' at first glance that A is ill-conditioned?
2. Let

   $$\tilde{x} = \begin{pmatrix} 0 \\ 3 \end{pmatrix}$$

   and calculate the residual vector $r = b - A\tilde{x}$. Which theorem tells us that an approximate solution $\tilde{x}$ may possess a large error $\tilde{x} - x$, even if its residual norm is small?

1.8.3 Determinant and Condition Number

Let $A \in \mathbb{R}^{20,20}$ with

$$A := \frac{1}{2} I = \frac{1}{2} \begin{pmatrix} 1 & & 0 \\ & \ddots & \\ 0 & & 1 \end{pmatrix}.$$

Calculate $\operatorname{cond}_\infty(A)$ and the determinant $\det(A)$. What is the message of this example?

1.8.4 Hilbert Matrices again

Show that Hilbert matrices

$$H_n := \begin{pmatrix} 1 & 1/2 & 1/3 & \cdots & 1/n \\ 1/2 & 1/3 & 1/4 & & \\ 1/3 & 1/4 & 1/5 & & \vdots \\ \vdots & & & \ddots & \\ 1/n & & \cdots & & 1/(2n-1) \end{pmatrix}$$

are symmetric positive-definite.
Hint: Show first that

$$x^T H_n y = \int_0^1 q_x(t)\, q_y(t)\, dt \quad \text{with} \quad q_z(t) := \sum_{i=1}^{n} z_i t^{i-1} \quad \forall z \in \mathbb{R}^n.$$

1.8.5 A Typical Positive-Definite Matrix

Show that the matrix

$$A = \begin{pmatrix} 2 & -1 & 0 & \cdots & 0 \\ -1 & 2 & -1 & 0 & \vdots \\ 0 & \ddots & \ddots & \ddots & 0 \\ \vdots & \ddots & \ddots & \ddots & -1 \\ 0 & \cdots & 0 & -1 & 2 \end{pmatrix}$$

is symmetric positive-definite.

1.8.6 Yet Another Algorithm for Sparse Systems

Let $A \in \mathbb{R}^{n,n}$ be symmetric positive-definite. Let F be defined as

$$F(x) := (Ax - b)^T A^{-1} (Ax - b) \quad \text{(see (1.29))}.$$

$F(x)$ can be minimized by the following algorithm:

Select some $x^{(0)} \in \mathbb{R}^n$ arbitrarily.

For i=1,2,...
  For k=1,2,..., n
    Calculate $x_k^{(i+1)}$ as minimum of
    $g(t) := F\left((x_1^{(i+1)}, \ldots, x_{k-1}^{(i+1)}, t, x_{k+1}^{(i+1)}, \ldots, x_n^{(i+1)})^T \right)$

Show that this method is identical with the Gauss–Seidel method.

2
Linear Optimization

Optimization is itself a large area within the field of applied mathematics. It deals with the minimization and maximization of functions with or without constraints and has many different applications. A large number of these applications belong to the area of *operations research* within the economic sciences. Not surprisingly, there are some software packages for optimization. But methods for the solution of simple optimization problems may also be found in most textbooks on numerical mathematics. Mathematical software libraries always contain a chapter on optimization. In the NAG Fortran Library there are more than 40 routines in Chapter E04 and some routines on integer programming in Chapter H. Chapter 8 of the IMSL Library contains more than 30 routines on optimization.

For operations research, large optimization problems, integer programming problems and special methods for these problems are of interest. To remain within the limits of our text we will consider only simple linear problems.

2.1 Introduction

The general problem of optimization is as follows.

Let the following functions be given:

$$\begin{aligned} f(x) : \mathbb{R}^n &\rightarrow \mathbb{R} \\ g(x) : \mathbb{R}^n &\rightarrow \mathbb{R}^s \\ h(x) : \mathbb{R}^n &\rightarrow \mathbb{R}^t. \end{aligned} \tag{2.1}$$

Search for a vector $\hat{x} \in \mathbb{R}^n$, for which

$$\begin{aligned} f(\hat{x}) &= \min_{x \in \mathbb{R}^n} f(x) \\ g(\hat{x}) &= 0 \\ h(\hat{x}) &\leq 0. \end{aligned} \tag{2.2}$$

f is called the *objective function*, $g(x) = 0$ are the *equality constraints*, $h(x) \leq 0$ the *inequality constraints*. The maximization problem does not need to be considered, because it will become a minimization problem simply by multiplying the objective function by '-1'. The same procedure leads from '$h(x) \geq 0$' to '$h(x) \leq 0$'.

If all functions f, g and h are *linear*, then we have a *linear optimization problem*. This is called an LP problem (linear programming problem), too. The LP problem can be written as follows.

Let there be given:

a vector of coefficients $b \in \mathbb{R}^n$,
two limit vectors $l, u \in \mathbb{R}^{m+n}$ and
the matrix of constraints $A \in \mathbb{R}^{m,n}$.

Search for a vector $\hat{x} \in \mathbb{R}^n$, for which

$$\begin{aligned} b^T\hat{x} &= \min_{x \in \mathbb{R}^n} b^T x \\ l &\leq \begin{bmatrix} \hat{x} \\ A\hat{x} \end{bmatrix} \leq u. \end{aligned} \tag{2.3}$$

Equality constraints are obtained by letting $l_i = u_i$. The values $l_i = -\infty$ and $u_i = +\infty$ are admissible. They cause the respective inequality constraint to become ineffective.

Production and transport problems are typical LP problems, see for example the exercise at the end of this chapter. We will take a biological problem as an example, namely a harvesting problem, which shows the influence of man on other species. Its model was first described in [27]; further examples for it can be found in [19] and in our application (see Section 2.3.1).

Example 2.1. We consider a natural population of plants or animals, which is influenced by 'harvesting'. The aim is to maximize the yields, but to preserve the species in a fixed size.

The population is counted regularly at certain points in time $t \in \mathbb{N}$, for example once per year. n classes of age $C_1, \ldots, C_n$ shall be distinguished. The respective values form the vector $x(t) \in \mathbb{R}^n$. For example $x_5(2)$ is the number of the species in the fifth class of age at the second point in time. The development of the population without human influence within one period (for example from t to $t+1$) can be described through a matrix model:

$$x(t+1) = Px(t) \tag{2.4}$$

with the *Leslie matrix*

$$P = \begin{pmatrix} F_1 & F_2 & F_3 & \cdots & \cdots & F_n \\ S_1 & 0 & 0 & \cdots & \cdots & 0 \\ 0 & S_2 & 0 & \ddots & \ddots & 0 \\ \vdots & \ddots & \ddots & \ddots & \ddots & \vdots \\ 0 & \cdots & 0 & S_{n-2} & 0 & 0 \\ 0 & 0 & \cdots & 0 & S_{n-1} & S_n \end{pmatrix}$$

Here F_i are the birth rates and S_i the natural survival rates within the respective age classes. It is assumed for the model that in each period of time a fixed number of species in each class are harvested. Let $(1 - H_k)$ be the harvesting rate within the class of age C_k. Then the development of the population changes to

$$x(t+1) = HPx(t), \tag{2.5}$$

where H is the diagonal matrix of non-natural survival rates.

$$H = \mathrm{diag}(H_1, H_2, \ldots, H_n).$$

The condition of preservation of species $x = x(t) = x(t+1)$ prescribes the strategy of harvesting. This means that after x has been computed, the matrix H is determined by

$$HPx = x \Longrightarrow H_k = \frac{x_k}{y_k}, \quad \text{with } y = Px. \tag{2.6}$$

The yields w_k of the species in different classes of age C_k may be different. Let w be the vector

$$w = (w_1, w_2, \ldots, w_n)^T$$

and I the identity matrix, then according to (2.6) the total yield T of one period is given as

$$T = w^T(I - H)Px = w^T(P - I)x. \tag{2.7}$$

So $-T$ is the objective function, because the yield will be maximized.

Constraints

1. Biologically trivial, mathematically necessary:

$$x \geq 0 \quad \text{(in terms of components).}$$

2. Growth necessary in all classes of age:

$$Px - x \geq 0 \quad \text{(in terms of components).}$$

3. Standardization of the total population:

$$\sum_{i=1}^{n} x_i = 1.$$

This means that the x_i are not absolute values, but represent the classes of age proportionally. This condition prevents an infinite solution.

We therefore get the following LP problem:

$$\text{Minimize} \qquad -w^T(P-I)x$$

with the constraints

$$\begin{aligned} x &\geq 0 \\ (P-I)x &\geq 0 \\ \sum_{i=1}^{n} x_i &= 1. \end{aligned}$$

This leads to the following parameters for the LP problem (2.3):

$$\begin{aligned} b &= (I-P^T)w \\ A &= \begin{pmatrix} P-I \\ 1\cdots 1 \end{pmatrix} \in \mathbb{R}^{m,n} \text{ with } m = n+1 \end{aligned} \tag{2.8}$$

$$\begin{aligned} l_1 = l_2 = \cdots = l_{2n} &= 0 \\ l_{2n+1} &= 1 \\ u_1 = u_2 = \cdots = u_{2n} &= \infty \\ u_{2n+1} &= 1. \end{aligned} \tag{2.9}$$

For $n = 1, 2, 3$ we can solve LP problems graphically. We will demonstrate that with a special population model. The development of the population shall depend on two classes of age only with the Leslie matrix

$$P = \begin{pmatrix} 0 & 5 \\ \frac{2}{3} & 0 \end{pmatrix}.$$

The yields are $w_1 = 1$ and $w_2 = 3$. Then the objective function is

$$\begin{aligned} b^T x &= -T = w^T(I-P)x = (1,3)\begin{pmatrix} 1 & -5 \\ -\frac{2}{3} & 1 \end{pmatrix}\begin{pmatrix} x_1 \\ x_2 \end{pmatrix} \\ &= -x_1 - 2x_2 \rightarrow \min. \end{aligned}$$

The set of admissible solutions is determined by the constraints

Constraint	Set
$x_1 \geq 0$	above the x_1-axis
$x_2 \geq 0$	right of the x_2-axis
$-x_1 + 5x_2 \geq 0$	above the straight line $x_2 = \frac{1}{5}x_1$
$\frac{2}{3}x_1 - x_2 \geq 0$	below the straight line $x_2 = \frac{2}{3}x_1$
$x_1 + x_2 = 1$	on the straight line $x_2 = 1 - x_1$

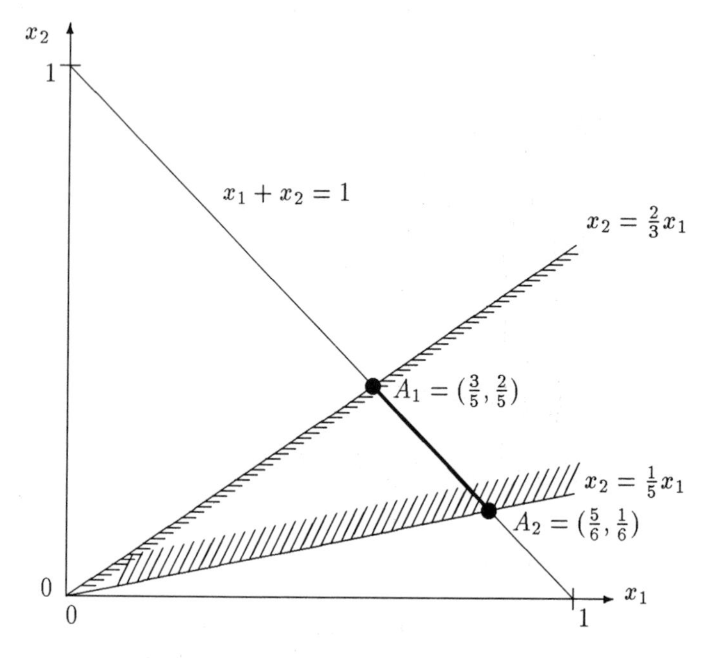

Figure 2.1. Constraints and admissible set

The conditions combine to yield the set of feasible solutions. In Figure 2.1 the set is represented by the boldface straight line. Now the theory tells us that only an edge or a vertex of the set of feasible points can be possible solutions. The vertices here are the points $A_1 = (\frac{3}{5}, \frac{2}{5})$ and $A_2 = (\frac{5}{6}, \frac{1}{6})$. For these points the objective function has the values

$$\begin{aligned} -T(A_1) &= -\tfrac{7}{5}, \\ -T(A_2) &= -\tfrac{7}{6}. \end{aligned}$$

So A_1 is the solution point. Hence population rates of $\frac{3}{5}$ and $\frac{2}{5}$ within the two classes of age yield the best harvest. Then the survival rates are

$H_1 = x_1/y_1 = 0.3$ and $H_2 = x_2/y_2 = 1$. For the harvesting strategy this means that 70% of the sexually not yet mature species are 'harvested'; the sexually mature are left in peace. In Section 2.3.1 we will look at a harvesting problem that considers the population of blue whales. ■

2.2 The Simplex Algorithm

The example has shown the main ideas for an algorithm for the numerical solution of LP problems (in terms of $\mathbb{R}^2$):

1. Determine the set of feasible points, i.e. all points which satisfy the constraints. If there are such points, then they form a convex polygon. That is a set limited by vertices and edges, which contains for every two points the straight line connecting them as well.
2. Determine the vertex (standard case) or the edge (exception) of this polygon for which the objective function takes on its minimal value.

Both tasks become complicated because of the necessity of dealing with several special cases. For this text we will give only the main steps without describing each special case separately. A program for the LP problem should naturally be able to solve the problem for all cases. A description of the Simplex algorithm may be found in many textbooks on (linear) programming, for example in [30].

First we consider the problem

$$\begin{array}{rcll} \tilde{b}^T\tilde{x} & \rightarrow & \min & \text{(objective function)} \\ \tilde{A}\tilde{x} & \leq & u & \text{(constraints)} \\ \tilde{x} & \geq & 0 & \text{(sign constraints).} \end{array} \tag{2.10}$$

All constraints can be transformed into equality constraints by introducing m new variables y_i, the so-called *slack variables*. They are determined together with $\tilde{x}$ such that

$$\tilde{A}\tilde{x} + y = u.$$

With the definitions

$$x := \begin{pmatrix} \tilde{x} \\ y \end{pmatrix}, \quad A := \begin{pmatrix} \tilde{A} & I \end{pmatrix}, \quad b := \begin{pmatrix} \tilde{b} \\ 0 \end{pmatrix}$$

this problem is transformed to

$$\begin{array}{rcl} b^T x & \rightarrow & \min \\ Ax & = & u \\ x & \geq & 0. \end{array} \tag{2.11}$$

For this form we will go back to the old dimensions:

$$x, b \in \mathbb{R}^n, \quad A \in \mathbb{R}^{m,n}, \quad u \in \mathbb{R}^m.$$

Problem (2.3) is easily transformed into this form. The non-square linear system $Ax = u$ may have one, none or infinitely many solutions. Its positive solutions form the set M of feasible solutions

$$M := \{x \in \mathbb{R}^n | Ax = u, x \geq 0\}. \tag{2.12}$$

We have to distinguish between three cases.

1. M is an empty set
   $\Longrightarrow$ The LP problem has got no solution at all.
2. M is an unbounded set. Then there are two possibilities:
   (a) $b^T x$ is not limited from below on M
   $\Longrightarrow$ There is no solution.
   (b) $b^T x$ is limited from below on M
   $\Longrightarrow$ There is a solution.
3. M is a non-empty, bounded set in $\mathbb{R}^n$
   $\Longrightarrow$ There is a solution.

These three cases are shown in the following figure. The hatching is at the feasible sides of the edges.

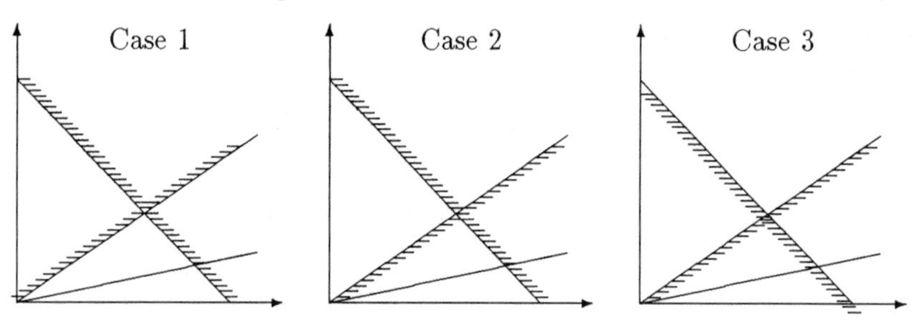

We consider only the standard case 3. Case 2(b) is easily transformed to case 3. For case 3 M is a bounded convex polyhedron. The minimum may lie at a vertex or on an edge. It lies on an edge if the objective function is parallel to this edge. For our introductory example this would have been the case for example for the objective function $x_1 + x_2$, see Figure 2.1.

If we leave aside the special cases (like unbounded set of feasible solutions, degenerate solutions or more rows than columns in A), we have to describe two algorithmic steps.

- Determine a starting vertex.
- Decrease the value of the objective function by exchanging two vertices.

If we avoid cycles, these steps lead to a solution after a finite number of operations, because a convex polyhedron can have only a finite number of vertices.

We now assume for $A \in \mathbb{R}^{m,n}$:

$$m \leq n \text{ and } \operatorname{rank}(A) = m.$$

Let $a^k \in \mathbb{R}^m$ be the columns of A. Let $x \in \mathbb{R}^n$ be a vector with components $x_k \geq 0$. Then we have the following.

1. x is a feasible point, if $\sum_{k=1}^{n} a^k x_k = u$.
2. x is a vertex of M, if for $\sum_{k=1}^{n} a^k x_k = u$ the vectors a^k belonging to *positive* components $x_k > 0$ are linearly independent. These vectors are called the *basis of the vertex* x.

From 2 we see that a vertex x can have at most m positive components x_k. We will assume that there are *exactly* m positive components x_k for each vertex. A vertex x with fewer than m positive components is called a *degenerate vertex*.

Determination of a Starting Vertex

Often a feasible vertex is known by the application, for example the origin. Then one may take that vertex as the starting vertex.

If there are only inequality constraints (as in problem (2.10)), and if the right-hand side of the constraints is non-negative ($u \geq 0$), then one introduces slack variables y and obtains as a starting vertex

$$\tilde{x} := 0, \quad y := u, \tag{2.13}$$

because the unit vectors are the basis of y and they are linearly independent.

Finally we come to problem (2.11) with equality constraints. Here we can construct a starting vertex through the solution of a new LP problem as follows.

1. All equations in $Ax = u$ with a negative right-hand side, $u_k < 0$, are multiplied by -1. So we can assume $u \geq 0$.
2. Solve the LP problem

$$\begin{aligned} \sum_{i=1}^{m} z_i &\rightarrow \min \\ Ax + z &= u \\ x &\geq 0 \\ z &\geq 0. \end{aligned} \tag{2.14}$$

Here we can take $x = 0$, $z = u$ as the starting vertex. Then we solve the problem. All z-components of the solution should be zero, because otherwise the original problem has no solution at all. The x-components of the solution then form a starting vertex for the original problem.

Exchange of Vertices

Let x^0 be a feasible vertex. For the sake of simplicity let the basis of this vertex consist of the vectors $a^k, k = 1, \ldots, m$. This means that $x_k^0 = 0$ for $k = m+1, \ldots, n$, and that

$$\sum_{k=1}^{m} x_k^0 a^k = u.$$

Because the $a^k, k = 1, \ldots, m$, are linearly independent, all columns of A can be represented by them:

$$a^i = \sum_{k=1}^{m} c_{ki} a^k, \quad i = 1, \ldots, n.$$

Now we try to replace the basis of the vertex x^0 by one for a new vertex x^1. To do that we look for a pair of indices k, j with $1 \leq k \leq m$ and $m+1 \leq j \leq n$ such that the new vertex x^1 is obtained by taking a^k off the basis and adding a^j to the basis. This must be done without increasing the value of the objective function. If there is no such pair of indices, then the vertex x^0 is a solution of the LP problem.

The details of this replacement are to be found in most textbooks on optimization or (linear) programming, see for example [17], [30], [39] or [92]. The Simplex method consists of repeating several such steps until a minimum has been found. Today there are computationally more efficient methods available, but they are extremely complex and their mathematical foundations are far from obvious.

2.3 Program and Example

Program LOPT, see B.2.1, is a frame program for NAG routine E04MBF. The components of the starting vector for E04MBF are preassigned by the program to be the average of the values l_i and u_i for the bounds with $\pm\infty$ replaced by zero. Input of a special starting vector is not possible.

The user has a choice between the more detailed output of NAG routine E04MBF and a shorter one, which merely gives the values of the variables and of the objective function at the minimum.

2.3.1 Application: Female Blue Whale Population

We will clarify the functionality of the program by a biological example which is a direct continuation of the harvesting example in Section 2.1 and is taken from [19]:

C_i	0–1	2–3	4–5	6–7	8–9	10–11	≥ 12
S_i	0.87	0.87	0.87	0.87	0.87	0.87	0.8
F_i	0	0	0.19	0.44	0.5	0.5	0.45
w_i	42	61	82	96	106	110	113.5

Table 2.1. Data for the female blue whale population

with $\left\{ \begin{array}{ll|ll} C_i: & \text{Classes of age} & S_i: & \text{Rates of survivors} \\ F_i: & \text{Rates of birth} & w_i: & \text{Yields} \end{array} \right.$

According to (2.8) the coefficients of the objective function and the matrix of constraints are given as:

$$b = (-11.07, -10.34, -9.5, -14.7, -10.7, -9.745, 3.8)^T,$$

$$A = \begin{pmatrix} -1 & 0 & 0.19 & 0.44 & 0.5 & 0.5 & 0.45 \\ 0.87 & -1 & 0 & 0 & 0 & 0 & 0 \\ 0 & 0.87 & -1 & 0 & 0 & 0 & 0 \\ 0 & 0 & 0.87 & -1 & 0 & 0 & 0 \\ 0 & 0 & 0 & 0.87 & -1 & 0 & 0 \\ 0 & 0 & 0 & 0 & 0.87 & -1 & 0 \\ 0 & 0 & 0 & 0 & 0 & 0.87 & -0.2 \\ 1 & 1 & 1 & 1 & 1 & 1 & 1 \end{pmatrix},$$

and we have $n = 7$ and $m = 8$. The values for l_i and u_i can be taken from (2.9), where $+\infty$ has to be replaced by 1.0D20.

Thus we come to the following input file, which has been created by the template of the PAN system for LOPT:

Input File for LOPT

```
Number of linear constraints:
8
Row and column dimension of matrix of constraints:
8 7
Matrix of constraints:
-1 0 0.19 0.44 0.5 0.5 0.45
0.87 -1 0 0 0 0 0
0 0.87 -1 0 0 0 0
0 0 0.87 -1 0 0 0
```

```
0 0 0 0.87 -1 0 0
0 0 0 0 0.87 -1 0
0 0 0 0 0 0.87 -0.2
1 1 1 1 1 1 1
Dimension of vector of lower bounds for the solution components:
7
Vector of lower bounds for the solution components:
0 0 0 0 0 0 0
Dimension of vector of upper bounds for the solution components:
7
Vector of upper bounds for the solution components:
1.0E20 1.0E20 1.0E20 1.0E20 1.0E20 1.0E20 1.0E20
Dimension of vector of lower bounds for the linear constraints:
8
Vector of lower bounds for the linear constraints:
0 0 0 0 0 0 0 1
Dimension of vector of upper bounds for the linear constraints:
8
Vector of upper bounds for the linear constraints:
1.0E20 1.0E20 1.0E20 1.0E20 1.0E20 1.0E20 1.0E20 1
Maximum number of iterations:
0
Compute a solution, if objective function is linear (no=0, yes=1):
1
Dimension of vector of coefficients of the objective function:
7
Vector of coefficients of the objective function:
-11.07 -10.34 -9.5 -14.7 -10.7 -9.745 3.8
Protocol of results (0=short, 1=long):
0
```

Results

Running the program LOPT with this input file gives the following results.

```
Output file for the PAN program LOPT:

Solution x:
  1          0.22598154
  2          0.19660394
  3          0.17104543
  4          0.14880952
  5          0.12946428
  6          0.11263393
  7          0.01546136

Value of the objective function at x:          -10.77106416
```

This means that we have an optimal 'harvesting' yield of 10.77 (which means 10.77 $\times N$ tons, if N is the size of the population) for the following rates for the age distribution:

Classes of age C_i	0–1	2–3	4–5	6–7	8–9	10–11	≥ 12
Distribution (%)	22.6	19.7	17.1	14.9	12.9	11.3	1.5
Harvesting rates (%)	0	0	0	0	0	0	86

Table 2.2. Optimal female blue whale population

The alternative more detailed output (slightly changed here) of NAG routine E04MBF gives (`LB = LOWER BOUND, UB=UPPER BOUND`):

```
WORKSPACE PROVIDED IS     IW(    80),  W(  3802).
TO SOLVE PROBLEM WE NEED  IW(    14),  W(   212).
EXIT LP PHASE.   INFORM =  0   ITER =   8

VARBL STATE    VALUE       LB    UB  LAGR MULT   RESIDUAL
V  1     FR  0.2259815    0.0  NONE      0.         0.2260
V  2     FR  0.1966039    0.0  NONE      0.         0.1966
V  3     FR  0.1710454    0.0  NONE      0.         0.1710
V  4     FR  0.1488095    0.0  NONE      0.         0.1488
V  5     FR  0.1294643    0.0  NONE      0.         0.1295
V  6     FR  0.1126339    0.0  NONE      0.         0.1126
V  7     FR  0.0154614    0.0  NONE      0.         0.0155

LNCON STATE    VALUE       LB    UB  LAGR MULT   RESIDUAL
L  1     LL -0.74593E-16  0.0  NONE    32.38    -0.7459E-16
L  2     LL -0.27755E-16  0.0  NONE    36.87    -0.2776E-16
L  3     LL  0.27755E-16  0.0  NONE    42.88     0.2776E-16
L  4     LL -0.55511E-16  0.0  NONE    43.68    -0.5551E-16
L  5     LL -0.27755E-16  0.0  NONE    29.31    -0.2776E-16
L  6     LL  0.27755E-16  0.0  NONE    15.16     0.2776E-16
L  7     FR  0.94899E-01  0.0  NONE     0.0      0.9490E-01
L  8     EQ    1.0000     1.0   1.0  -10.77      0.0

EXIT E04MBF - OPTIMAL LP SOLUTION FOUND.

LP OBJECTIVE FUNCTION =  -1.077106D+01

NO. OF ITERATIONS =     8
```

This output gives the additional information that the constraints 'growth in all age classes' take on the lower bound zero for the first six inequalities. Only the seventh inequality is inactive (FREE).

Naturally the simple linear programming model cannot consider all biological influences and side effects, for example those of killing mother animals. Taking into account all biological circumstances mathematically would complicate the model heavily and for example lead to a system of differential equations or a control theory problem (i.e. a coordinate optimization and differential equation problem).

2.4 Linear Optimization with IMSL

NAG routine E04MBF corresponds to IMSL routine DLPRS. DLPRS does not have the option of searching only for a feasible but not necessarily optimal point. The workspace for DLPRS is about 1.5 times larger than that for E04MBF. DLPRS gives two optimal results: the solution of the original problem as well as the solution of the dual problem.

2.5 Exercise: Optimal Mine Production

A mine is mining n different ores A_i, $i = 1, \ldots, n$. From ore A_i a maximum of t_i tons can be mined per day, and the total mining capacity is T tons per day. There are m clients, who are living d_k, $k = 1, \ldots, m$, miles away from the mine. There are contracts between the mine and the clients about fixed minimal output of m_{ik}, $i = 1, \ldots, n$, $k = 1, \ldots, m$, tons of ore A_i for client k per day. The mine has a transport capacity of W ton-miles per day. The yield per ton of ore A_i is g_i.

Write a program which computes the maximum yield under the conditions given. You may use the program LOPT, copy and change it.

Test example: Apply your program to the following situation. A mine has two clients and produces two ores. The maximal mining capacities are 30 and 15 tons per day for ore A_1 and A_2, respectively. Client 1 lives 10 miles away and has to be delivered with at least 10 tons of ore A_2. Client 2 lives 25 miles away and has to be delivered with at least 16 tons of ore A_1. The transport capacity of the mine is 750 ton-miles per day. The yield per ton is £25 for A_1 and £20 for A_2. Construct a graphical solution according to the introductory example by hand on grid paper and compare it with the solution your program gives.

3

Interpolation and Approximation

This chapter will show the different possibilities of approximating a given function or set of data points by a linear combination of (simple) functions taken from a certain function system.

For approximation this function is determined such that a certain norm of the difference between the function looked for and the given set of data points or function is rendered minimal.

For interpolation the approximating function has to match prescribed values at data points or the values of a given function at these points. Furthermore, interpolation is a tool for the development of numerical methods for integration and differentiation and for the solution of differential equations.

The quality of the approximation or interpolation depends on the method chosen as well as on the approximating function system. Therefore we consider the different methods in one chapter.

Within this introductory section we mention all the methods briefly. In the remaining sections the different methods are considered in more detail. All methods are transferable from the one-dimensional to the multidimensional case.

Interpolation or Approximation?

Given a table of data (x_i, f_i), $i = 0, 1, \ldots, n$, we will point out two extreme, but typical, situations.

1. There are a large number of data points given, i.e. n is very large. Then one should not normally choose interpolation, especially if the data are underlying data or measurement errors. So one is looking for a smooth curve through the data 'cloud' like that in Figure 3.1.
2. Only a few data points are given and it may be important that the approximating function matches the given points. Then one will choose an interpolation method resulting in a curve like that in Figure 3.2. Both plots have been prepared with MATLAB.

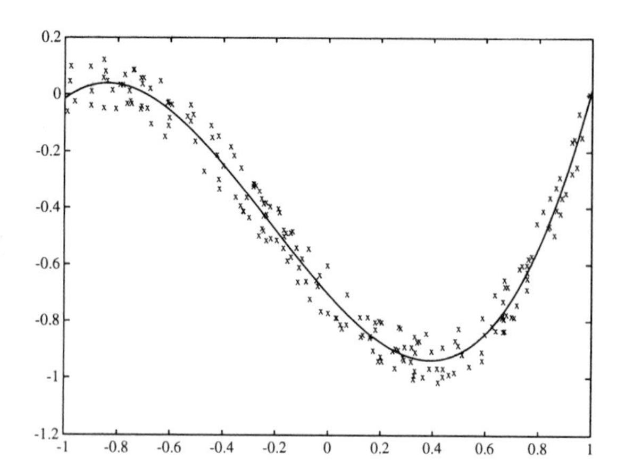

Figure 3.1. Polynomial least-squares approximation

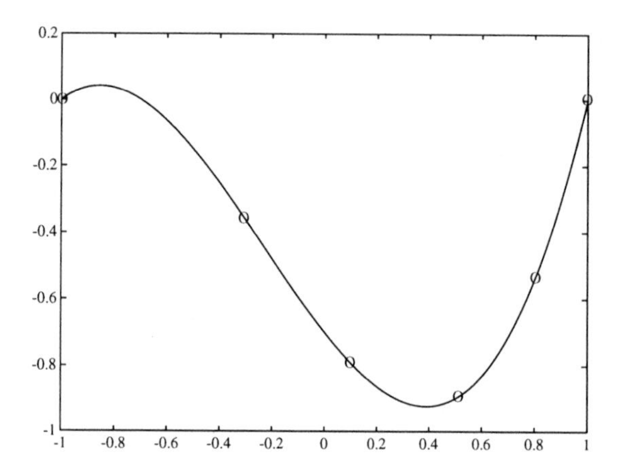

Figure 3.2. Cubic spline interpolation

The case of a function to be interpolated is simple, because we only have to choose the points x_i and to evaluate $f_i := f(x_i)$. Approximation problems are formulated in different mathematical terms, if a function is given, see below.

Polynomial Interpolation

This classical method of determining the coefficients a_i of a polynomial

$$P(x) := a_0 + a_1 x + \cdots + a_n x^n$$

such that

$$P(x_i) = y_i\,, \quad i = 0, 1, \ldots, n,$$

makes sense only for few data points and small polynomial degrees.

Spline Interpolation

With more than about 8 data points the problem may become numerically unstable, i.e. the interpolating polynomial will oscillate; see Example 3.4.

Nevertheless there may be reasons to interpolate. Then this can be done with piecewise polynomials of small degree. But the resulting curve will have break-points, i.e. it is not differentiable. It is possible to construct an interpolation function which also fulfills differentiability conditions. Such functions are called spline functions or simply splines.

Least-Squares Approximation

The approximating function is represented as a linear combination of functions of a function system $\{\varphi_i\}$:

$$g(x) := g(x; \alpha_0, \alpha_1, \ldots, \alpha_n) := \sum_{i=0}^{n} \alpha_i \varphi_i(x)\ ,$$

the coefficients α_i of which have to be determined.

For approximation it is more important than for interpolation to distinguish between

- the continuous case – i.e. the approximation of a given function

and

- the discrete case – i.e. the approximation of a set of data points.

For both cases a norm is defined for measuring the distance between the data and the approximating function. Then the coefficients α_i are determined such that this distance becomes minimal.

The most convenient choice for the norm is the Euclidian or 2-norm for the discrete case, to which the choice of the L_2-norm for square-integrable functions is the continuous analogue. For both cases the approximation method is called Gaussian or least-squares approximation.

If we choose splines as the function system, then we have to solve a linear system, which is similar to the one to be solved for spline interpolation: it is an easy-to-be-solved banded system.

For other functions systems – for example polynomials – it is important for numerical stability to choose systems of orthogonal functions (see Section 1.2.3). We will consider two possibilities for such systems: approximation with Chebyshev polynomials and with trigonometric function.

Choice of a Method for One-Dimensional Problems

If we take B-splines as basis functions, we get a method for interpolation *and* approximation with generally good approximation characteristics. The interpolating spline is a piecewise polynomial. Therefore approximation is necessary only for a large number of data points or for data with errors. But for example cubic splines are only twice differentiable.

If the existence of higher derivatives is important or if the approximating function must be of a special form, then we may choose approximation with orthogonal polynomials or with trigonometric functions.

Polynomial interpolation can only be recommended for small amounts of data or for cases when derivative values have to be interpolated too.

Multi-Dimensional Interpolation or Approximation

All one-dimensional methods can be extended to the multi-dimensional case. Sometimes it is preferable to use special multi-dimensional methods. We will only consider two-dimensional splines.

Interpolation of Curves

The drawing of curves is an important task for graphical software and CAD. At the end of this chapter we therefore consider the interpolation of multi-dimensional data with smooth curves. Once more we choose splines as interpolating functions.

3.1 Polynomial Interpolation

3.1.1 Introduction

Theorem 3.1. *Let there be given $n+1$ distinct real interpolation nodes $x_0, x_1, \ldots, x_n$, with corresponding values $y_0, y_1, \ldots, y_n$.*

Then there is exactly one real polynomial P of maximum degree n

$$P(x) := a_0 + a_1 x + \cdots + a_n x^n \tag{3.1}$$

with

$$P(x_i) = y_i, \quad i = 0, 1, \ldots, n. \tag{3.2}$$

The given values y_i can be values of a given function f, which shall be interpolated at the interpolation nodes x_i:

$$f(x_i) = y_i, \quad i = 0, 1, \ldots, n, \tag{3.3}$$

or data points (x_i, y_i), $i = 0, 1, \ldots, n$, for example from discrete measurements.

In the first case it is possible to represent the interpolation error and estimate it. This estimate can be carried over to the data point case only with some uncertainty and if some assumptions about the data can be made.

The distribution of the interpolation nodes has some influence on the error. If a given function is to be interpolated, and the function can be evaluated everywhere in the interval $[a, b]$, then an optimal distribution of the interpolation nodes can be chosen.

Theorem 3.2. *Let all interpolation nodes be within the interval* $[a, b]$. *Let* f *in* $[a, b]$ *be* $(n+1)$ *times continuously differentiable:* $f \in C^{n+1}[a, b]$ *with* $y_i := f(x_i)$, $i = 0, 1, \ldots, n$. *Let* P *be the polynomial which interpolates the set of data* (x_i, y_i) *and*

$$\omega(x) := (x - x_0)(x - x_1) \cdots (x - x_n) . \tag{3.4}$$

Then for each $\tilde{x} \in [a, b]$ *there is a* $\xi \in [a, b]$ *with*

$$f(\tilde{x}) - P(\tilde{x}) = \frac{\omega(\tilde{x}) f^{n+1}(\xi)}{(n+1)!} , \tag{3.5}$$

and it follows that

$$|f(\tilde{x}) - P(\tilde{x})| \leq \frac{|\omega(\tilde{x})|}{(n+1)!} \max_{\xi \in [a,b]} |f^{n+1}(\xi)| . \tag{3.6}$$

Theorem 3.2 shows that the error distribution in $[a, b]$ is mainly influenced by the distribution of $\omega(x)$. For equidistant interpolation nodes $\omega(x)$ is much bigger near the end-points than in the middle of the interval.

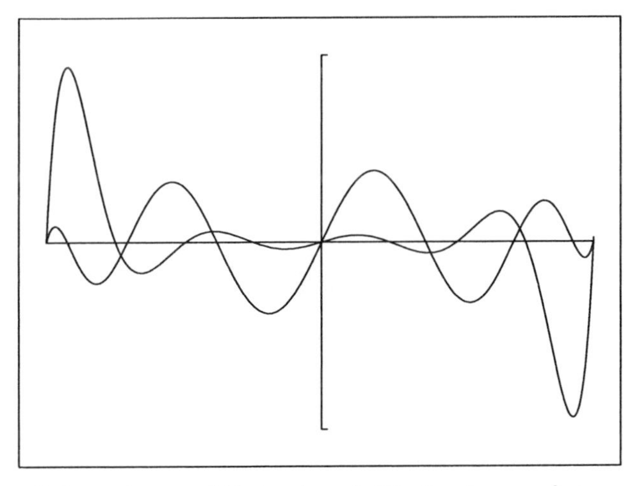

Figure 3.3. ω for equidistant and Chebyshev points, $n = 8$

The error curve becomes more regular if instead of equidistant nodes we take the so-called *Chebyshev points* as interpolation nodes:

$$x_i = \frac{a+b}{2} + \frac{b-a}{2}\cos\left(\frac{i}{n}\pi\right), \quad i = 0, 1, \ldots, n. \tag{3.7}$$

These are the positions of the extrema of the Chebyshev polynomials, see Section 3.4.3. The Chebyshev points are more densely distributed near the end-points of the interval $[a, b]$ than in the middle, see Figure 3.3.

3.1.2 Numerical Methods

We will consider four methods:

- *Lagrange interpolation*, which easily makes obvious that a polynomial can always be constructed according to Theorem 3.1. Lagrange interpolation can not to be recommended for practical purposes.
- *Newton interpolation*, which is a numerically stable and economical method of constructing an interpolating polynomial. Newton's method is implemented in NAG routine E01AEF, which itself is used in our program INTERDIM1.
- *Aitken–Neville interpolation*, which is suitable for the calculation of a *single* interpolated value. Aitken's method is implemented in NAG routine E01AAF.
- *Hermite interpolation*, which is the analogue of Newton interpolation for the case when derivative values have also to be interpolated. Hermite's method is implemented in NAG routine E01AEF, which itself is used in our program INTERDIM1.

Lagrange Interpolation

With the $n+1$ Lagrangian polynomials

$$\begin{aligned} L_i(x) &:= \prod_{\substack{j=0 \\ j\neq i}}^{n} \frac{(x-x_j)}{(x_i-x_j)} \qquad (i = 0, 1, \ldots, n) \\ &= \frac{(x-x_0)\cdots(x-x_{i-1})(x-x_{i+1})\cdots(x-x_n)}{(x_i-x_0)\cdots(x_i-x_{i-1})(x_i-x_{i+1})\cdots(x_i-x_n)} \end{aligned} \tag{3.8}$$

the interpolating polynomial P is easily constructed as

$$P(x) := \sum_{i=0}^{n} y_i L_i(x). \tag{3.9}$$

It fulfills (3.2), because for the Lagrangian polynomials

$$L_i(x_k) = \delta_{ik} = \begin{cases} 1, & \text{if } i = k, \\ 0, & \text{if } i \neq k. \end{cases} \tag{3.10}$$

Lagrange interpolation is not suited to the practical calculation of polynomial values, because it needs too much effort and is not as stable as Newton's method.

Newton Interpolation

For this method the interpolating polynomial is represented differently:

$$\begin{aligned} P(x) := c_0 &+ c_1(x - x_0) + c_2(x - x_0)(x - x_1) + \cdots \\ &+ c_n(x - x_0)(x - x_1) \cdots (x - x_{n-1}). \end{aligned} \tag{3.11}$$

The coefficients c_i of this representation are easily calculated, if the data points are put in one after the other:

$$\begin{aligned} P(x_0) = y_0 &\implies c_0 = y_0 \\ P(x_1) = y_1 &\implies c_1 = \frac{y_1 - y_0}{x_1 - x_0} \\ &\ldots. \end{aligned}$$

Following this process recursively always gives fractions of differences of the fractions calculated in the previous step. These expressions are called *divided differences*:

(a) the trivial divided difference

$$[y_k] := y_k, \qquad k = 0, 1, \ldots, n,$$

(b) the first divided difference

$$\begin{aligned} [y_k, y_{k+1}] &:= \frac{y_{k+1} - y_k}{x_{k+1} - x_k} \\ &= \frac{[y_{k+1}] - [y_k]}{x_{k+1} - x_k}, \qquad k = 0, 1, \ldots, n-1, \end{aligned}$$

(c) the general divided difference

$$[y_{i_0}, y_{i_1}, \ldots, y_{i_j}] := \frac{[y_{i_1}, \ldots, y_{i_j}] - [y_{i_0}, \ldots, y_{i_{j-1}}]}{x_{i_j} - x_{i_0}}. \tag{3.12}$$

With this definition one arrives at a recursive scheme for the calculation of the coefficients, which we will give for $n = 3$ only:

$$
\begin{array}{c|lllll}
x_0 & y_0 = c_0 & & & \\
 & & [y_0, y_1] = c_1 & & \\
x_1 & y_1 & & [y_0, y_1, y_2] = c_2 & \\
 & & [y_1, y_2] & & [y_0, y_1, y_2, y_3] = c_3 \\
x_2 & y_2 & & [y_1, y_2, y_3] & \\
 & & [y_2, y_3] & & \\
x_3 & y_3 & & &
\end{array}
$$

This recursive calculation of the coefficients c_i is the first step of the algorithm. The computation steps forward from left to right and within each column from bottom to top. For this sequence one may overwrite the divided differences of the last column not subsequently used.

For the calculation of polynomial values the representation (3.11) is very suitable as well, because common factors can be placed outside the brackets:

$$
\begin{aligned}
P(x) &= c_0 + c_1(x - x_0) + \cdots \\
&\quad + c_n(x - x_0)(x - x_1)\cdots(x - x_{n-1}) \\
&= c_0 + (x - x_0) \\
&\quad \{c_1 + (x - x_1)\,[c_2 + (x - x_2)\,(\cdots c_{n-1} + (x - x_{n-1})c_n)]\}\,.
\end{aligned}
\tag{3.13}
$$

The calculation of polynomial values in this way by starting within the innermost brackets is traditionally called the *Horner scheme*. The Horner scheme is the second step of the algorithm:

1. Newton interpolation: Calculation of the coefficients

for $k = 0, 1, \ldots, n$:

$c_k := y_k$.

for $k = 1, 2, \ldots, n$:

for $i = n, n-1, \ldots, k$:

$$c_i := \frac{c_i - c_{i-1}}{x_i - x_{i-k}}$$

2. Horner scheme: Calculation of polynomial values

$p := c_n$.

for $k = n-1, n-2, \ldots, 0$:

$p := c_k + (x - x_k)p$.

Then we have $p = P(x)$.

Example 3.3. We want to calculate $p := \ln(1.57)$[†]. We are given a table for the natural logarithm with following values:

[†]All calculations have been done in double precision, but we only show up to six significant digits in the results given.

x	1.4	1.5	1.6	1.7
$\ln(x)$	0.336472	0.405465	0.470004	0.530628

From this table we obtain the following Newton scheme:

$$\begin{array}{c|cccc}
1.4 & \underline{\underline{0.336472}} & & & \\
 & & \underline{\underline{0.689929}} & & \\
1.5 & \boxed{0.405465} & & \underline{\underline{-0.222718}} & \\
 & & \boxed{0.645385} & & \underline{0.0900752} \\
1.6 & 0.470004 & & -0.195695 & \\
 & & 0.606246 & & \\
1.7 & 0.530628 & & &
\end{array}$$

From this scheme we can pick up the coefficients of several interpolating polynomials with interpolation nodes around the point 1.57. First there is the linear polynomial with the framed coefficients

$$P_1(x) = 0.405465 + 0.645385(x - 1.5).$$

Taking into consideration the first three data points we get the quadratic polynomial with the doubly underlined coefficients

$$\begin{aligned}
P_2(x) &= 0.336472 + 0.689929(x - 1.4) - 0.222718(x - 1.4)(x - 1.5) \\
&= 0.336472 + (x - 1.4)[0.689929 - (x - 1.5)0.222718].
\end{aligned}$$

Finally we obtain the coefficients of the interpolating cubic polynomial, if we take all the underlined values:

$$\begin{aligned}
P_3(x) = &\quad 0.336472 + 0.689929(x - 1.4) \\
- &\quad 0.222718(x - 1.4)(x - 1.5) \\
+ &\quad 0.0900752(x - 1.4)(x - 1.5)(x - 1.6) \\
= &\quad 0.336472 + (x - 1.4)\{0.689929 \\
+ &\quad (x - 1.5)[-0.222718 + 0.0900752(x - 1.6)]\} \\
= &\quad P_2(x) + 0.0900752(x - 1.4)(x - 1.5)(x - 1.6).
\end{aligned}$$

This example shows one important advantage of the representation (3.11): on adding one data point we only have additionally to calculate the bottom oblique line of the Newton scheme values.

We now will compare the errors of the three polynomials, and give the respective estimates from Theorem 3.2. With $\ln(1.57) = 0.451076\ldots$ we get:

i	$P_i(1.57)$	$\|P_i(1.57) - \ln(1.57)\|$	Using (3.6)
1	0.450642	4.3×10^{-4}	5.4×10^{-4}
2	0.451110	3.4×10^{-5}	4.3×10^{-5}
3	0.451078	2.0×10^{-6}	3.0×10^{-6}

■

Hermite interpolation

The problem is now slightly changed. The polynomial shall not only have specified values, but also derivative values of different orders. We will consider this problem briefly and only in the following special form.

Let there be $n+1$ distinct interpolation nodes $x_0, x_1, \ldots, x_n$, corresponding function values $y_0, y_1, \ldots, y_n$ and corresponding derivative values $y'_0, y'_1, \ldots, y'_n$.

We look for a polynomial P_{2n+1} of maximum degree $2n+1$ with

$$\begin{aligned} P_{2n+1}(x_i) &= y_i, \quad i = 0, 1, \ldots, n, \\ P'_{2n+1}(x_i) &= y'_i, \quad i = 0, 1, \ldots, n. \end{aligned} \tag{3.14}$$

For the solution of this problem the Newton scheme is expanded as follows. Each interpolation node is written down twice and the respective first divided difference is replaced by the derivative value, i.e. $[y_i, y_i] := y'_i$. For $n = 1$ we come to the following scheme:

$$\begin{array}{c|llll} x_0 & y_0 = c_0 & & & \\ & & [y_0, y_0] := y'_0 = c_1 & & \\ x_0 & y_0 & & [y_0, y_0, y_1] = c_2 & \\ & & [y_0, y_1] & & [y_0, y_0, y_1, y_1] = c_3. \\ x_1 & y_1 & & [y_0, y_1, y_1] & \\ & & [y_1, y_1] := y'_1 & & \\ x_1 & y_1 & & & \end{array}$$

The interpolating polynomial now contains quadratic terms:

$$P_3(x) = c_0 + c_1(x - x_0) + c_2(x - x_0)^2 + c_3(x - x_0)^2(x - x_1)$$

or in general

$$\begin{aligned} P_{2n+1}(x) = \quad & c_0 + c_1(x - x_0) + c_2(x - x_0)^2 + c_3(x - x_0)^2(x - x_1) \\ + \quad & c_4(x - x_0)^2(x - x_1)^2 + \cdots \\ + \quad & c_{2n+1}(x - x_0)^2(x - x_1)^2 \cdots (x - x_{n-1})^2(x - x_n). \end{aligned}$$

Aitken–Neville Interpolation

If we have a given set of data points and are looking only for one interpolated value, the explicit calculation of the polynomial coefficients is not necessary. The two algorithmic steps of Newton interpolation are put together instead such that the calculation of the single value takes as little effort and storage as necessary. For this process Aitken proposed a scheme, which was subsequently refined by Neville. We will give this scheme algorithmically only:

Neville interpolation

For $k = 0, 1, \ldots, n$:

$p_k := y_k$.

For $k = 1, 2, \ldots, n$:

For $i = n, n-1, \ldots, k$:

$$p_i := p_i + (x - x_i)\frac{p_i - p_{i-1}}{x_i - x_{i-k}}$$

Then we have $P(x) = p_n$.

3.1.3 Program and Example

Polynomial and spline interpolation are to be found together in the program INTERDIM1 (see Section 3.3). There you will also find an example comparing the two interpolating methods.

Here we will give an example for the instability of polynomial interpolation with too many interpolation points.

Example 3.4. If we take nine interpolation points from a polynomial of degree three, which means we have a table of data (x_i, y_i) with $y_i := p_3(x_i)$, then the solution of the interpolation problem with $n = 9$ is identical to the polynomial $p_3(x)$. If we change the table of data slightly to get $(\tilde{x}_i, \tilde{y}_i)$ we will get an oscillating polynomial of degree eight as the solution with heavily varying values. To demonstrate this we prepared two input files for INTERDIM1 with the following point and value vectors:

x_i	$y_i := p_3(x_i)$	$\tilde{x}_i$	$\tilde{y}_i$
0.0	−0.2669	0.00333	-0.37925637
0.2	−0.0679	0.165	-0.191389
0.3	−0.0164	0.302	-0.0317
0.4	0.0111	0.452	0.069275
0.5	0.0206	0.637	0.050489
0.6	0.0181	0.757	-0.052838
0.7	0.0097	0.868	0.10215
0.8	0.0012	0.917	0.3018
1.0	0.0082	0.9286	0.43092

The results are shown in Figure 3.4 for the polynomial of degree 3 values and in Figure 3.5 for the arbitrary values. ■

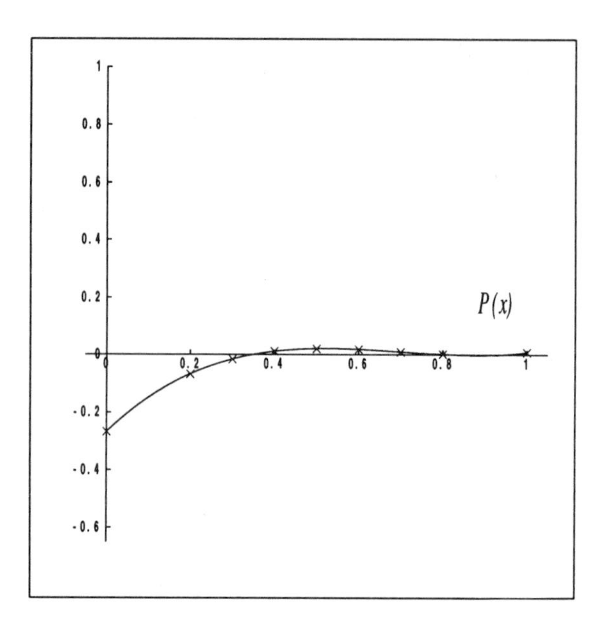

Figure 3.4. Polynomial interpolation; similar data ...

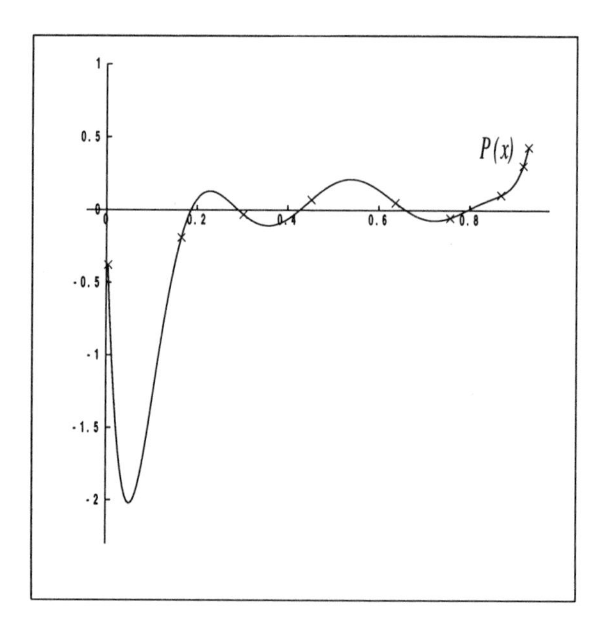

Figure 3.5. ...different results

3.2 Splines

Polynomial interpolation is a classical method for small problems. If the polynomial degree is greater than about 7, the interpolating polynomials underly unstable oscillations.

We may change to rational interpolation, which shows good approximation properties and is especially suitable for extrapolation. But rational interpolation is a nonlinear mathematical problem and has the detrimental possibility of poles in the vicinity of which we get overflow instead of good interpolated values.

Another possibility of improving the interpolation quality is to link together several simple interpolating polynomials from interval to interval. But a function calculated in this way has break points at the inner interval end-points; it is not differentiable (see the solid curve in Figure 3.6).

A better possibility is to impose additional differentiability conditions to the low-degree polynomials linked together. This leads to the construction of the so-called *spline functions*, known as splines (see the broken curve in Figure 3.6).

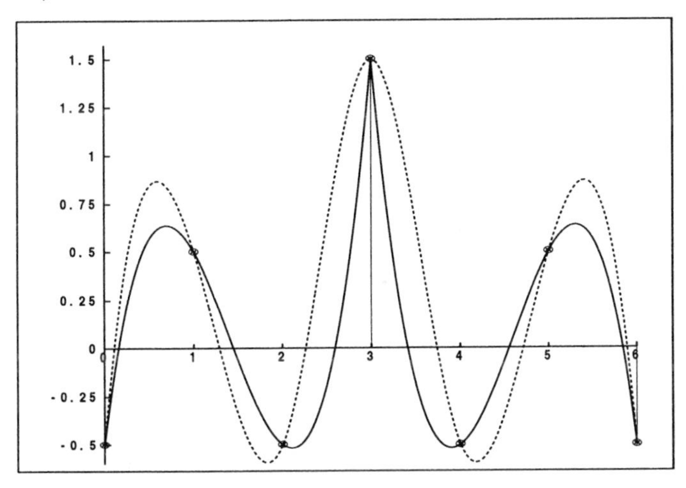

Figure 3.6. Differently linked functions: polynomials versus splines

Spline functions can be characterized by three properties:

- They are piecewise polynomials of maximum degree k.
- They are p times continuously differentiable.
- Within the class of functions which fulfill these two conditions and which interpolate a given set of data points, the interpolating spline is the smoothest one.

Most often used are the cubic splines, for which $k = 3$ and $p = 2$. The concept of so-called B-splines is introduced, first in the simple form of linear B-splines, then in the usual form of cubic B-splines.

3.2.1 Cubic Splines

Since the 1920s small wooden rods have been used for the construction of aircrafts to get smooth lines and surfaces for the body and the wings. Such rods are called *splines*. These rods are fixed with nails at certain points (the interpolation nodes). The line of the rod then naturally obeys an energy minimum principle. This means that it follows the curve of minimal curvature.

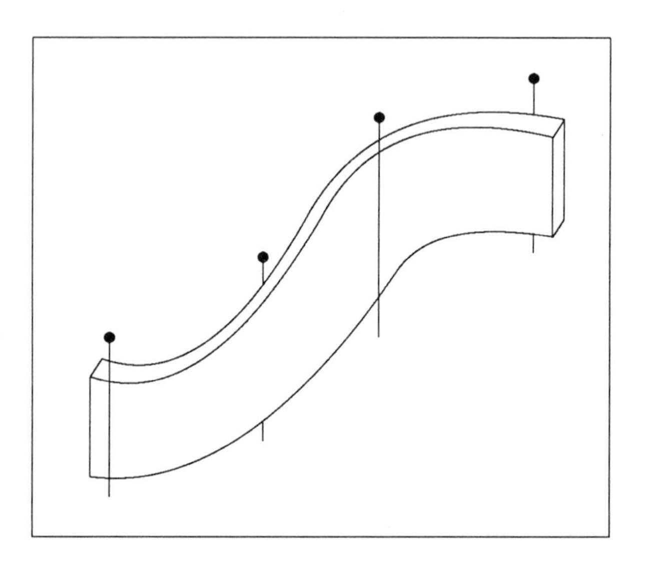

Figure 3.7. A wooden spline

We will try to formulate this physical problem in mathematical terms. Let

$$x_0 < x_1 < \cdots < x_n \tag{3.15}$$

be $n + 1$ distinct points and $y_0, y_1, \ldots, y_n$ corresponding values. Look for a function s which fulfills the following conditions.

1. s interpolates the given set of data points:

$$s(x_i) = y_i \quad i = 0, 1, \ldots, n. \tag{3.16}$$

2. Within each subinterval s is a polynomial of maximum degree 3:

$$s(x) \quad = \quad s_i(x) \quad \text{for } x \in [x_i, x_{i+1}] \quad (i = 0, 1, \ldots, n-1)$$

with

$$s_i(x) = a_i(x-x_i)^3 + b_i(x-x_i)^2 + c_i(x-x_i) + d_i. \quad (3.17)$$

3. s is a twice continuously differentiable function:

$$s \in C^2[x_0, x_n]. \quad (3.18)$$

These conditions do not define the cubic interpolating spline function completely, because there are still two degrees of freedom. The spline is uniquely defined with two additional boundary conditions. Most frequently used are the following three possibilities:

Natural spline interpolation:	$s''(x_0) = 0,$	$s''(x_n) = 0.$
Complete spline interpolation:	$s'(x_0) = y'_0,$	$s'(x_n) = y'_n.$
Periodic spline interpolation:	$s'(x_0) = s'(x_n),$	$s''(x_0) = s''(x_n).$

For the solution of the spline interpolation problem one has to solve a tridiagonal system of linear equation (with two additional 'corner' elements for the periodic case). Details of the solution of these systems are to be found in most textbooks on numerical analysis, for example in [92].

The prescription of additional boundary conditions can be avoided by using B-splines, which we will consider below.

We will point out again the remarkable minimal property of a spline function: Within the class of functions, which are twice continuously differentiable, interpolate the given set of data points, and fulfill the additional boundary conditions, the interpolating spline is the one with the smallest mean second derivative or curvature:

$$\int_{x_0}^{x_n} (s''(x))^2\, dx = \min_{f \in C^2[x_0,x_n]} \int_{x_0}^{x_n} (f''(x))^2\, dx. \quad (3.19)$$

3.2.2 Linear B-splines

For computational applications it is easier to built up a spline function of simple basis functions, the *B-splines*. On the other hand the theoretical development of general B-splines is quite complicated. Therefore we will first consider a very special and simple example, the B-splines of degree 1, more vividly called hat functions. These are continuous piecewise linear functions.

Example 3.5. For the four points

$$x_0 < x_1 < x_2 < x_3 \quad (3.20)$$

we are looking for four functions, which

- are linear polynomials in each of the intervals $[x_0, x_1]$, $[x_1, x_2]$ and $[x_2, x_3]$,

- take on the value 1 at exactly one of these points, and
- are zero at the other points.

Obviously we get the four functions of Figure 3.8.

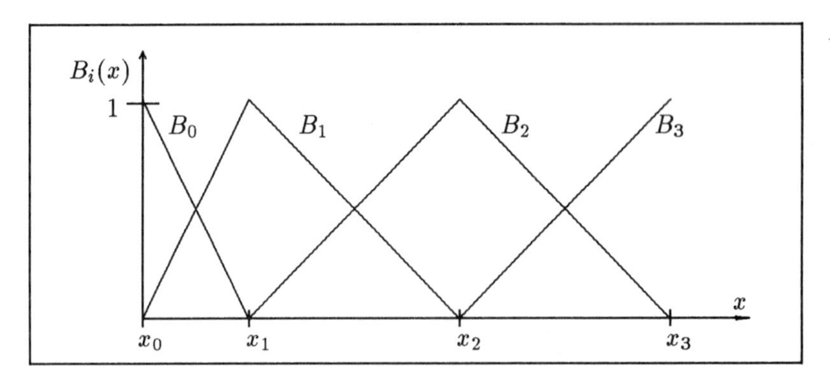

Figure 3.8. Linear B-splines for four nodes

We will develop the expression for these four simple functions in the same (complicated) way in which general B-splines can be constructed. Firstly we will mention that B-splines are defined on a set of knots which is not necessarily identical with the set of interpolation nodes. If $\{x_0, x_1, x_2, x_3\}$ is the set of interpolation points we will define the B-spline knots here as

$$t_0 := t_1 := x_0, \quad t_2 := x_1, \quad t_3 := x_2, \quad t_4 := t_5 := x_3 \ . \tag{3.21}$$

Now we define the truncated power function (of degree 1 here)

$$F_x(t) := \begin{cases} (t-x) & \text{if} \quad x \le t \\ 0 & \text{if} \quad x > t \ . \end{cases} \tag{3.22}$$

$F_x(t)$ is a function of t for each (fixed) x. Let its values for the knots t_i be called $f_i := F_x(t_i)$. Now the B-splines are defined with divided differences (see Section 3.1.2):

$$B_i(x) := (t_{i+2} - t_i)[f_i, f_{i+1}, f_{i+2}] \ . \tag{3.23}$$

For B_0 we get $B_0(x) := (t_2 - t_0)[f_0, f_1, f_2]$.

Now we have to observe two peculiarities. Firstly, the divided differences are functions of x, because they are formed with t-values. This means that the f_i depend on x, and so do the B_i. Therefore we need different Newton schemes for the different intervals $[x_i, x_{i+1}]$. Secondly, we have to form divided differences for equal points as with Hermite interpolation; but here they are replaced by zeros. For B_0 and $x_0 \le x \le x_1$ we get the following Newton scheme:

t_i	$F_x(t_i)$		
x_0	0		
		0	
			$\dfrac{\frac{x_1-x}{x_1-x_0}}{x_1-x_0}$
x_0	0		
		$\dfrac{x_1-x}{x_1-x_0}$	
x_1	x_1-x		

Outside $[x_0, x_1]$ we have $B_0(x) = 0$ according to (3.22), because here $F_x(t) = 0$ for $x > t_2$. This leads to

$$B_0(x) = \begin{cases} \dfrac{x_1 - x}{x_1 - x_0} & \text{if} \quad x \in [x_0, x_1] \\ 0 & \text{otherwise.} \end{cases}$$

Even more long-winded is the determination of $B_1(x) := (t_3 - t_1)[f_1, f_2, f_3]$, because now we have to distinguish between two Newton schemes. On the other hand it is easy to see that

$$B_1(x) = \begin{cases} 1 - \dfrac{x_1 - x}{x_1 - x_0} & \text{if} \quad x \in [x_0, x_1] \\ \dfrac{x_2 - x}{x_2 - x_1} & \text{if} \quad x \in [x_1, x_2] \\ 0 & \text{otherwise.} \end{cases}$$

The somewhat awkward definition is useful for general purposes, as we will see later. ■

After having calculated the B-splines for given sets of interpolation nodes and knots, the interpolation problem can easily be solved.

Determine coefficients $\alpha_i, \; i = 0, \ldots, n$, such that

$$\sum_{i=0}^{n} \alpha_i B_i(x_k) = y_k, \quad k = 0, 1 \ldots, n \, . \tag{3.24}$$

With

$$B_i(x_k) = \begin{cases} 1 & \text{if} \quad i = k \\ 0 & \text{if} \quad i \neq k \end{cases}$$

we have

$$\alpha_i := y_i \, .$$

3.2.3 Cubic B-splines

B-splines of higher order than one are too complicated to be calculated by hand, but they are suitable for automatic computation, because they can be built up recursively from B-splines of lower degrees.

We will now consider twice continuously differentiable cubic B-splines. Let the interpolation nodes be $x_0, x_1, \ldots, x_n$. Then the knots $\{t_i\}$ are defined as

$$\begin{array}{l} (t_0,\ t_1,\ t_2,\ t_3,\ t_4,\ t_5,\ \ldots,\ t_n,\quad t_{n+1},\ t_{n+2},\ t_{n+3},\ t_{n+4}) := \\ (x_0,\ x_0,\ x_0,\ x_0,\ x_2,\ x_3,\ \ldots,\ x_{n-2},\ x_n,\quad x_n,\quad x_n,\quad x_n). \end{array}$$

For these knots $t_0, t_1, \ldots, t_{n+4}$ we define the B-spline of degree 0 as

$$B_{i,0}(x) = \begin{cases} 1 & \text{if} \quad t_i < x \le t_{i+1} \\ 0 & \text{otherwise} \end{cases} \tag{3.25}$$

$$B_{i,0}(x) \equiv 0 \quad \text{if} \quad t_i = t_{i+1}. \tag{3.26}$$

Then the B-splines of higher degrees can be calculated recursively as

$$B_{i,k}(x) =^* \frac{x - t_i}{t_{i+k} - t_i} B_{i,k-1}(x) + \frac{t_{i+k+1} - x}{t_{i+k+1} - t_{i+1}} B_{i+1,k-1}(x). \tag{3.27}$$

($=^*$ means that terms with a zero denominator are set to zero.)

For the set of knots given and the B-splines of degree zero to three we get

$$\begin{array}{cccccccccccccc} B_{0,0} & & B_{1,0} & & B_{2,0} & \cdots & & B_{n,0} & & B_{n+1,0} & & B_{n+2,0} & & B_{n+3,0} \\ \downarrow & \swarrow & \downarrow & \swarrow & & \cdots & & \downarrow & \swarrow & \downarrow & \swarrow & \downarrow & \swarrow & \\ B_{0,1} & & B_{1,1} & & \cdots & \cdots & & B_{n,1} & & B_{n+1,1} & & B_{n+2,1} & & \\ \downarrow & \swarrow & \downarrow & & \cdots & \cdots & & \downarrow & \swarrow & \downarrow & \swarrow & & & \\ B_{0,2} & & B_{1,2} & & \cdots & \cdots & & B_{n,2} & & B_{n+1,2} & & & & \\ \downarrow & \swarrow & \downarrow & & \cdots & \cdots & & \downarrow & \swarrow & & & & & \\ B_{0,3} & & B_{1,3} & & \cdots & \cdots & & B_{n,3} & & & & & & \end{array} \tag{3.28}$$

Now let x be a point in $[t_i, t_{i+1}]$ with $t_i < t_{i+1}$. For the calculation of a series of B-splines at the point x we then need all B-spline values $B_{i,k}(x) \neq 0$ for $k = 0, 1, 2, 3$. These are the B-splines from the following triangle scheme:

$$\begin{array}{cccc} & & & 0 \\ & & 0 & \\ & 0 & & B_{i-3,3} \\ 0 & & B_{i-2,2} & \\ & B_{i-1,1} & & B_{i-2,3} \\ B_{i,0} & & B_{i-1,2} & \\ & B_{i,1} & & B_{i-1,3} \\ 0 & & B_{i,2} & \\ & 0 & & B_{i,3} \\ & & 0 & \\ & & & 0 \end{array} \tag{3.29}$$

Such a calculation of the B-spline values can be carried out recursively, too.

Because the construction of the B-splines is not very obvious, we will characterize these functions with some of their properties.

1. They have the smallest possible number of support intervals:

$$B_0(x) > 0 \quad \text{if} \quad x \in [x_0, x_2)$$

$$
\begin{aligned}
B_1(x) > 0 \quad &\text{if} \quad x \in (x_0, x_3) \\
B_2(x) > 0 \quad &\text{if} \quad x \in (x_0, x_4) \\
B_3(x) > 0 \quad &\text{if} \quad x \in (x_0, x_5) \\
B_i(x) > 0 \quad &\text{if} \quad x \in (x_{i-2}, x_{i+2}) \quad (i = 4, \ldots, n-4) \\
B_{n-3}(x) > 0 \quad &\text{if} \quad x \in (x_{n-5}, x_n) \\
B_{n-2}(x) > 0 \quad &\text{if} \quad x \in (x_{n-4}, x_n) \\
B_{n-1}(x) > 0 \quad &\text{if} \quad x \in (x_{n-3}, x_n) \\
B_n(x) > 0 \quad &\text{if} \quad x \in (x_{n-2}, x_n]
\end{aligned}
$$

Outside these intervals the functions are zero.

2. $$\sum_{i=0}^{n} B_i(x) = 1 \quad \forall x \in [x_0, x_n]$$

3. The matrix B with the coefficients

$$b_{ik} := B_k(x_i) \tag{3.30}$$

is a positive-definite banded matrix with bandwidth seven, but it is not necessarily a symmetric matrix.

4. Derivative values and integrals of B-splines are also calculated recursively.

These nice properties make B-splines convenient function systems for the solution of integral and differential equations.

The first four and the last four B-splines have different shapes from the other ones because of their vicinity to the interval end-points. We give the formula for an 'inner' B-spline for equidistant interpolation nodes:

$$x_i := x_0 + ih, \quad i = 0, 1, \ldots, n.$$

Then with $I_k := [x_k, x_{k+1}]$ we have for $i = 4, 5, \ldots, n-4$:

$$
B_i(x) = \frac{1}{6h^3} \begin{cases}
(x - x_{i-2})^3, & x \in I_{i-2} \\
h^3 + 3h^2(x - x_{i-1}) + 3h(x - x_{i-1})^2 - 3(x - x_{i-1})^3, & x \in I_{i-1} \\
h^3 + 3h^2(x_{i+1} - x) + 3h(x_{i+1} - x)^2 - 3(x_{i+1} - x)^3, & x \in I_i \\
(x_{i+2} - x)^3, & x \in I_{i+1} \\
0, & \text{otherwise.}
\end{cases}
$$

In Figure 3.9 the first five cubic B-splines are shown for at least nine interpolation nodes. The knots are marked by '×'. The only inner B-spline with a support of four intervals and symmetric shape is B_4. These inner B-splines are interpolating natural cubic splines (see Section 3.2.1) with equidistant interpolation nodes $\{x_{i-2}, \ldots, x_{i+2}\}$ and corresponding values

$$y_i = \{0, 1/6, 2/3, 1/6, 0\}.$$

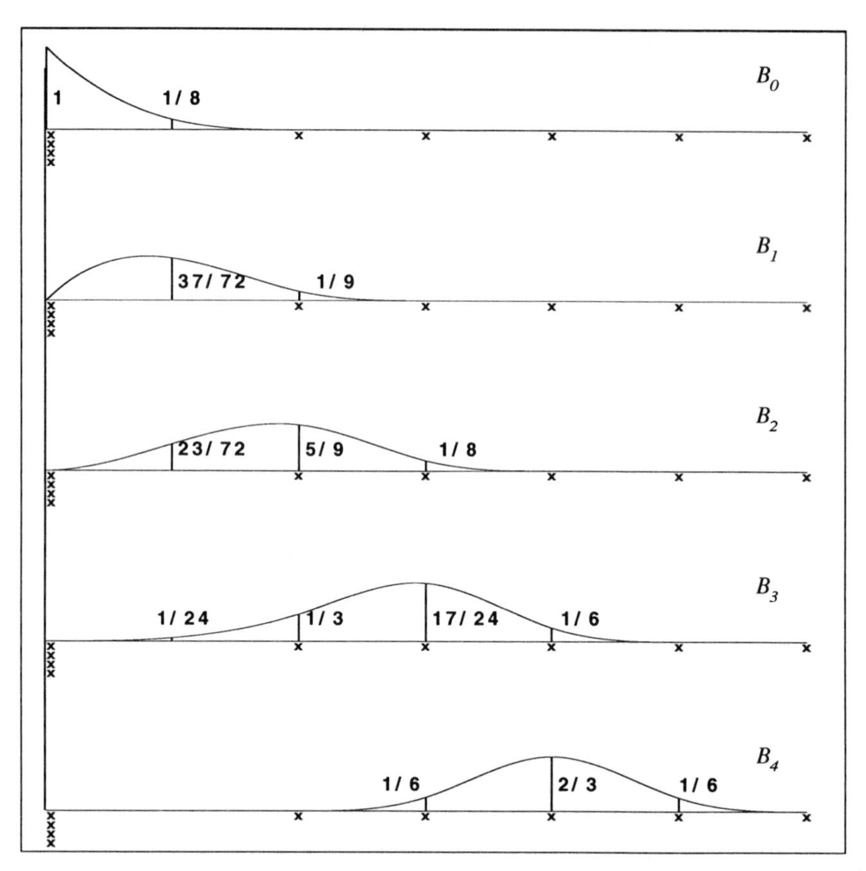

Figure 3.9. Cubic B-splines with equidistant interpolation nodes

Interpolation

The interpolating spline s is represented as a linear combination of B-splines:

$$s(x) = \sum_{k=0}^{n} \alpha_k B_k(x). \tag{3.31}$$

With these $n+1$ coefficients we can meet $n+1$ interpolation conditions to get a unique solution. Thus there are no boundary conditions necessary.

The unknown coefficients are calculated by the solution of a linear systems $B\alpha = y$, where α and y are vectors with α_i and y_i as components, respectively, or

$$\sum_{k=0}^{n} \alpha_k B_k(x_i) = y_i, \quad i = 0, 1, \ldots, n. \tag{3.32}$$

A numerically stable solution can be calculated with little effort. Spline interpolation is carried out by our program INTERDIM1 (see Section 3.3 for an example) using NAG routine E01BAF. Details of the algorithm of NAG routine E01BAF are to be found in [18].

Approximation

If m data points are to be fitted by a linear combination of $n+1$ B-splines with $m > n+1$, then we have to solve a discrete least-squares approximation problem getting the respective linear system $B\alpha = y$ now with a rectangular m by $n+1$ matrix B. This can be solved by the method of least squares (see Section 1.5):

$$\sum_{i=1}^{m} \left(\sum_{k=0}^{n} \alpha_k B_k(x_i) - y_i \right)^2 \rightarrow \min_{\alpha_k} . \tag{3.33}$$

Spline approximation is carried out by NAG routine E02BAF.

For further details about B-splines we refer the reader to [21].

3.3 One-Dimensional Interpolation: INTERDIM1

Program INTERDIM1, see B.3.1, interpolates a table of data points with a polynomial or a spline function. For a given function users have to prepare such a table themselves. The program uses the methods of the preceding sections. Additionally the program is suitable for rational interpolation, which is not described here, but the use of which is obvious from the comment lines in program and input file. In general it is preferable to create the necessary input file with the template program of the PAN system. One of the following methods can be selected:

1 Newton interpolation with polynomials.
2 Interpolation with B-splines.
3 Rational interpolation.

For cases 1 and 2 the following different evaluations are possible:

Polynomial P	Spline S
Evaluate P(x) Replace: $P \rightarrow P'$ Replace: $P \rightarrow \int P$	Evaluate S(x) Evaluate $S'(x)$ Evaluate $S''(x)$ Evaluate $S'''(x)$ Evaluate $\int_{x_0}^{x_n} S(x)\,dx$

This means that for polynomial interpolation the function to be evaluated changes if we want to calculate derivative or integral values, and the integration is an analytical one. For example if we want to calculate the integral

$$\int_a^b P(x)\,dx,$$

we have to replace P by $\int P$ first, then to evaluate the new function 'P' – we will call it Q now – at the point b to get

$$\int_a^b P(x)\,dx = Q(b) \quad (Q(a) = 0).$$

INTERDIM1 has a graphical interface which allows all possibilities of the program to be used graphically as there is input of the points to be interpolated, output of resulting evaluations, drawing of the interpolating function and its derivatives.

We will clarify these possibilities by two examples.

Firstly consider the function

$$f(x) = \sqrt{x},$$

which will be interpolated with the following points:

x	0	1	4	9	16	25	36	49	64
y	0	1	2	3	4	5	6	7	8

This is an artificial example, but it shows the difference in quality of polynomial and spline interpolation.

We are mainly interested in interpolated values near the interval endpoints. Therefore we prepare the following input files.

Input Files

```
Input file for the PAN program INTERDIM1:
Method (1 = polynomial , 2 = spline , 3 = rational interp.):
1                           (2 for the second run)
Graphical input of table of values (no = 0, yes = 1):
0
Dimension of vector of points:
9
Vector of points:
0 1 4 9 16 25 36 49 64
Dimension of vector of values:
9
```

```
Vector of values:
0 1 2 3 4 5 6 7 8
Left margin:
0.0
Right margin:
64.0
Lower margin:
0.0
Upper margin:
20.0
Input of evaluation points (no = 0, yes = 1):
1
Dimension of vector of evaluation points:
10
Vector of evaluation points:
0.1 0.3 0.5 0.7 0.9 55 57 59 61 63
```

With these input files we get two sets of results which we show in the following table together with the values of $f(x) = \sqrt{x}$:

x	$P(x)$	$S(x)$	$f(x) = \sqrt{x}$
0.1	0.128820	0.124346	0.316227
0.3	0.364987	0.355691	0.547722
0.5	0.574768	0.564967	0.707106
0.7	0.760697	0.753454	0.836660
0.9	0.925141	0.922429	0.948683
55.0	56.773501	7.417362	7.416198
57.0	78.917289	7.551075	7.549834
59.0	94.778908	7.682272	7.681145
61.0	92.051716	7.811066	7.810249
63.0	51.276268	7.937570	7.937253

In the upper plot of Figure 3.10 we see the function f and the interpolating polynomial P. P is heavily oscillating and it disappears from the plot at $x \approx 50$, taking on values of up to 100. In the lower plot f is shown again together with the interpolating spline, which is a much better approximation to f, with only slight oscillations at the left end-point of the interval.

To demonstrate the graphical use of the program, we will show the input and resulting output file for a purely graphical session.

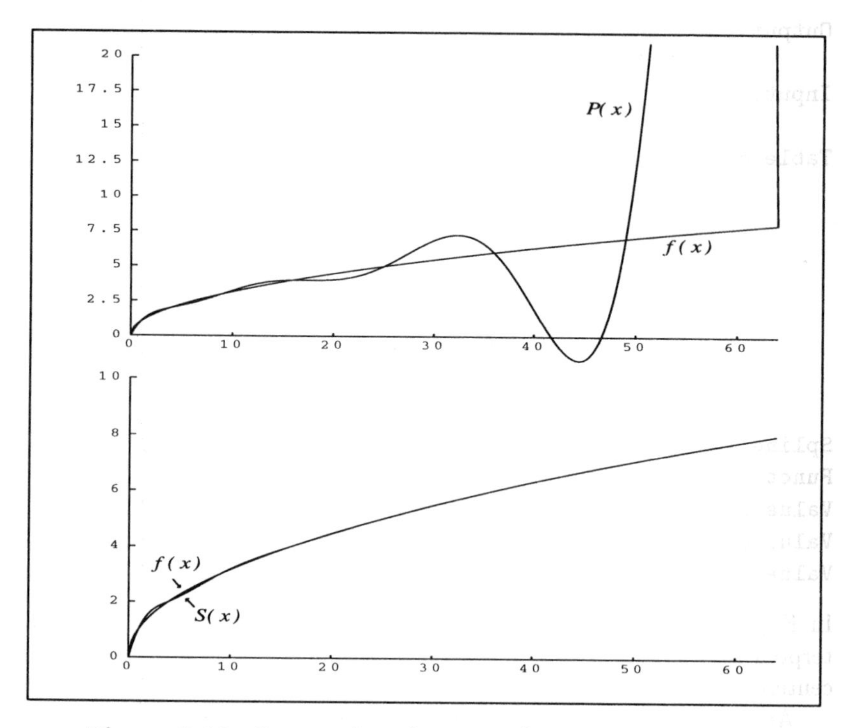

Figure 3.10. Comparison between interpolation with polynomials and with splines

```
Input file for the PAN program INTERDIM1:
Method (1 = polynomial , 2 = spline , 3 = rational interp.):
2
Graphical input of table of values (no = 0, yes = 1):
1
Left margin:
0.0
Right margin:
10.0
Lower margin:
0.0
Upper margin:
10.0
Input of evaluation points (no = 0, yes = 1):
0
```

The mouse input is recorded in the output file as well as the evaluation of function and derivative values at points defined with mouse-input:

```
Output file for the PAN program INTERDIM1:

Input: Template

Table of values with  7 points:
       0.54990208        5.99804258
       1.77103722        8.06457901
       3.25048900        4.61252403
       4.80039120        2.61643839
       7.80626249        2.92172194
       9.09784698        5.62230873
       9.82583141        5.85714293

Spline interpolation:
Function value at               7.33659458:     2.23335888
Value of 1. derivative at       7.38356209:     1.17044522
Value of 2. derivative at       7.38356209:     1.49071217
Value of 3. derivative at       7.36007881:     0.56404193
```

In Figure 3.11 the input points (+) are to be seen together with the interpolating function ($\times$) and derivative values (one, two and three small centered circles) near the point $x = 7.35$.

Additionally the method can be changed and the interpolating functions of different methods shown in one figure.

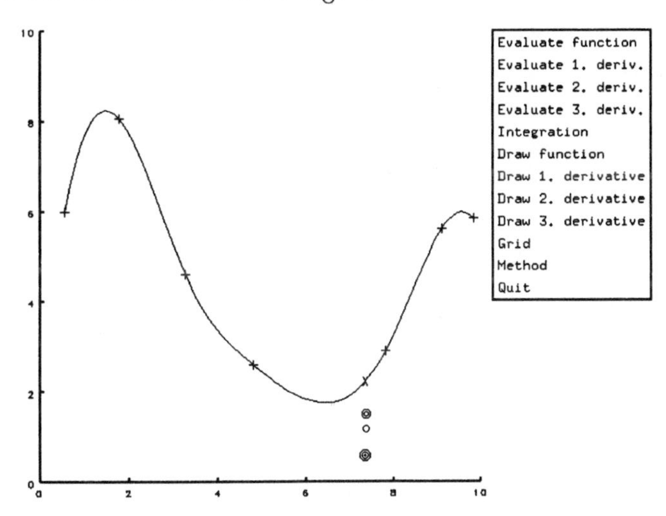

Figure 3.11. Graphics window with input and results

3.4 Curve Fitting with Polynomials

We now give a brief theoretical introduction to the curve fitting or least-squares approximation problem. In this section we discuss the problem of curve fitting with Chebyshev polynomials. The following section is devoted to curve fitting with trigonometric functions, where the method of fast Fourier transform plays a dominant role. Because of different applications there are different names in use for this method, such as Gaussian approximation, data fitting or regression analysis.

3.4.1 Least-Squares Approximation of Functions

Let f be a piecewise continuous function, which is square-integrable in the interval (a, b). Let ω be a positive, integrable *weight function* with $0 < \int_a^b \omega(x)\, dx < \infty$.

Let $\{\varphi_i\}_{i=0}^{\infty}$ be a system of linearly independent functions, defined and square-integrable on (a, b). Determine the coefficients α_i, $i = 0, 1, \ldots, n$, of

$$g(x) := g(x; \alpha_0, \alpha_1, \ldots, \alpha_n) := \sum_{i=0}^{n} \alpha_i\, \varphi_i(x), \tag{3.34}$$

such that

$$F(\alpha_0, \alpha_1, \ldots, \alpha_n) := \int_a^b (f(x) - g(x))^2 \omega(x)\, dx \to \min_{\alpha_i}. \tag{3.35}$$

If we define an inner product of two functions as

$$(r, s) := \int_a^b r(x) s(x) \omega(x)\, dx, \tag{3.36}$$

then the necessary conditions for minimizing F lead to the system of linear equations

$$\begin{pmatrix} (\varphi_0, \varphi_0) & (\varphi_0, \varphi_1) & \cdots & (\varphi_0, \varphi_n) \\ (\varphi_1, \varphi_0) & \ddots & & \vdots \\ \vdots & & \ddots & \vdots \\ (\varphi_n, \varphi_0) & \cdots & \cdots & (\varphi_n, \varphi_n) \end{pmatrix} \begin{pmatrix} \alpha_0 \\ \alpha_1 \\ \vdots \\ \alpha_n \end{pmatrix} = \begin{pmatrix} (f, \varphi_0) \\ (f, \varphi_1) \\ \vdots \\ (f, \varphi_n) \end{pmatrix}. \tag{3.37}$$

This is called a *system of normal equations*. If the functions φ_i are linearly independent, the system has a unique solution. If its matrix of coefficients is a full matrix, its solution may be spoiled by rounding errors because it is usually ill-conditioned. An example for ill-conditioned normal systems can be found in Section 1.2.3. Therefore it is important to work with function systems, which lead to sparse systems. Mostly one chooses *orthogonal* function systems, for which

$$(\varphi_i, \varphi_j) = 0, \quad \text{if} \quad i \neq j. \tag{3.38}$$

Then the system of normal equations is diagonal and its solution obviously is given as

$$\alpha_i = \frac{(f, \varphi_i)}{(\varphi_i, \varphi_i)}, \quad i = 0, 1, \ldots, n. \tag{3.39}$$

Because the inner products are integrals, they normally have to be calculated by numerical integration. In this way the problem is discretized. This means that there is no significant difference from the discrete least-squares approximation, which is also preferred by most software libraries. Then an inner product has to be defined for vectors and associated properties such as orthogonality have to be carried over. This is possible for trigonometric functions, see Section 3.5, as well as for Chebyshev polynomials, see Section 3.4.3.

3.4.2 Least-Squares Approximation of Data

Now instead of the function f a set of data points (x_i, y_i) and a vector of *weights* with positive components $\omega_i > 0$ are given, i.e. we have the data

$$(x_i, y_i, \omega_i), \quad i = 1, \ldots, m.$$

The problem is now to determine the coefficients α_k, $k = 0, 1, \ldots, n$, of

$$g(x) := g(x; \alpha_0, \alpha_1, \ldots, \alpha_n) := \sum_{k=0}^{n} \alpha_k \, \varphi_k(x), \tag{3.40}$$

such that

$$F(\alpha_0, \alpha_1, \ldots, \alpha_n) := \sum_{i=1}^{m} (y_i - g(x_i))^2 \omega_i \to \min_{\alpha_k}. \tag{3.41}$$

Let us now for two functions $r(x)$, $s(x)$ define a positive semi-definite symmetric bilinear form in $\mathbb{R}^m$ as

$$(r, s) := \sum_{i=1}^{m} r_i \, s_i \, \omega_i, \tag{3.42}$$

where $r_i = r(x_i)$, $s_i = s(x_i)$. Then we have the discrete analogue to (3.36), and we get formally the same system (3.37) of normal equations, where now the integrals are replaced by bilinear forms of vectors.

This discrete least-squares approach can be applied to a given function f, too, if we simply take the values y_i to be

$$y_i = f(x_i), \quad i = 1, \ldots, m.$$

To get a diagonal system of normal equations again, we now have to choose the points $\{x_i\}$ such that the orthogonality remains valid for the inner product in $\mathbb{R}^m$:

$$\begin{aligned}\sum_{i=1}^{m} \varphi_j(x_i)\,\varphi_k(x_i)\,\omega_i &= 0 \quad \text{for all} \quad j,k=0,\ldots,n \quad \text{with} \quad j\neq k.\\ \sum_{i=1}^{m} \varphi_j(x_i)\,\varphi_j(x_i)\,\omega_i &\neq 0 \quad \text{for all} \quad j=0,\ldots,n.\end{aligned}$$

3.4.3 Approximation with Chebyshev Polynomials

Chebyshev polynomials

According to the addition formulas of trigonometry the function $\cos(k\varphi)$ is representable for each $k \in \mathbb{N}_0$ as a polynomial in $\cos(\varphi)$.

Thus for $k \in \mathbb{N}_0$ the definition

$$T_k(x) := T_k(\cos(\varphi)) := \cos(k\varphi) \quad \text{with} \quad x := \cos(\varphi), \quad x \in [-1,1] \tag{3.43}$$

yields a system of polynomials $\{T_k\}$ of degree k called *Chebyshev polynomials*. The first seven of them are

$$\begin{aligned}\cos(0\,\varphi) = 1 \quad &\Longrightarrow \quad T_0(x) = 1,\\ \cos(1\,\varphi) = \cos(\varphi) \quad &\Longrightarrow \quad T_1(x) = x,\\ \cos(2\,\varphi) = 2\cos^2(\varphi) - 1 \quad &\Longrightarrow \quad T_2(x) = 2x^2 - 1,\\ \cos(3\,\varphi) = 2\cos(\varphi)\cos(2\,\varphi) - \cos(\varphi) &\\ = 4\cos^3(\varphi) - 3\cos(\varphi) \quad &\Longrightarrow \quad T_3(x) = 4x^3 - 3x.\end{aligned}$$

$$\begin{aligned}T_4(x) &= 8x^4 - 8x^2 + 1,\\ T_5(x) &= 16x^5 - 20x^3 + 5x,\\ T_6(x) &= 32x^6 - 48x^4 + 18x^2 - 1.\end{aligned}$$

From this definition we easily obtain the following characteristics:

1. $$|T_k(x)| \leq 1 \quad \text{for} \quad x \in [-1,1], \quad k \in \mathbb{N}_0. \tag{3.44}$$

2. Chebyshev polynomials can recursively be defined:

$$\begin{aligned}T_0(x) = 1, \qquad & T_1(x) = x\\ T_{k+1}(x) \quad = \quad & 2\,x\,T_k(x) - T_{k-1}(x), \quad k \geq 1.\end{aligned} \tag{3.45}$$

3. $$T_k(-x) = (-1)^k\,T_k(x).$$

4. $T_k(x)$ has its extremal points for $k\varphi = l\pi$. This means that

$$x_l^{(e)} = \cos\left(\frac{l\pi}{k}\right), \quad l = 0, 1, \ldots, k, \quad k \geq 1. \tag{3.46}$$

They are symmetric with respect to the origin and have alternating signs. They correspond to an equidistant partition of a circle. Therefore they are more densely distributed near the interval end-points.

5. From $\cos(\varphi) = 0$ we have $k\varphi = (2l-1)\pi/2$. Therefore $T_k(x)$ becomes zero at the points

$$x_l = \cos\left(\frac{(2l-1)\pi}{2k}\right), \quad l = 1, \ldots, k, \quad k \geq 1. \tag{3.47}$$

They also are symmetric with respect to the origin and more densely distributed near the interval end-points. They all lie within the interval $[-1, 1]$.

The most important property of Chebyshev polynomials for for use in least-squares approximation is that they form an orthogonal system.

Theorem 3.6. *The Chebyshev polynomials T_k, $k = 0, 1, 2, \ldots$, are orthogonal with respect to the weight function $\omega(x) = 1/\sqrt{1-x^2}$:*

$$\int_{-1}^{1} T_l(x)\, T_j(x) \frac{1}{\sqrt{1-x^2}}\, dx = \left\{ \begin{array}{ll} 0, & \text{if } l \neq j \\ \pi/2, & \text{if } l = j > 0 \\ \pi, & \text{if } l = j = 0 \end{array} \right\}, \quad l, j \in \mathbb{N}_0. \tag{3.48}$$

Least-Squares Approximation of Functions

Because of the orthogonality of Chebyshev polynomials the continuous least-squares approximation problem is mainly a matter of numerical integration of the inner products in (3.39).

Let f be a function piecewise continuous in $[-1, 1]$. Then

$$g_n(x) = \frac{1}{2} c_0 T_0(x) + \sum_{l=1}^{n} c_l T_l(x) \tag{3.49}$$

with

$$c_l = \frac{2}{\pi} \int_{-1}^{1} f(x)\, T_l(x) \frac{1}{\sqrt{1-x^2}}\, dx, \quad l = 0, 1, \ldots, n, \tag{3.50}$$

is the least-squares approximation for f.

Least-Squares Approximation of Data

If we have a given set of data points $(x_j, y_j), j = 1, \ldots, m$, instead of a function f, then we can only calculate approximations c_l^* to the coefficients c_l.

In order to maintain an orthogonal system, one should take the Chebyshev points

$$x_j = \cos(\phi_j) \quad \text{with} \quad \phi_j := \frac{(j-1)\pi}{m-1}, \quad j = 1, \dots, m, \tag{3.51}$$

whenever possible. Then we have

$$c_l^* = \frac{2}{m} \sum_{j=1}^{m} y_j \cos(l\phi_j), \quad l = 0, 1, \dots, k; \; k < n. \tag{3.52}$$

This is a trapezoidal sum for the integrals (3.50). Even if we cannot take the Chebyshev points, it is possible to generate a set of polynomials that is orthogonal with respect to the summation over the special data set (see [31]). These polynomials can then be represented in their Chebyshev series form and the summation can be carried out by fast algorithms (see [15] and [38]) as for the trigonometric functions (see Section 3.5.3).

3.4.4 Program and Example

The program APPROX solves approximation problems with trigonometric functions or with Chebyshev polynomials. It is presented in Section 3.6 together with examples.

3.5 Trigonometric Approximation

Least-squares approximation with trigonometric functions is simply a matter of summation and discrete Fourier transform, which we will briefly describe in the next few sections.

3.5.1 Harmonic Analysis

Strictly *harmonic analysis* is a continuous least-squares approximation problem, but it can be treated as a discrete least-squares approximation problem.

Let f be a piecewise continuous function defined in the interval $[0, 2\pi]$, which is periodic with period 2π:

$$f(x + 2\pi) = f(x) \quad \forall\, x \in \mathbb{R}. \tag{3.53}$$

At discontinuities f must have unique limit values:

$$\lim_{h \to +0} f(x_0 - h) = y_0^-, \qquad \lim_{h \to +0} f(x_0 + h) = y_0^+. \tag{3.54}$$

The function f will be approximated in the least-squares sense by its trigonometric Fourier series

$$g_n(x) = \frac{1}{2}a_0 + \sum_{k=1}^{n-1}(a_k\cos(kx) + b_k\sin(kx)) + a_n\cos(nx). \tag{3.55}$$

This means that the function system chosen is the set of trigonometric functions[†]

$$\{1,\ \cos(x),\ \sin(x),\ \cos(2x),\ \sin(2x),\ \ldots,\ \sin((n-1)x),\ \cos(nx)\}. \tag{3.56}$$

With the inner product

$$(f,g) = \int_0^{2\pi} f(x)g(x)\,dx,$$

i.e. for the weight function $\omega \equiv 1$, this functions system is orthogonal:

$$\begin{aligned}
(\cos(jx),\cos(kx)) &= \begin{cases} 0 & \text{if } \ j \neq k \\ 2\pi & \text{if } \ j = k = 0 \\ \pi & \text{if } \ j = k \neq 0 \end{cases} \\
(\sin(jx),\sin(kx)) &= \begin{cases} 0 & \text{if } \ j \neq k \\ \pi & \text{if } \ j = k \neq 0 \end{cases} \\
(\cos(jx),\sin(kx)) &= 0 \ \ \forall \ \ j \geq 0,\ k > 0.
\end{aligned} \tag{3.57}$$

Because of the orthogonality the coefficients can be calculated according to (3.39), which gives

$$\begin{aligned}
a_k &= \frac{1}{\pi}\int_0^{2\pi} f(x)\cos(kx)\,dx, \quad k = 0,1,\ldots,n, \\
b_k &= \frac{1}{\pi}\int_0^{2\pi} f(x)\sin(kx)\,dx, \quad k = 1,2,\ldots,n-1.
\end{aligned} \tag{3.58}$$

These coefficients are called *Fourier coefficients*, and (3.55) is called a *finite Fourier series*. It is an approximation to the *Fourier series* of f:

$$g(x) = \frac{1}{2}a_0 + \sum_{k=1}^{\infty}(a_k\cos(kx) + b_k\sin(kx)). \tag{3.59}$$

[†] Instead of $[0, 2\pi]$ one can choose any other finite interval $[a, b]$. The linear transformation $t = \dfrac{x-a}{b-a}2\pi$ then leads to the function system

$$1, \quad \cos(kt), \quad \sin(kt), \quad \ldots.$$

3.5.2 Calculation of the Fourier Coefficients

We now come to a discrete problem by integrating the Fourier coefficients a_k and b_k, (3.58), with the trapezoidal rule at equidistant points:

$$h = \frac{2\pi}{N}, \quad x_j = jh = \frac{2\pi j}{N}, \quad j = 0, \ldots, N. \tag{3.60}$$

Taking the periodicity of f into account we obtain for a_k and b_k the approximations

$$\begin{aligned} a_k^* &= \frac{2}{N}\sum_{j=1}^{N} f(x_j)\cos(kx_j), \quad k = 0, 1, \ldots, n, \\ b_k^* &= \frac{2}{N}\sum_{j=1}^{N} f(x_j)\sin(kx_j), \quad k = 1, \ldots, n-1. \end{aligned} \tag{3.61}$$

(In general $n \neq N$.)

This is equivalent to a discrete least-squares approximation with the weights

$$\omega_0 = \omega_N = 1/2, \quad \omega_j = 1, \quad j = 1, \ldots, N-1, \tag{3.62}$$

if we assume that for the trigonometric functions with these points and weights a discrete orthogonality is valid. The following theorems will show that this is the case. Proofs for these theorems may be found in [10].

Theorem 3.7. *If x_j are the points (3.60), then*

$$\begin{aligned} \sum_{j=1}^{N}\cos(kx_j) &= \begin{cases} N & \text{if } k = 0, N, 2N, 3N, \ldots \\ 0 & \text{otherwise,} \end{cases} \\ \sum_{j=1}^{N}\sin(kx_j) &= 0, \quad k = 1, \ldots, n-1. \end{aligned} \tag{3.63}$$

Theorem 3.8. *Taking the trigonometric function system (3.56) and the equidistant points (3.60) we obtain the following discrete orthogonality relations ($\mathbb{Z}$ is the set of integer numbers):*

$$\begin{aligned} \sum_{j=1}^{N}\cos(kx_j)\cos(lx_j) &= \begin{cases} N & \text{if } \frac{k+l}{N} \in \mathbb{Z} \text{ and } \frac{k-l}{N} \in \mathbb{Z} \\ \frac{N}{2} & \text{if either } \frac{k+l}{N} \in \mathbb{Z} \text{ or } \frac{k-l}{N} \in \mathbb{Z} \\ 0 & \text{otherwise,} \end{cases} \\ \sum_{j=1}^{N}\sin(kx_j)\sin(lx_j) &= \begin{cases} \frac{N}{2} & \text{if } \frac{k+l}{N} \notin \mathbb{Z} \text{ and } \frac{k-l}{N} \in \mathbb{Z} \\ -\frac{N}{2} & \text{if } \frac{k+l}{N} \in \mathbb{Z} \text{ and } \frac{k-l}{N} \notin \mathbb{Z} \\ 0 & \text{otherwise,} \end{cases} \end{aligned} \tag{3.64}$$

$$\sum_{j=1}^{N} \cos(kx_j)\sin(lx_j) \quad = \quad 0 \quad \forall\, k,l \in \mathbb{N}_0.$$

Theorem 3.9. *If $N = 2n, \quad n \in \mathbb{N}$, then the finite Fourier series*

$$g_n^*(x) := \frac{1}{2}a_0^* + \sum_{k=1}^{n-1} [a_k^* \cos(kx) + b_k^* \sin(kx)] + \frac{1}{2}a_n^* \cos(nx) \qquad (3.65)$$

with the coefficients (3.61) interpolates the function f with interpolation nodes x_j. This interpolation is unique.

Theorem 3.10. *If $N = 2n, \quad n \in \mathbb{N}$ and $m < n$, then the finite Fourier series*

$$g_m^*(x) := \frac{1}{2}a_0^* + \sum_{k=1}^{m} [a_k^* \cos(kx) + b_k^* \sin(kx)] \qquad (3.66)$$

with coefficients (3.61) is the discrete least-squares approximation of the function f, i.e. it minimizes the root-mean-square error

$$\left[\sum_{j=1}^{N} (g_m^*(x_j) - f(x_j))^2\right]^{1/2} . \qquad (3.67)$$

3.5.3 Fast Fourier Transform

To calculate the coefficients a_k^*, b_k^* one uses fast Fourier transform. If we leave the factor $2/N$ in (3.61) aside, then we need an efficient algorithm for calculating the numbers

$$\begin{aligned} a_k' &= \sum_{j=0}^{N-1} f(x_j)\cos(kx_j), \quad k = 0,1,\ldots,n \\ b_k' &= \sum_{j=0}^{N-1} f(x_j)\sin(kx_j), \quad k = 1,\ldots,n-1. \end{aligned} \qquad (3.68)$$

As early as 1903, Runge ([84], [85]) found an efficient algorithm for $N = 4m$, $m \in \mathbb{N}$, which splits the Fourier series to reduce the amount of computational work. In 1965 Cooley and Tukey looked again at an idea of Gauss, which they could make this splitting method even more efficient with. This method is called *fast Fourier transform* or briefly FFT.

We will consider this method for the easiest case of N being a power of 2:

$$N = 2^\gamma. \qquad (3.69)$$

Using complex notation we can write:

$$\begin{aligned} w_n &:= \mathrm{e}^{-\mathrm{i}2\pi/n} = \cos\left(\frac{2\pi}{n}\right) - \mathrm{i}\sin\left(\frac{2\pi}{n}\right), \quad \mathrm{i} := \sqrt{-1} \\ \text{and} & \\ y_j &:= f(x_{2j}) + \mathrm{i}\, f(x_{2j+1}), \quad j = 0, 1, \ldots, n-1, \quad n := \frac{N}{2}. \end{aligned} \tag{3.70}$$

This gives us instead of (3.68)

$$c_k = \sum_{j=0}^{n-1} y_j \mathrm{e}^{-\mathrm{i}jk2\pi/n} = \sum_{j=0}^{n-1} y_j w_n^{jk}, \quad k = 0, 1, \ldots, n-1. \tag{3.71}$$

We will not give the connections between a'_k, b'_k and c_k here. Instead we come to the *discrete Fourier transform* or DFT of the y_j to the c_k. If n is even, $n = 2m$, the indices k and j in (3.71) can be represented as

$$\begin{aligned} k &= p_1 + 2p, \quad \text{with} \quad 0 \le p_1 < 2, \quad 0 \le p < m, \\ j &= q_1 m + q, \quad \text{with} \quad 0 \le q_1 < 2, \quad 0 \le q < m. \end{aligned}$$

Hence

$$\begin{aligned} c_k = c_{p_1+2p} &= \sum_{q_1=0}^{1} \sum_{q=0}^{m-1} y_{q_1 m+q} w_n^{(p_1+2p)(q_1 m+q)} \\ &= \sum_{q=0}^{m-1} y_q w_n^{(p_1+2p)q} + \sum_{q=0}^{m-1} y_{m+q} w_n^{(p_1+2p)(m+q)}. \end{aligned}$$

The powers w_n^l, $l = 0, \ldots, n-1$, of the roots of unity form a regular n-gon on the unit circle in the complex number plane. Therefore we have

$$\begin{aligned} w_n^{2m} &= w_n^n = 1 \\ \text{and} \quad w_{2m}^{q+m} &= -w_{2m}^q \\ \text{and} \quad w_n^{(p_1+2p)q} &= (w_n^2)^{pq}\, w_n^{p_1 q} \\ w_n^{(p_1+2p)(m+q)} &= (-1)^{p_1} (w_n^2)^{pq}\, w_n^{p_1 q}. \end{aligned} \tag{3.72}$$

For $p_1 = 0$ and $p_1 = 1$ we obtain the two cases

$$\begin{aligned} c_{2p} &= \sum_{q=0}^{m-1} (y_q + y_{m+q}) w_m^{pq}, \\ c_{2p+1} &= \sum_{q=0}^{m-1} \left[(y_q - y_{m+q}) w_n^q\right] w_m^{pq}, \end{aligned} \tag{3.73}$$

or with the auxiliary variables

$$\left.\begin{array}{lll} z_q & := & y_q + y_{m+q} \\ z_{m+q} & := & (y_q - y_{m+q})\, w_n^q \end{array}\right\} \quad q = 0, 1, \ldots, m-1, \tag{3.74}$$

$$\left.\begin{array}{lll} c_{2p} & := & \sum_{q=0}^{m-1} z_q\, w_m^{pq} \\ c_{2p+1} & := & \sum_{q=0}^{m-1} z_{m+q}\, w_m^{pq} \end{array}\right\} \quad p = 0, 1, \ldots, m-1. \tag{3.75}$$

Each equation in (3.75) is a DFT like (3.71) on its own. So, if m is even, we can repeat the reduction steps. For $N = 2^\gamma$ we can split the sums up until we have $m = 1$.

The effort involved in obtaining the sum is then $O(n \log_2 n)$ complex operations to be compared with $O(n^2)$ operations for direct summation. This means that, if $n = 32$, we have an acceleration by a factor 12.8; for $n = 4096$ the factor is 683. So the algorithm deserves the name FFT.

We will not give details of the most commonly used algorithms, which are the algorithms of Cooley–Tukey and of Sande–Tukey. These two algorithms are well described in Chapters 11 and 12 of [10], for example. Software libraries allow nearly all numbers for N. The effort depends on the prime factor decomposition of N. NAG prefers multiples of 2, 3 and 5 for N, but allows up to 20 factors of prime numbers not greater than 19.

3.6 One-Dimensional Approximation: APPROX

The program APPROX, see B.3.2, solves the least-squares approximation problem with trigonometric functions using FFT or with orthogonal polynomials in their Chebyshev series representation form. It uses NAG routines C06FAF, E02DAF, E02AFF, and E02AKF. At present it can only approximate functions. In the near future it will be able to approximate a set of data points as well. We will give an examples for the use of APPROX.

We will approximate a function with discontinuities, see Figure 3.12, which is a typical application for harmonic analysis:

$$f(x) = \begin{cases} x^2 & \text{if } 0 \le x < 1 \\ 2 & \text{if } 1 \le x < 2 \\ 4 - x & \text{if } 2 \le x < 3 \\ 0 & \text{if } 3 \le x \le 4. \end{cases}$$

We approximate f in $[0, 4]$ with 8, 32 and 128 trigonometric functions.

Input File for APPROX

```
Input file for the PAN program APPROX:
Method (1 = Fourier , 2 = Chebyshev approximation):
```

```
1
Output of coefficients (no = 0, yes = 1):
0
Number of data points:
512
First data point:
0
Last data point:
4
Polynomial degree:
128
Left margin:
0
Right margin:
4
Lower margin:
0
Upper margin:
2
Input of evaluation points (no = 0, yes = 1):
0
```

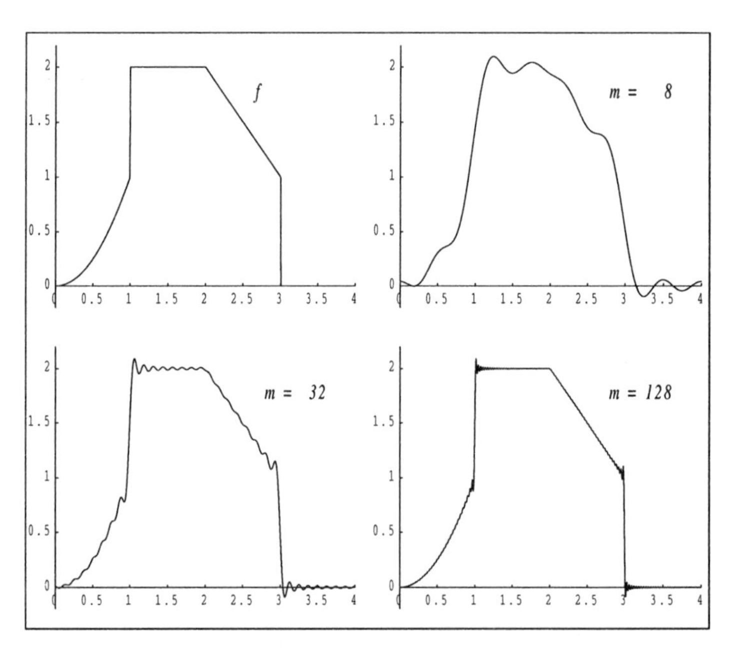

Figure 3.12. f **and its Fourier approximations with** $m = 8, 32, 128$

3.7 Two-Dimensional Splines

We consider two problems. Let

$$(x_k, y_k, f_k) \quad k = 1, 2, \ldots, m \tag{3.76}$$

be a set of two-dimensional data points. Determine a function

$$S(x, y) := \sum_{i=1}^{n} \sum_{j=1}^{l} c_{ij} s_i(x) t_j(y) \tag{3.77}$$

with

- **Interpolation**

$$S(x_k, y_k) = f_k, \quad k = 1, 2, \ldots, m, \tag{3.78}$$

or

- **Approximation**

$$\sum_{k=1}^{m} (S(x_k, y_k) - f_k)^2 \to \min_{c_{ij}}. \tag{3.79}$$

s_i and t_j are functions from an one-dimensional functions system. The solution of these problems becomes much simpler if one only chooses points on a *rectangular grid*: Let

$$x_i, \quad i = 1, 2, \ldots, n \quad \text{and} \quad y_j, \quad j = 1, 2, \ldots, l \tag{3.80}$$

be two one-dimensional coordinate vectors and

$$f_{ij}, \quad i = 1, 2, \ldots, n, \quad j = 1, 2, \ldots, l$$

the respective function values.

Determine a function

$$S(x, y) := \sum_{k=1}^{n} \sum_{\mu=1}^{l} c_{k\mu} s_k(x) t_\mu(y) \tag{3.81}$$

with

- **Interpolation**

$$S(x_i, y_j) = f_{ij}, \quad i = 1, 2, \ldots, n, \quad j = 1, 2, \ldots, l \tag{3.82}$$

or

- **Approximation**

$$\sum_{i=1}^{n}\sum_{j=1}^{l}(S(x_i,y_j)-f_{ij})^2 \to \min_{c_{k\mu}}. \tag{3.83}$$

3.7.1 Solution with Bicubic B-Splines

We now choose the functions s and t in (3.78), (3.79) and (3.80) to be one-dimensional cubic B-splines

$$\begin{aligned} s_k(x) &:= B_k(x), \quad k=1,2,\ldots,n \\ t_\mu(y) &:= B_\mu(y), \quad \mu=1,2,\ldots,l \end{aligned} \tag{3.84}$$

which have been introduced in Section 3.2.3. The knots necessary for defining the B-splines are defined in accordance with the one-dimensional case for the set of interpolation nodes $\{x_i\}$ and $\{y_j\}$, respectively, and with numbering starting with 1 instead of 0. If all inner knots are distinct the B-splines are twice continuously differentiable. A typical two-dimensional B-spline on a 10×10-grid, which was drawn by NAG routine J06HEF (see Appendix D), can be seen in Figure 3.13.

Figure 3.13. Two-dimensional B-spline

With (3.84) we obtain the two-dimensional B-splines

$$\hat{B}_{k\mu}(x,y) := B_k(x)B_\mu(y). \tag{3.85}$$

In each rectangle R_ν of the grid these functions are polynomials of maximum degree 6:

$$\hat{B}_{k\mu}(x,y) = \sum_{i=0}^{3}\sum_{j=0}^{3} a_{ij}^{k\mu\nu} x^i y^j, \quad \text{if } (x,y) \in R_\nu. \tag{3.86}$$

$\{\hat{B}_{k\mu}\}$ spans the two-dimensional function space looked for (see [21], or [79] for the similar case of Lagrange polynomials). The application of this two-dimensional approach with the finite element method to partial differential equations is represented in [79] in such a way that is easily understood. B-splines are defined slightly differently in [79] and in [21].

To solve the interpolation problem (3.78) on the rectangular grid (3.80) one has to solve the linear system

$$\sum_{k=1}^{n}\sum_{\mu=1}^{l} c_{k\mu} s_k(x_i) t_\mu(y_j) = f_{ij}, \quad i = 1,2,\ldots,n, \quad j = 1,2,\ldots,l \tag{3.87}$$

for the coefficients $c_{k\mu}$ of (3.81). This is a square regular system with nl unknowns. Furthermore it is sparse and has a block banded matrix of coefficients. So one must apply special methods for its solution, for example a special QR-decomposition with minimal fill-in or a relaxation method.

For the more general problem of arbitrarily distributed data points we get a linear system as well. If the data are given as in (3.76) then (3.87) becomes

$$\sum_{k=1}^{n}\sum_{\mu=1}^{l} c_{k\mu} B_k(x_i) B_\mu(y_i) = f_i, \quad i = 1,2,\ldots,m. \tag{3.88}$$

Now the B-splines will no longer be constructed with the interpolation nodes $\{x_i\}$ and $\{y_i\}$, but with knots depending only on the area covered by the nodes and perhaps on the density of the nodes in different sub-areas. Details of how to manage this can be found in the routine documents which describe NAG routine E02DAF. For this more general problem both the amount of time and storage necessary will grow. The linear system (3.88) is square only if $m = n\,l$. In order not to be bound to this condition the system is best solved with the method of least squares. Thus one can obtain a solution to the approximation problem if $m > n\,l$.

3.7.2 Program and Examples

INTERDIM2, see B.3.3, uses NAG routines E01DAF, E02DAF, E02DEF and E02ZAF to interpolate or approximate two-dimensional data on a rectangular grid or arbitrarily distributed two-dimensional data. To show how it works we will run one example using both possibilities in turn.

In Figure 3.14 (right) we see the two-dimensional function $F(X,Y)$, which represents the meeting of two shallow-water waves. Mathematically it represents the solution of a Korteweg–de Vries equation. This is a partial differential equation which in some special cases is solvable analytically (see [35]). The solution (with y as the time-variable) is

$$F(x,y) \quad := \quad -2\frac{(a_1 + a_2 + \frac{5}{168}a_{12})^2}{(1 + \frac{5}{8}a_1 + \frac{5}{6}a_2 + \frac{25}{2352}a_{12})^2} + 2\frac{\frac{8}{5}a_1 + \frac{6}{5}a_2 + \frac{1}{12}a_{12}}{1 + \frac{5}{8}a_1 + \frac{5}{6}a_2 + \frac{25}{2352}a_{12}}$$

with

$$a_1 \quad := \quad \exp\left(\frac{128}{125}y - \frac{8}{5}x\right)$$

$$a_2 \quad := \quad \exp\left(\frac{54}{125}y - \frac{6}{5}x\right)$$

$$a_{12} \quad := \quad \exp\left(\frac{182}{125}y - \frac{14}{5}x\right).$$

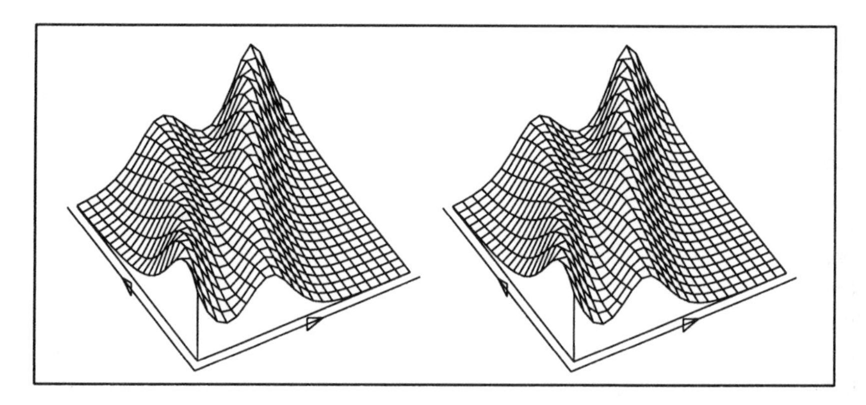

Figure 3.14. Spline approximation(left) of a 2Soliton(right)

We will approximate this function in the rectangle $[-8,7] \times [-8,7]$ with a bicubic spline. As set of data points we take the integer grid points with the respective function values. As knots for the construction of the bicubic spline we choose

$$LAMDA(4+i) = MU(4+i) := -7+i \quad \text{for} \quad i = 1, 2, \ldots, 12.$$

Taking into consideration the fourfold boundary knots, this gives rise to $20 \times 20 = 400$ knots. We evaluate the approximating spline surface at 31×31 points and draw it together with the function itself with NAG routine J06HDF (see Appendix D). The maximal error of this approximation is about 10% at values in the lower left corner. The root-mean-square error is much smaller. We will give two input files for this example. The first uses the possibility of input data on a rectangular grid; the second uses the same data but not from a grid, to show the difference. The resulting spline approximation is the same. In both cases we use a function for evaluating the input data. For the input of a set of data points the function values have to be added at the appropriate place.

Input Files for Two Cases

To economize on space we show the numerical data at the end of the text lines instead of starting a new line as it must be in the actual input files.

Case 1: Rectangular grid.

```
Input file for the PAN program INTERDIM2:
Rectangular grid (no = 0, yes = 1):                                1
Input of values at data points (no = 0, yes = 1):                  0
Is the rectangular grid equidistant (no = 0, yes = 1):             1
Number of data points in x direction:                             12
Number of data points in y direction:                             12
Dummy dimension of next vector:                                    2
Vector of first and last x co-ordinate of data points:          -6 5
Dummy dimension of next vector:                                    2
Vector of first and last y co-ordinate of data points:          -6 5
Evaluation at single points (no = 0, yes = 1):                     0
Evaluation on an equidistant grid (no = 0, yes = 1):               1
Number of points in x direction:                                  12
Number of points in y direction:                                  12
Graphical output (no = 0, X Window = 1, PostScript = 2):           1
Number of grid lines in x direction:                              23
Number of grid lines in y direction:                              23
Angle of elevation:                                               45
Rotation:                                                        240
```

Case 2: Non-rectangular grid.

```
Input file for the PAN program INTERDIM2:
Rectangular grid (no = 0, yes = 1):                                0
Input of values at data points (no = 0, yes = 1):                  0
Number of parallels to the y axis:                                12
Number of parallels to the x axis:                                12
Number of x coordinates of parallels to the y axis:               12
Vector of x coordinates:              -6 -5 -4 -3 -2 -1 0 1 2 3 4 5
Number of y coordinates of parallels to the x axis:               12
Vector of y coordinates:              -6 -5 -4 -3 -2 -1 0 1 2 3 4 5
```

```
Number of x coordinates of data points:                      256
Vector of x coordinates:
-8 -7 -6 -5 -4 -3 -2 -1 0 1 2 3 4 5 6 7
-8 -7 -6 -5 -4 -3 -2 -1 0 1 2 3 4 5 6 7
                ...
-8 -7 -6 -5 -4 -3 -2 -1 0 1 2 3 4 5 6 7
Number of y coordinates of data points:                      256
Vector of y coordinates:
-8 -8 -8 -8 -8 -8 -8 -8 -8 -8 -8 -8 -8 -8 -8 -8
-7 -7 -7 -7 -7 -7 -7 -7 -7 -7 -7 -7 -7 -7 -7 -7
                ...
7 7 7 7 7 7 7 7 7 7 7 7 7 7 7 7
Epsilon for determining the rank:                        1.0e-10
Evaluation at single points (no = 0, yes = 1):                 0
Evaluation on an equidistant grid (no = 0, yes = 1):           1
Number of points in x direction:                              12
Number of points in y direction:                              12
Graphical output (no = 0, yes = 1):                            1
Number of grid lines in x direction:                          23
Number of grid lines in y direction:                          23
Angle of elevation:                                           45
Rotation:                                                    240
```

3.8 Interpolation of Curves

3.8.1 Introduction

Let

$$\left(x_1^{(i)}, x_2^{(i)}, \ldots, x_m^{(i)}\right), \quad i = 0, 1, \ldots, n \tag{3.89}$$

be $n+1$ points in $\mathbb{R}^m$. They will be interpolated by a smooth curve.

A parameter $t \in \mathbb{R}$ is introduced, for example the arc length along the curve, normalized to the interval $[0,1]$. We are looking for m one-dimensional functions

$$x_1(t), x_2(t), \ldots, x_m(t), \quad t \in [0,1], \tag{3.90}$$

which interpolate the components of the given points for the parameter values t_i:

$$x_k(t_i) = x_k^{(i)}, \quad i = 0, 1, \ldots, n, \quad k = 1, 2, \ldots, m. \tag{3.91}$$

As parameter values we can take

$$t_0 \quad := \quad 0$$

$$t_i \quad := \quad t_{i-1} + \sqrt{\sum_{k=1}^{m} (x_k^{(i)} - x_k^{(i-1)})^2}, \quad i = 1, \ldots, n. \qquad (3.92)$$

They can be normalized by

$$t_i := \frac{t_i}{t_n}, \quad i = 1, \ldots, n. \qquad (3.93)$$

3.8.2 Numerical Solution by Splines

We solve m one-dimensional spline interpolation problems. As interpolation nodes we always have the parameter values t_i; function values for the kth spline $x_k(t)$ are the values $x_k^{(i)}$. Then equation (3.91) is fulfilled. The parametric curve in $\mathbb{R}^m$ is determined as

$$(x_1(t), x_2(t), \ldots, x_m(t)) \quad t \in [0, 1]. \qquad (3.94)$$

3.8.3 Program and Examples

The program INTERPAR, see B.3.4, solves the curve interpolation problem in m dimensions with cubic splines. It uses NAG routines E01BAF and E02BBF. As input it only needs the values m for the dimension of the problem and n for the number of data points, and then the points $x^{(i)}$, for $i = 0, 1, \ldots, n$, row by row.

Parametric or curve interpolation is important for many graphical purposes, for example in Computer Aided Design (CAD). Details of other methods like Beta-splines or Bezier curves, which use more sophisticated methods for controlling the shape of the curve, can be found in several textbooks on computer graphics and related mathematics (see for example [7], [8], [24]). Even the scientific text publishing system LaTeX, [65], which this text was written with, has a simple option to draw smooth curves (quadratic Bezier ones). We have used this possibility for example for the wooden spline in Figure 3.3.

An example will show how efficient or inefficient curve interpolation can be. We wish to draw a distorted Archimedean spiral:

$$r(\phi) = c\,\phi \left(1 + \frac{bc\phi}{d} \sin(a\,\phi)\right). \qquad (3.95)$$

(r, ϕ) are polar coordinates. We take the following values:

$d = 500$:	distortion	$c = 2$:	opening of the spiral
$b = 0.2$:	superposition factor	$a = 5.03$:	superposition frequency

Here we obtain the parameter representation as

$$(x(t), y(t)) = (r\cos(\phi), r\sin(\phi)) \qquad (3.96)$$

with $t = \phi$. We will draw 32 turns of the spiral and try this with 503 and 1006 points. The results can be seen in Figure 3.15. The lower curves interpolate 503 points, the upper ones 1006 points. The left-hand-side curves were interpolated with a piecewise linear polynomial (broken line), the right-hand-side ones with cubic splines. The difference makes obvious how important reasonable software is for graphics.

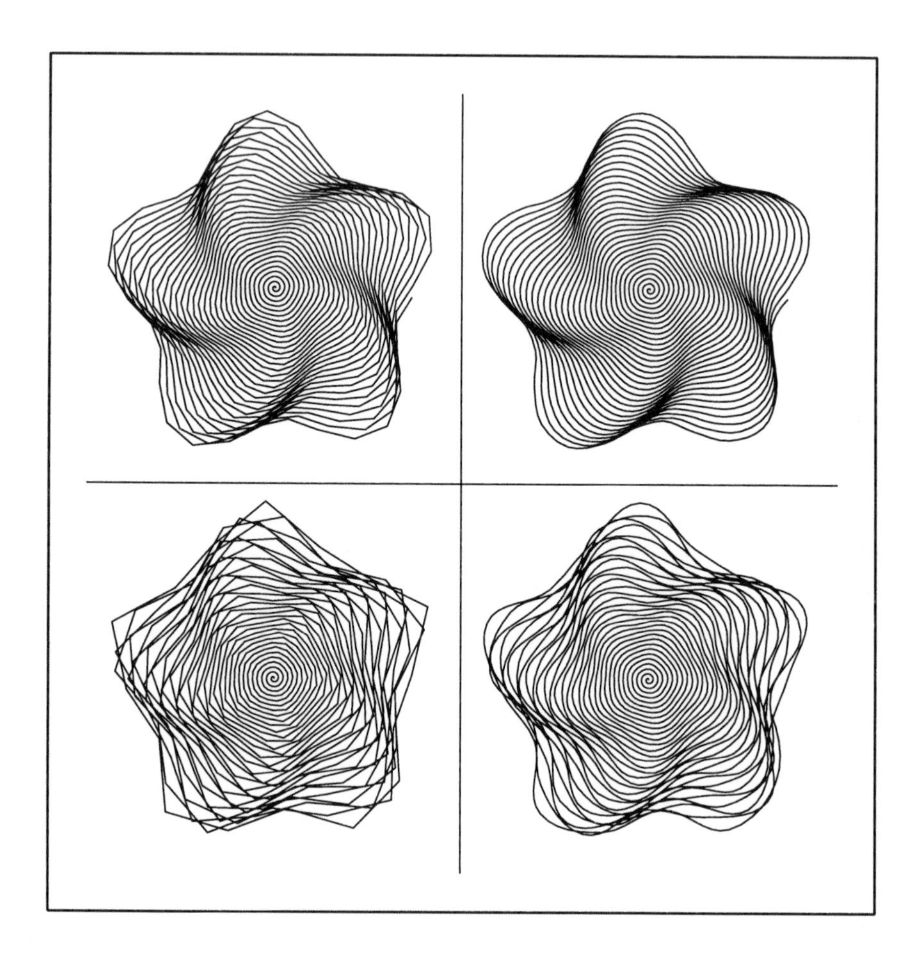

Figure 3.15. Broken line and spline interpolation with 503 and 1006 curve points

3.9 Interpolation and Approximation using IMSL

3.9.1 Polynomial Interpolation and Approximation

The IMSL library contains no routine which merely interpolates like NAG routines E01AAF, E01AEF or E02AFF. IMSL routines RCURV and RLINE calculate interpolating or approximating polynomials. RCURV approximates a set of points in $\mathbb{R}^m$ with the method of least squares. The polynomial approximation in IMSL is carried out as a regression analysis, i.e. mean value, variance and the root-mean-square error are calculated for the approximation. RLINE is equivalent to RCURV, but it uses only linear polynomials. The polynomials are given in the form $p(x) = a_0 + a_1 x + \ldots + a_n x^n$, not as a Chebyshev series as in NAG routines E01AEF and E02AFF. The results of RCURV tend to oscillate if the polynomial degree becomes too great, say greater than 10. RCURV is similar to NAG routines E02ADF and E02AEF, but without the possibility of weighting the input data. The IMSL routine for evaluating the polynomial is PPVAL. PPVAL can evaluate piecewise polynomials as well. Integrals and derivatives are calculated by IMSL routines PPITG and PPDER, respectively. These routines only calculate single values for the integral or the derivative and not some series for them like NAG routines E02AJF and E02AHF. PPITG and PPDER can evaluate piecewise polynomials as well.

3.9.2 One-Dimensional Spline Interpolation

In contrast to the NAG Fortran library, IMSL is able to calculate splines of different degrees. IMSL holds the following routines for one-dimensional interpolation with cubic splines:

CSINT Find a spline, the third derivative of which is continuous at the second and the second last interpolation node.

CSDEC Find a spline, the first or second derivative of which is prescribed at the interval end-points.

CSHER Hermite interpolation.

CSAKM Akima interpolation.

CSCON Find a spline with the same convexity properties as the given data.

CSPER Find a periodic splines.

There are two IMSL routines for least-squares approximation: BSLSQ, which is equivalent to NAG routine E02BAF, and BSVLS, which searches for the optimal knots when the number of knots is given. NAG routine E01BAF for the calculation of a one-dimensional B-spline corresponds to IMSL routine BSINT. NAG routines E02BBF, E02BCF and E02BDF for the evaluation of the spline, its derivatives and its integral correspond to IMSL routines BSVAL, BSDER and BSITG, respectively. E02BCF always calculates the first, second and third derivatives; BSDER has to be given the number wanted. BSITG can integrate the spline for a given interval, E02BDF always integrates over the interval formed by the interpolation nodes.

3.9.3 Trigonometric Approximation

NAG routine C06FAF corresponds to IMSL routine FFTRF. FFTRF has no limitation so far as the prime factors of n are concerned.

3.9.4 Multi-Dimensional Splines

The NAG library contains routines only for one- and two-dimensional interpolation with B-splines. IMSL contains one routine, BS3IN, for three-dimensional interpolation as well. There is no IMSL routine for interpolation at one point only, like NAG routine E01ACF. NAG routine E02DAF corresponds to IMSL routine BS2IN, but BS2IN can not approximate the given data like E02DAF. BS2IN only works on grids, it does not work on arbitrarily distributed data points like E02DAF. When using cubic B-splines BS2IN needs much less workspace than E02DAF, namely $116\max(n_x, n_y)+8n_x+4n_y$ bytes, if n_x and n_y are the numbers of data points in the x- and the y-direction, respectively. The workspace necessary for E02DAF grows quadratically with the number of interpolation points in the y-direction and linearly with the number of interpolation points in the x-direction. NAG routine E02DBF for evaluating the interpolating function corresponds to IMSL routine BS2VL (or BS3VL for three-dimensional splines). BS2VL and BS3VL can evaluate the spline only at one point. IMSL contains routines for the evaluation of derivatives of two- and three-dimensional splines (BS2DR and BS3DR) and for integrating them over a rectangle and a cuboid (BS2IG and BS3IG).

3.10 Exercises

3.10.1 Beloved B-Splines

Hat Functions

Let a data table (x_i, y_i), $i = 0, \ldots, n$, be given with $x_0 < \cdots < x_n$.
Let $s : [x_0, x_n] \to \mathbb{R}$ be the piecewise linear function interpolating the values y_i. (s is a linear combination of hat functions).

Show that s is the only function fulfilling the following conditions:

1. s is continuous in $[x_0, x_n]$ and continuously differentiable in (x_i, x_{i+1}) for $0 \le i < n$.

2. $s(x_i) = y_i, \quad i = 0, \ldots, n.$

3. Under all functions r fulfilling the first two conditions s uniquely minimizes the functional

$$F(r) := \int_{x_0}^{x_n} (r'(t))^2 \, dt.$$

Cubic B-Splines

Show that for the cubic B-splines $B_i(x) := B_{i,3}(x)$ defined in Section 3.2.3 the following characteristic positivity inequalities (equivalent to those on Page 92) hold:

$$\begin{aligned} B_i(x) &> 0 \text{ for } t_i < x < t_{i+k} \\ B_i(x) &= 0 \text{ if } x < t_i \text{ or } t_{i+k} < x. \end{aligned}$$

3.10.2 Square-Wave Impulse

Consider the 2π-periodic square-wave function f defined by

$$f(x) := \begin{cases} -2 & \text{if} \quad 0 \leq x < \pi \\ 2 & \text{if} \quad \pi \leq x < 2\pi. \end{cases}$$

1. Calculate all coefficients of the Fourier series (3.59) of f.

2. Use the series found in 1 to calculate the finite Fourier series (3.55) $g_n(x)$ for $x = 4.5$ and $x = 5.9$ and for $n = 4, 16, 256$. (A pocket calculator with a few programming steps will do.)
   Alternatively you may use the PAN program APPROX to do the work for you. Even better, use APPROX to draw $g_n(x)$ for $n = 4, 16, 256$.

3.10.3 Design a Car

Take grid paper and your hand or a simple plotting program at your computer to draw a simple car silhouette like this one:

Measure as many points as you like from your grid paper, interpolate them by a spline curve and compare the plots. You may use the PAN program INTERPAR or write your own program according to Section 3.8 setting $m = 2$. Where are the critical points? How should you distribute your measurements to get a better approximation? Did your car approximation look like mine:

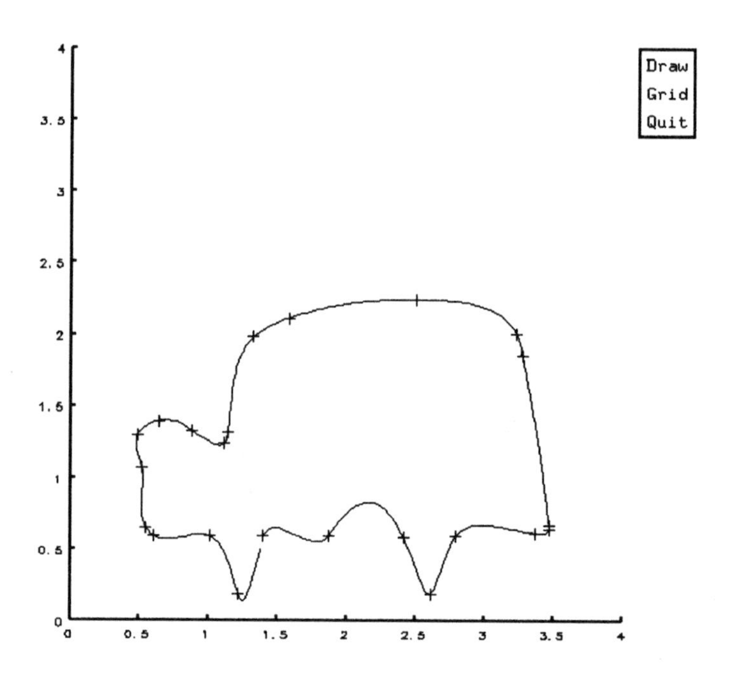

Figure 3.16. Poorly approximated car

4
Nonlinear Equations

The determination of roots of a given function is one of the classical mathematical problems. In most applications this problem is part of a larger one. Often it is a multi-dimensional problem. This means that we have to solve a system of nonlinear equations. Normally it is not possible to find analytical solutions. Therefore we need reliable numerical methods.

For simple problems a scaled drawing may yield good approximations for the roots, which can be improved with one of the well known numerical methods like Newton's method. Software libraries should contain reliable Black Box routines as well as general purpose routines with certain control possibilities for the user.

We will look first at simple methods, which are in part the basis for the more sophisticated methods. Later we consider methods of more importance for specific use in software libraries. They are not as straightforward as the simple methods, but they have to achieve the aim of guaranteeing convergence to a root or a fixed point even when used as Black Box routines. Therefore we often find combined methods: a simple find-algorithm to give a value close to the root, and a complicated method to reach the root with as few steps as possible and with high precision.

4.1 Introduction

4.1.1 The Problem

Let f be a continuous vector function: $f : \mathbb{R}^n \to \mathbb{R}^n$.
Find a vector $\bar{x} \in \mathbb{R}^n$ with

$$f(\bar{x}) = 0. \tag{4.1}$$

This is a system of n nonlinear equations with n unknowns:

$$\begin{aligned} f_1(x_1, x_2, \ldots, x_n) &= 0, \\ &\vdots \\ f_n(x_1, x_2, \ldots, x_n) &= 0. \end{aligned} \tag{4.2}$$

Example 4.1. We look for a solution of the system

$$\begin{aligned} x_2 \, e^{x_1} - 2 &= 0, \\ x_1^2 + x_2 - 4 &= 0. \end{aligned}$$

This means that

$$\begin{aligned} f_1(x_1, x_2) &= x_2 \, e^{x_1} - 2, \\ f_2(x_1, x_2) &= x_1^2 + x_2 - 4. \end{aligned}$$

These equations can be written as:

$$\begin{aligned} x_2 &= 2 \, e^{-x_1}, \\ x_2 &= 4 - x_1^2. \end{aligned} \tag{4.3}$$

Therefore we are looking for intersection points of these functions $x_2(x_1)$. For these intersection points we can get approximate values from a plot, see Figure 4.1. We see that there are two intersection points. Figure 4.1 yields good approximate values for both of them. ■

The example shows that even simple nonlinear problems often have several solutions. The problem of existence and uniqueness of solutions is not as easy as with linear systems.

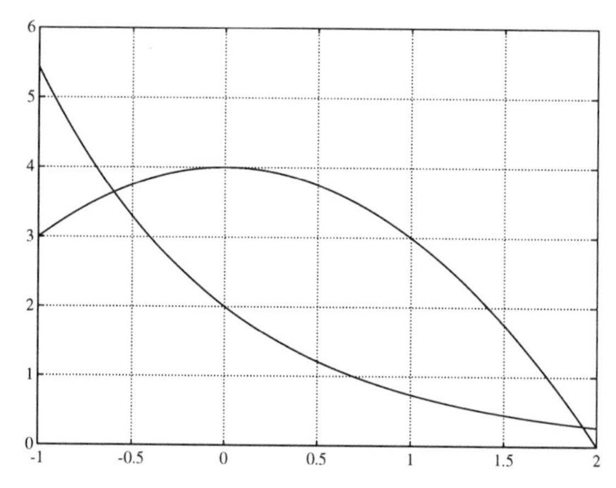

Figure 4.1. Intersection points of two functions

For the numerical solution of nonlinear systems we will mainly apply iterative methods, which can be written as

$$\begin{aligned} &\text{Choose:} && x_0 \in \mathbb{R}^n. \\ &\text{Calculate:} && x_{i+1} := \Phi(x_i), \quad i = 0, 1, 2, \ldots. \end{aligned} \tag{4.4}$$

The convergence of this general method can be discussed by considering special characteristics of the function Φ. But first the problem $f(x) = 0$ has to be transformed to an equivalent *fixed point problem.*

Let there be a continuous vector function $f : \mathbb{R}^n \to \mathbb{R}^n$.
Find a vector function $\Phi : \mathbb{R}^n \to \mathbb{R}^n$ with

$$f(\bar{x}) = 0 \iff \bar{x} = \Phi(\bar{x}). \tag{4.5}$$

$\bar{x}$ is called *fixed point* of Φ. Such a transformation is always possible; the simplest construction is

$$\Phi(x) := f(x) + x.$$

But not every transformation yields convergence for method (4.4). Therefore one looks for more systematic constructions of functions Φ from f.

4.1.2 General Convergence Theory

Definition 4.2. *The sequence $\{x_i\}$ with limit $\bar{x}$ is convergent with order p if a constant $K > 0$ exists, which is independent of i, such that*

$$||x_{i+1} - \bar{x}|| \leq K\, ||x_i - \bar{x}||^p \quad \forall i \geq 0 \tag{4.6}$$

with $K < 1$, if $p = 1$. K is called the convergence rate.

Remark 4.3.

1. *Small rates K and large orders p are desirable for fast convergence, but a large order is by far more important than a small rate. $p = 1$ means linear convergence, $p = 2$ quadratic convergence and $p = 3$ cubic convergence. One will not very often find higher convergence orders than $p = 3$.*
2. *There are inclusion methods such as bisection, which do not yield a sequence of real (vector) values, but one of (multi-dimensional) intervals instead, which include the solution. For considering the convergence of such methods one should rather look at the interval width than at the sequence of mid-points.*

Theorem 4.4. *Each method of pth order, $p \geq 1$, for determining a fixed point is locally convergent: There is a neighbourhood $U(\bar{x})$ such that for all starting values $x_0 \in U(\bar{x})$ the sequence $\{x_i\}$ generated by Φ converges with $\bar{x}$ as limit.*

Definition 4.5. *A function $\Phi : U \to U$, $U \subset \mathbb{R}^n$, is called contracting in a set $U \subset \mathbb{R}^n$, if for some norm in $\mathbb{R}^n$ we have:*

$$||\Phi(x) - \Phi(y)|| \leq K||x - y|| \;\; \text{with} \;\; K < 1 \;\; \forall x, y \in U. \tag{4.7}$$

Theorem 4.6. Fixed Point Theorem by Banach
Let $\Phi : \mathbb{R}^n \to \mathbb{R}^n$ have a fixed point $\bar{x} : \Phi(\bar{x}) = \bar{x}$ and let Φ be contracting in $U_r(\bar{x}) := \{x \in \mathbb{R}^n \mid ||x - \bar{x}|| < r\}$.
Then the method $x_{i+1} = \Phi(x_i)$, $i = 0, 1, 2, \ldots$, is at least linearly convergent for all starting values $x_0 \in U_r(\bar{x})$ and we have for $i = 1, 2, \ldots$.

1. $$x_i \in U_r(\bar{x}).$$
2. $$||x_i - \bar{x}|| \leq K^i ||x_0 - \bar{x}||. \tag{4.8}$$
3. *A priori error estimate:*
$$||x_i - \bar{x}|| \leq \frac{K^i}{1-K} ||x_1 - x_0||. \tag{4.9}$$
4. *A posteriori error estimate:*
$$||x_i - \bar{x}|| \leq \frac{K}{1-K} ||x_i - x_{i-1}||. \tag{4.10}$$

The contraction condition can be replaced by a condition on the derivatives of Φ, if the respective derivatives exist. We will consider this condition only for a single nonlinear equation ($n = 1$).

Theorem 4.7. *Let Φ: $\mathbb{R} \to \mathbb{R}$ be $p + 1$ times continuously differentiable:*

$$\Phi \in C^{p+1}(\mathbb{R}) \quad or \quad \Phi \in C^{p+1}(U_r(\bar{x})).$$

Then the iteration method (4.4) is at least linearly convergent for all $x_0 \in U_r(\bar{x})$ if

$$|\Phi'(x)| < 1 \quad for\ x \in U_r(\bar{x}). \tag{4.11}$$

The method is of pth order if

$$\begin{aligned} \Phi^{(k)}(\bar{x}) &= 0 \quad for \quad k = 1, 2, \ldots, p-1, \\ but \quad \Phi^{(p)}(\bar{x}) &\neq 0. \end{aligned} \tag{4.12}$$

Example 4.8. We are looking for a root of the function

$$f(x) = x - \exp\left(1 - \frac{1}{x^2}\right).$$

The easiest transformation to a fixed point equation is

$$x = \exp\left(1 - \frac{1}{x^2}\right) \Leftrightarrow f(x) = 0.$$

$$\text{This yields:} \quad \Phi(x) = \exp\left(1 - \frac{1}{x^2}\right).$$

For the starting value $x_0 = 2$ we obtain the sequence

$$\begin{array}{lllll} x_1 &=& \Phi(x_0) \quad = \exp(1-1/4) &=& 2.117 \\ x_2 &=& 2.1746 \qquad\qquad\qquad x_3 &=& 2.2002 \\ x_4 &=& 2.2110 \qquad\qquad\qquad x_5 &=& 2.2154 \\ \ldots && \ldots \qquad\qquad\qquad\quad x_{10} &=& 2.2184. \end{array}$$

The following iterates are equal to x_{10} within the five digits shown. We will consider this slowly convergent method more detailed with the given theorems:

$$\Phi'(x) = \frac{2}{x^3} \exp\left(1 - \frac{1}{x^2}\right).$$

It is

$$|\Phi'(x)| < 1\,, \quad \text{if } x > 1.523,$$

and at the fixed point

$$\Phi'(\bar{x}) \approx 0.42.$$

Hence the method is linearly convergent for a neighbourhood $U_r(2.2184)$ with $r < 0.69$. As contraction constant we take the absolute maximum of the derivative of Φ in $U_{0.2184}(\bar{x})$, which is $K = \Phi'(2) \leq 0.53$. The a priori and a posteriori error estimates for x_5 are:

$$\begin{array}{lllll} |x_5 - \bar{x}| &\leq& \dfrac{0.53^5}{0.47}\, 0.117 = 0.0104 & & \text{a priori,} \\[2ex] |x_5 - \bar{x}| &\leq& \dfrac{0.53}{0.47}\, 0.0044 = 0.00496 & & \text{a posteriori,} \\[2ex] |x_5 - \bar{x}| &=& 0.003 & & \text{exactly.} \end{array}$$

As is shown in this case the a posteriori estimate is generally sharper than the a priori estimate, because it contains more information about the sequence $\{x_i\}$. ■

4.1.3 Stability and Condition

With some one-dimensional examples we will try to see typical stability situations for the root finding problem. In Figure 4.2 we see four quadratic curves of identical shape, but with very different root positions. We have the following situations from left to right.

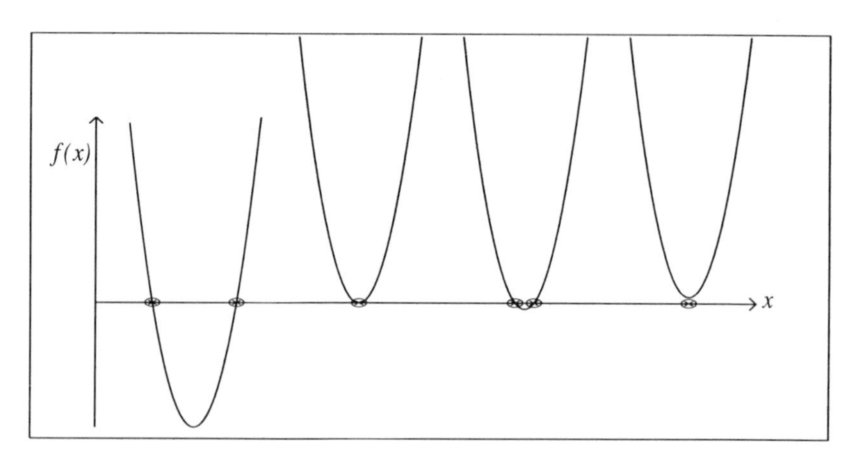

Figure 4.2. Different conditions for root finding

1. Stable situation with two well separated and easy-to-determine roots.
2. A zero of multiplicity two, which means

$$f(\bar{x}) = f'(\bar{x}) = 0. \tag{4.13}$$

   The numerical computation of a double root is always badly conditioned, because a small displacement by rounding errors may yield two or no real roots (see the next two situations).
3. Two or more different but very close roots may lead to numerical instability especially if we are looking for all roots.
4. This function only has complex roots. So either we calculate with complex numbers or we choose a method with which we obtain quadratic elementary factors $Ax^2 + Bx + C$. With these factors the complex roots are easily calculated. For polynomials there are such methods like Bairstow's method.

For more detailed considerations on stability, condition numbers and errors we draw the reader's attention to Wilkinson's book [97]. There we learn for example that even the root finding problem for polynomials with well separated roots can be badly conditioned, see Figure 4.3. The roots of the functions shown have the same distance from each other, but if we define an interval $[x_-, x_+]$ for each of the roots, such that

$$f(x) < \varepsilon \;\; \forall x \in [x_-, x_+] \tag{4.14}$$

with ε small, then we obtain for the function at the left-hand side two small and well separated intervals, while the respective intervals for the function at the right-hand side are much wider or even unite to one interval.

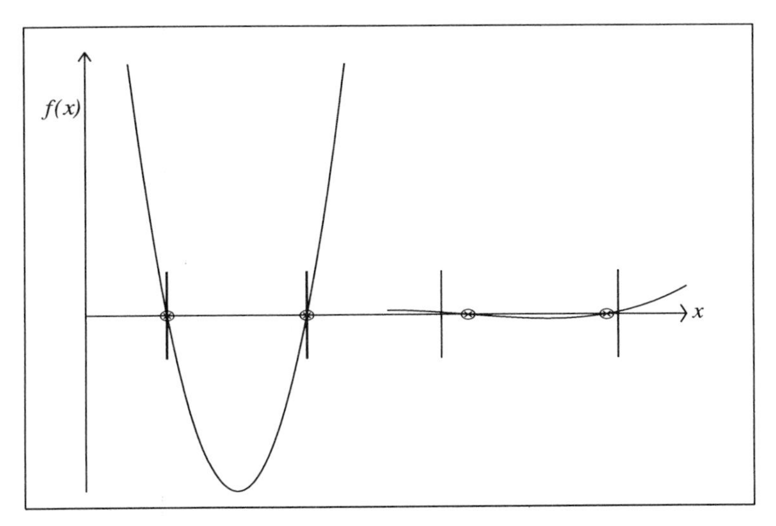

Figure 4.3. Differently conditioned situations

4.2 A Single Nonlinear Equation

Let f be a continuous function on $[a,b]$. Find a point $\bar{x} \in [a,b]$ with

$$f(\bar{x}) = 0. \qquad (4.15)$$

For solving this problem we firstly will consider the two best known methods, Newton's method with fast convergence, and the method of bisection with slow, but guaranteed, convergence of intervals, which include the root $\bar{x}$. Afterwards we will consider more complicated Black Box methods, which normally are used in software libraries.

4.2.1 Newton's Method

Newton's method uses the derivative of f to find successive approximations to the root $\bar{x}$. Therefore we assume that f' exists in $[a,b]$. Newton's method needs a good initial guess x_0 of the solution, because it has no guaranteed convergence for all x_0. Then it takes as approximation x_{i+1} the root of the line tangent to the graph of f at the last approximation x_i, see Figure 4.4.

$$\begin{aligned} &\text{Initial guess:} && x_0. \\ &\text{For } \ i = 0,1,2,\ldots, && x_{i+1} = x_i - \frac{f(x_i)}{f'(x_i)}, \quad f'(x_i) \neq 0. \end{aligned} \qquad (4.16)$$

If Newton's method converges, then it converges quite quickly; it yields quadratic convergence for simple roots as is shown in Figure 4.4.

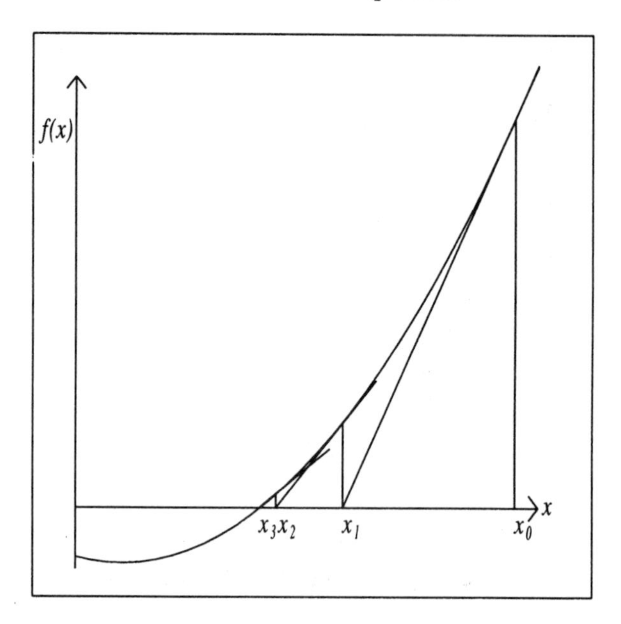

Figure 4.4. Convergence with Newton's method

If the initial guess x_0 is not sufficiently near to the root, we may have divergence, convergence to a root further away, or alternating divergence as shown in Figure 4.5.

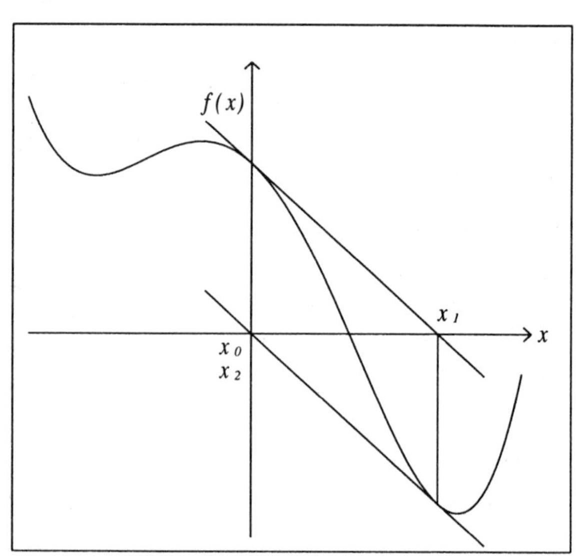

Figure 4.5. Divergence with Newton's method

4.2.2 Bisection Method

If we know an interval $[a_0, b_0]$ with $f(a_0) \cdot f(b_0) < 0$ and if f is a continuous function, then bisection is a simple inclusion method, with which we can find an arbitrarily narrow inclusion for a root in $[a_0, b_0]$ simply by interval bisection and looking for the sign of f at the interval mid-points.

(1) Initial interval $[a_0, b_0]$ with $f(a_0) \cdot f(b_0) < 0$.
(2) Let $i := 0$.
(3) Interval mid-point: $m := 0.5\,(a_i + b_i)$.
(4) Function evaluation: $fm := f(m)$.
(5) If $fm = 0 \rightarrow$ STOP, m is a root.
(6) If $f(a_i) \cdot fm < 0$,then

$$\text{let } a_{i+1} := a_i, \quad b_{i+1} := m,$$

otherwise

$$\text{let } a_{i+1} := m, \quad b_{i+1} := b_i.$$

(7) If $|b_{i+1} - a_{i+1}| < \varepsilon \rightarrow (9)$.
(8) Let $i := i + 1$ and continue with (3).
(9) Take $\tilde{x} := 0.5(a_{i+1} + b_{i+1})$ as root approximation.

For the approximation $\tilde{x}$ we have $|\tilde{x} - \bar{x}| < 0.5\,|b_{i+1} - a_{i+1}|$. This means that the maximum error for the bisection method converges linearly to zero with convergence rate 0.5.

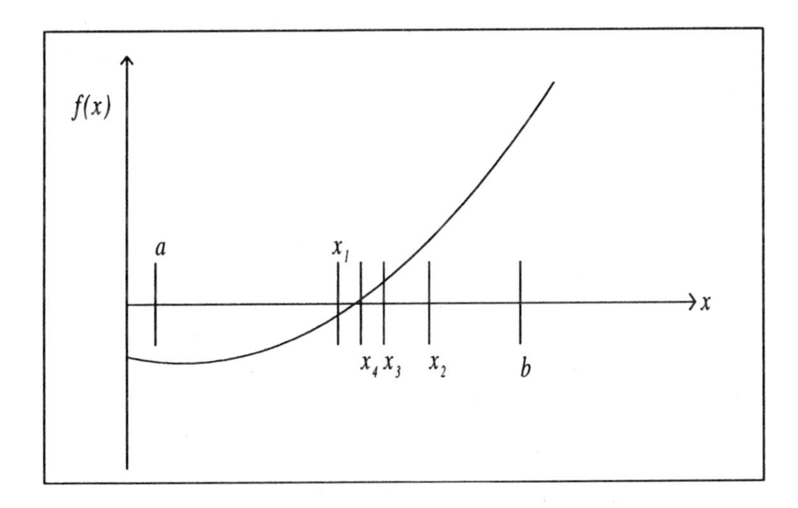

Figure 4.6. Bisection in $[a, b]$

We will demonstrate the convergence of the different methods by Example 4.8, which we will treat again with different methods.

Example 4.9.

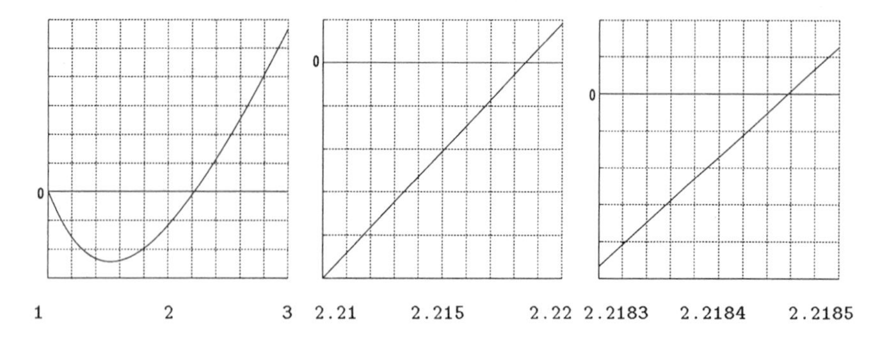

Figure 4.7. Graphical method for finding a root

Let f again be defined as

$$f(x) = x - \exp\left(1 - \frac{1}{x^2}\right),$$

with

$$f'(x) = 1 - \frac{2}{x^3}\exp\left(1 - \frac{1}{x^2}\right).$$

f obviously has a root at $\bar{x} = 1.0$ and one in $[2, 3]$. We will find an approximation for the latter one. In Section 4.1.2 we applied an intuitive method of successive approximations to f. Now we will apply a graphical and some more systematic methods to the same problem.

First we will play a graphical game. We draw the function with a graphical tool in $[1, 3]$ (see the left-hand plot in Figure 4.7). With the help of a grid it is easy to see that the root is included in $[2.21, 2.22]$, therefore we now draw f for that interval (see the middle plot in Figure 4.7). Again with the grid and a ruler we can reduce the size of the interval to $[2.2182, 2.2187]$ to get the right-hand plot in Figure 4.7. From this figure we read an approximate value $\tilde{x} = 2.21846$ for the root, which is correct to six digits. This shows that this funny man/machine method is not too bad.

Newton's method yields seven correct digits for every initial guess in $[2, 3]$ with only four iteration steps, whereas the bisection method with the initial interval $[a_0, b_0] = [2, 3]$ needs eighteen steps for a similar precision.

Brent's method, which we will consider in the next section, only needs one step more than Newton's method for a similar precision. The main results of all these methods are shown in the table below. ■

Step No.	Newton	Bisection $0.5(a_i + b_i)$	Brent	Graphics	Exact Root
0	3.0	2.5	2.5	2.2	
1	2.3077	2.25	2.1709	2.2185	
2	2.2212	2.125	2.2214	2.21846	2.218457490
3	2.2184605	2.1875	2.21840	—	
4	2.2184575	2.221875	2.2184574	—	
5	2.2184575	2.203125	2.2184576	—	2.218457490
⋮	—	⋮	—	—	
18	—	2.2184561	—	—	2.218457490

4.2.3 Brent's Method

There are some methods which try to combine the guaranteed convergence of a method like bisection with a faster convergence speed. Such methods are important as Black Box routines for software libraries.

An example for such a combination is the *Regula Falsi method*. But it may converge extremely slowly, as Figure 4.8 shows. Here the Regula Falsi method converges even more slowly than the bisection method.

Brent's method, [9], tries to maintain the inclusion property and minimum convergence of the bisection method, but to be faster if possible. This aim is reached with inverse quadratic interpolation. Three points $a := x_{i-2}$, $b := x_{i-1}$ and $c := x_i$ are used; they are interpolated with a quadratic polynomial in y. Then one takes the polynomial value at $y = 0$ as the new value $x := x_{i+1}$.

$$
\begin{aligned}
\text{With } R &:= \frac{f(c)}{f(b)}, \quad S := \frac{f(c)}{f(a)}, \quad T := \frac{f(a)}{f(b)} \quad \text{and} \\
P &:= S\{T(R-T)(b-c) - (1-R)(c-a)\}, \\
Q &:= (T-1)(R-1)(S-1) \\
\text{we get } x &:= c + \frac{P}{Q}.
\end{aligned} \tag{4.17}
$$

In most cases the approximation c of the root is improved by adding P/Q. Then the convergence of the method is *superlinear*, which means that (4.6) is valid for some p with $1 < p < 2$. If the new iterate x_{i+1} yields no inclusion or if the correction P/Q yields a value outside the last inclusion interval, then a Regula Falsi method step or, if that is too slow, a step of the bisection method is made instead. This way the method guarantees convergence, which in most cases is superlinear.

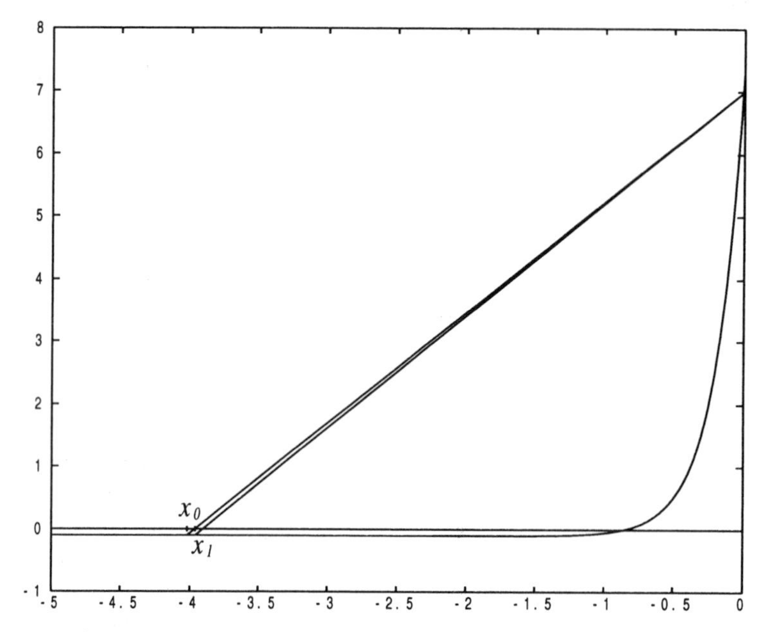

Figure 4.8. Slow convergence with the regula falsi

4.2.4 Determination of an Including Interval

For starting Brent's method, one needs an interval $[a, b]$ with $f(a) \cdot f(b) < 0$. Often such an interval is not known to the user, for example if the root finding method is only one part of a larger problem. Then one needs a 'find'-routine. We will look at a simple, but useful method, [93], which is one of many known methods. It needs a starting point a and a step length h, and has the ability of backtracking. The method is only weakly dependent of the values of the parameters a and h. Therefore one may consider it as a reliable Black Box routine.

Find-algorithm

(1) Choose a and h.
    Set $b := a$.
(2) $b := b + h$, $fa := f(a)$, $fb := f(b)$.
(3) If $fa \cdot fb \leq 0 \longrightarrow$ STOP.
(4) Is $\text{abs}(fa) < \text{abs}(fb)$?
    Yes: Set $h := -2h$.
    No: Set $h := 2h$, $a := b$, $fa := fb$.
(5) $\longrightarrow$ (2).

4.2.5 Broyden's Method

Broyden, [12], proposed a very successful method, which gives up the inclusion of the root. It starts with an arbitrary point x_0 and solves a sequence of problems, the first of which has x_0 as a root:

$$\begin{aligned} h(x,\alpha) &:= f(x) - \alpha f(x_0) \\ \text{with} \quad h(x_0,1) &= 0 \\ \text{and} \quad h(\bar{x},0) &= f(\bar{x}) = 0. \end{aligned} \tag{4.18}$$

This means that we can progress from a problem with known root to one with the root searched for. To achieve that we have to select a sequence of values

$$\alpha_1 = 1 > \alpha_2 > \cdots > \alpha_m = 0$$

such that each problem can be solved easily by using the solution to the preceding problem. One possible realization of this idea (according to [93]) is the following algorithm:

(1) **Initial phase**
Let $\alpha_1 := 1$, $\alpha_2 := 0.995$, $\alpha_3 := 0.99$
$x_1 := x_0$
Solve for x_2 and x_3:
$h(x, \alpha_r) = 0, \quad r = 2,3$ with the secant method.
Let $r := 3$.

(2) **Determination of α_{r+1}**
Interpolate the values
$(\alpha_{r-2}, x_{r-2}), (\alpha_{r-1}, x_{r-1}), (\alpha_r, x_r)$
by a quadratic polynomial
$q_r(\alpha) = x_r + a_r(\alpha - \alpha_r) + b_r(\alpha - \alpha_r)^2$ and let
$v_r(\alpha) := x_r + a_r(\alpha - \alpha_r)$
Let $\lambda := 1/2$ and determine α_{r+1} as the solution of
$\lambda |v_r(\alpha) - x_r| = |q_r(\alpha) - v_r(\alpha)|$
which means (note $\alpha_{r+1} < \alpha_r$):
$\alpha_{r+1} = \alpha_r - \lambda |a_r / b_r|$
Is $\alpha_{r+1} < 0$, then let
$\alpha_{r+1} = 0$

(3) **Secant method**
Determine x_{r+1} as the solution of
$h(x, \alpha_{r+1}) = 0$ with the secant method.

(4) Let $r := r + 1$.
If $\alpha_r > 0$, then go to (2).

(5) $\bar{x} := x_r$.

This algorithm can be improved by variations. For more safety and efficiency the values for λ may vary with the number of steps of the secant method. If the secant method fails, then α_{r+1} may be set back to the mean value

$$\alpha_{r+1} := \frac{\alpha_{r+1} + \alpha_r}{2}$$

and then the $(r+1)$th step is repeated. More details and a benchmark test is to be found in [93]. Obviously for this method one may replace the secant method by another one.

4.3 Systems of Nonlinear Equations

We now come back to the general problem (4.1).
Let $f : \mathbb{R}^n \to \mathbb{R}^n$. Determine a vector $\bar{x} \in \mathbb{R}^n$ such that

$$f(\bar{x}) = 0. \tag{4.19}$$

For this general problem we will reformulate Newton's method and describe the Powell hybrid method, which is more suitable as a 'Black Box' algorithm. Broyden's method from Section 4.2.5 is easily transferred to the multi-dimensional case, see [20]. Finally we like to refer to the recent ABS methods, [1]. They mainly coincide for nonlinear equations with the quasi-Newton methods of Brent/Broyden/Powell.

4.3.1 Newton's Method

For Newton's method the derivative $f'(x_i)$ has to be replaced for the multi-dimensional case by the functional or Jacobian matrix Df (often simply called *Jacobian*), which contains all partial derivatives of the functions f_i:

$$Df := \begin{pmatrix} \dfrac{\partial f_1}{\partial x_1} & \dfrac{\partial f_1}{\partial x_2} & \cdots & \dfrac{\partial f_1}{\partial x_n} \\ \dfrac{\partial f_2}{\partial x_1} & \dfrac{\partial f_2}{\partial x_2} & & \vdots \\ \vdots & & \ddots & \vdots \\ \dfrac{\partial f_n}{\partial x_1} & \dfrac{\partial f_n}{\partial x_2} & \cdots & \dfrac{\partial f_n}{\partial x_n} \end{pmatrix}. \tag{4.20}$$

Equation (4.16) then becomes

$$x^{(i+1)} = x^{(i)} - \left(Df(x^{(i)})\right)^{-1} f(x^{(i)}). \tag{4.21}$$

This can be written as a system of linear equations:

$$\begin{aligned} Df(x^{(i)})\,\Delta x^{(i)} &= -f(x^{(i)}) \\ \text{with } \Delta x^{(i)} &:= x^{(i+1)} - x^{(i)}. \end{aligned} \tag{4.22}$$

This system must be solved once per iteration step. As a new approximation one obtains:

$$x^{(i+1)} = x^{(i)} + \Delta x^{(i)}. \tag{4.23}$$

Example 4.10. We will apply Newton's method to Example 4.1. It is $n = 2$,

$$f(x) = \begin{pmatrix} x_2\, e^{x_1} - 2 \\ x_1^2 + x_2 - 4 \end{pmatrix}$$

and

$$Df = \begin{pmatrix} x_2\, e^{x_1} & e^{x_1} \\ 2\,x_1 & 1 \end{pmatrix}.$$

From Figure 4.1 we get as a starting value for the right root the approximation

$$(x_1^{(0)}, x_2^{(0)}) = (1.9, 0.3).$$

For this value the function f and the Jacobian take up the values

$$f(x^{(0)}) = (0.00577, -0.09)^T$$

and

$$Df(x^{(0)}) = \begin{pmatrix} 2.00577 & 6.68589 \\ 3.8 & 1 \end{pmatrix}.$$

The solution of (4.22) for $i = 0$ gives

$$\Delta x^{(0)} = (0.026, -0.0087)^T,$$

and therefore

$$x^{(1)} = (1.926, 0.2913)^T.$$

Now the values for the functions f_i and the matrix Df have to be calculated again for the next step. The results may be found in the following table with $f_1 := f_1(x_1^{(i)}, x_2^{(i)})$, $f_2 := f_2(x_1^{(i)}, x_2^{(i)})$ and $\varepsilon := \|x^{(i)} - x^{(i-1)}\|$.

i	$x_1^{(i)}$	$x_2^{(i)}$	f_1	f_2	ε
0	1.9	0.3	0.006	-0.09	—
1	1.925961	0.291349	-0.0008	0.0007	0.03
2	1.92573714	0.291536495	$-2_{10}-7$	$5_{10}-8$	$3_{10}-4$
3	1.92573712	0.291536536	$-5_{10}-15$	$8_{10}-16$	$5_{10}-8$

■

4.3.2 The Powell Hybrid Method

In [81] Powell describes a method for the solution of a system of nonlinear equations, which is a typical example for an algorithm suitable for a reliable software library routine. It is composed of different methods so that in most of the possible cases convergence can be shown. On the other hand the numbers of operations necessary for the different parts of the algorithm is minimized.

We will consider one step $i \to i+1$ of the algorithm, omitting the index (i) at all step-dependent terms except for the vectors x. The mathematical details may be found in [81].

The starting point for the method is the system of linear equations from Newton's method:

$$Df\,\Delta x = -f(x^{(i)})$$
$$x^{(i+1)} = x^{(i)} + \Delta x.$$

This will be modified for convergence reasons to

$$x^{(i+1)} = x^{(i)} + \lambda\,\Delta x. \tag{4.24}$$

The values $\lambda := \lambda_i$ are determined in each step such that

$$F(x^{(i+1)}) \; < \; F(x^{(i)}) \tag{4.25}$$

with the function

$$F(x) = \sum_{k=1}^{n} (f_k(x))^2 . \tag{4.26}$$

F is equal to zero if and only if all f_i are equal to zero and each local minimum of F with value zero is a global minimum. If Df is not a singular matrix, then this method yields a monotonically decreasing sequence of values $F(x^{(i)})$ with limit zero. However, if Df is singular at the root and F has a local minimum with a value greater than zero, then this method can converge to a wrong limit value; an example can be found in [81].

Another possibility of modifying Newton's method is the following. Minimizing the function $F(x)$ is equivalent to calculating the least-squares solution (lss) of the system $Df\,\Delta x = -f(x^{(i)})$. But the lss is the solution of the normal equations (see (1.38))

$$(Df)^T Df\,\Delta x = -(Df)^T f(x^{(i)}). \tag{4.27}$$

This can be modified to

$$\begin{aligned} \left((Df)^T Df + \mu I\right)\eta &= -(Df)^T f(x^{(i)}), \\ x^{(i+1)} &= x^{(i)} + \eta. \end{aligned} \tag{4.28}$$

with a small parameter $\mu := \mu_i$[†]. The solution η depends on μ: $\eta = \eta(\mu)$, and we have

$$\eta(0) = \Delta x .$$

On the other hand we have

$$\mu\eta \to -(Df)^T f(x^{(i)}) \quad \text{for} \quad \mu \to \infty. \tag{4.29}$$

This means that the method becomes the classical method of steepest descent with a length factor $1/\mu$ going to zero with $\mu \to \infty$. With these considerations one can show that there is always a value μ with

$$F(x^{(i+1)}) < F(x^{(i)}) . \tag{4.30}$$

This method can fail only if the function f has no bounded first derivatives or in extreme cases through the influence of rounding errors. For implementation in a software library one needs tests on too many loops or function evaluations anyway. One such test is

$$\begin{aligned} \text{Is} \quad F(x^{(i)}) \quad &> \quad M \, \| g^{(i)} \| \\ \text{with} \qquad \qquad & \\ g_j^{(i)} \quad &:= \quad \frac{\partial}{\partial x_j} F(x) \bigg|_{x=x^{(i)}} ? \end{aligned} \tag{4.31}$$

If this is the case for a sufficiently large M, then probably the method converges against a local minimum of F which is not a root of f. When designing a method for the solution of nonlinear equations one always has to think of the case that there may not exist a solution at all. For this case a test on the number of function evaluations is necessary as well.

For the case that the Jacobian Df is not given explicitly, the algorithm must contain an approximate method, which normally is the application of finite divided differences.

The number of operations needed in each step for solving the linear system is $O(n^3)$ but for standard cases the matrix Df only varies slightly. Therefore the method may be modified such that the inverse of Df is only calculated once at the beginning:

$$H^{(0)} := \left(Df(x^{(0)})\right)^{-1} \tag{4.32}$$

and then this matrix is updated only after each step. This can be done with Broyden's rank-one modification, see [11] or [92]. With this method the number of operations needed per step is only $O(n^2)$ instead of $O(n^3)$, while the convergence is superlinear instead of quadratic.

[†] This idea is applied to the regularized solutions of incorrectly posed problems or of ill-conditioned systems of equations, too.

4.4 Programs and Examples

The NAG library contains six routines for determining roots of a function $f : \mathbb{R} \to \mathbb{R}$, three written in 'direct communication' form and three in 'reverse communication' form. The 'direct communication' routines are designed for the unexperienced user. They call one or more of the 'reverse communication' routines, which permit more control of calculation. Direct and reverse communication routines differ in the following features.

- Function f: For 'direct communication' routines f can be coded as a user-supplied routine while 'reverse communication' routines only call single function values through their parameter lists.
- For 'direct communication' routines certain control parameters are fixed, while they can be set by the user for 'reverse communication' routines.
- The 'direct communication' routines are called once per root finding process, the 'reverse communication' routines are called once per step. That allows the user for example to change some of the control parameters from step to step depending on the interim results.

For this chapter we wrote two programs. DIRECT uses 'direct communication' routines for the solution of a single nonlinear equation only, while NLIN uses both 'direct' and 'reverse communication' routines. Therefore NLIN is an easy-to-use program for the solution of a system of nonlinear equations as well as one for the more sophisticated user with control of the iteration process from step to step.

4.4.1 A Single Equation

The program DIRECT, see B.4.1, uses NAG routines C05ADF, C05AGF or C05AJF for determining the root of a real-valued function f. They all are Black Box routines. The program is easy to understand, therefore we only look at Example 4.8, Page 126, again.

DIRECT finds the root with or without a start interval including the root using the template within the PAN system or preparing an input file. We will give input files for three different cases.

Case 1: Interval including the root is given

```
Input file for the PAN program DIRECT:
Epsilon (0 = 100*tau):
0.0
Eta:
0.0
Interval with change of sign (no = 0, yes = 1):
```

```
1
Left margin:
2.0
Right margin:
3.0
```

Case 2: Good starting value is known

```
Input file for the PAN program DIRECT:
Epsilon (0 = 100*tau):
0.0
Eta:
0.0
Interval with change of sign (no = 0, yes = 1):
0
Input of a good starting value (no = 0, yes = 1):
1
Starting value:
4.0
Maximal number of function evaluations:
100
```

Case 3: General case

For this case the program has to search for a good starting value itself:

```
Input file for the PAN program DIRECT:
Epsilon (0 = 100*tau):
0.0
Eta:
0.0
Interval with change of sign (no = 0, yes = 1):
0
Input of a good starting value (no = 0, yes = 1):
0
Rough estimate for the root:
5
Step length:
0.1
```

The resulting output file is the same in all these three cases:

```
Output file for the PAN program DIRECT:

Zero of the function:                    2.21845749
Value of the function at the zero:        0.00E+00
```

4.4.2 Systems of Nonlinear Equations

The program NLIN, see B.4.2, mainly uses the possibilities of NAG routines C05NBF, C05NCF, C05PBF and C05PCF. One should mention the fact that detailed control is worth the effort only for extreme cases or if one wishes to have interim results.

We will become better acquainted with the possibilities of the program by treating an extreme example. It is a linear transformation of Powell's example for a non-convergent case of the modified Newton method. Let

$$\begin{aligned} f_1(x) &= x_1 - 5, \\ f_2(x) &= \frac{10x_1 - 50}{x_1 - 4.9} + 2(x_2 - 5)^2. \end{aligned}$$

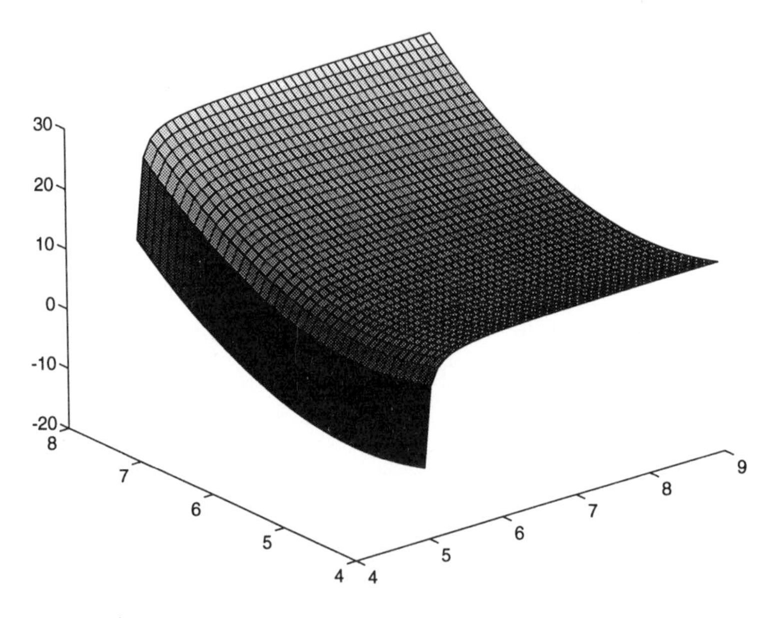

Figure 4.9. The trough function

This system has only one root at $\bar{x} = (5, 5)$. For this root the Jacobian

$$Df = \begin{pmatrix} 1 & 0 \\ 1/(x_1 - 4.9)^2 & 4(x_2 - 5) \end{pmatrix}$$

is singular. f_1 does not depend on x_2, f_2 is according to the variable x_2 a parabola with minimum in $x_2 = 5$, which is the x_2-value of the root. According to x_1 we find a pole of f_2 in $x_1 = 4.9$, which is very near to the root. On the right of the root the shape of $f_2(x_1, \bar{x}_2)$ becomes more and more flat for fixed $\bar{x}_2$ and the derivative $\partial f_2/\partial x_1$ has limit zero with

$x_1 \to \infty$. This means that we are in a long flat trough. The situation becomes clearer with Figure 4.9, where the function $f_2(x_1, x_2)$ is shown for $(x_1, x_2) \in [4.95, 8] \times [4, 8]$.

We found that for all starting values from the rectangle $[6.5, 7.5] \times [4.5, 5.7]$ the modified Newton method remains within the trough; see [81] for a more detailed discussion. For example (6.8016, 5) is an attraction point for the starting point (7.9, 5.6) and many similar pairs of starting values, but obviously no root.

For this example we have to write two subroutines for the calculation of the function and the derivative values. `fo` must return the values of the functions f_i at a point x. Depending upon the value of **`iflag`**, `fm` must either return the values of the functions f_i at a point x or return the Jacobian at x. If the Jacobian is computed by the program, i.e. it is not supplied by the user, `fm` can be declared as a dummy routine. Otherwise `fo` can be declared as a dummy routine.

We here will only show `FM`.

```
      SUBROUTINE FM(N,X,FVEC,FJAC,LDFJAC,IFLAG)
      INTEGER N, IFLAG, LDFJAC
      DOUBLE PRECISION FVEC(N), X(N), FJAC(LDFJAC,N)
      IF (IFLAG.EQ.0) THEN
        PRINT*,X(1), X(2),FVEC(1), FVEC(2)
      END IF
      IF (IFLAG.EQ.1) THEN
        FVEC(1) = X(1) - 5.0D0
        FVEC(2) = (1.0D1*X(1)-5.0D1)/(X(1)-4.9D0)
     *            +2.0D0*(X(2)-5.0)*(X(2)-5.0)
      ENDIF
      IF (IFLAG.EQ.2) THEN
        FJAC(1,1) = 1.0D0
        FJAC(1,2) = 0.0D0
        FJAC(2,1) = 1.0D0/((X(1)-4.9D0)*(X(1)-4.9D0))
        FJAC(2,2) = 4.0D0*(X(2)-5.0)
      ENDIF
      RETURN
      END
```

Now we will look at different calculations with NLIN for this example. We carried these calculations out with the template for NLIN within the PAN system. We here will give the input files produced by the template, which can be used for batch calculations.

1. Firstly we assume the knowledge of a good starting value near the solution.

```
Input file for the PAN program NLIN:
Number of functions:
2
Dimension of starting vector:
2
Starting vector:
6 6
Accuracy (0 = sqrt(tau)):
0.0
Supply of the Jacobian by the user (no = 0, yes = 1):
1
Checking of the Jacobian (no = 0, yes = 1):
0
Simple version of the program (no = 0, yes = 1):
1
```

It produces the following result:

```
Output file for the PAN program NLIN:

Input: Template

Root vector:

  1          5.00000000
  2          5.00000000

Values of the functions at the root vector:

  1            0.00E+00
  2            0.63E-29
```

2. If we now start near the attraction point (6.8016, 5) of the modified Newton method, and if we additionally let the routine calculate the derivative values, we get the input file

```
Input file for the PAN program NLIN:
Number of functions:
2
Dimension of starting vector:
2
starting vector:
6.8016 5
Accuracy (0 = sqrt(tau)):
0.0
```

```
Supply of the Jacobian by the user (no = 0, yes = 1):
0
Simple version of the program (no = 0, yes = 1):
1
```

and exactly the same results as above.

3. The dependence of control parameters will be shown next using the same starting point, but the sophisticated version of the program. The input file now reads

```
Input file for the PAN program NLIN:
Number of functions:
2
Dimension of starting vector:
2
starting vector:
6.8016 5
Accuracy (0 = sqrt(tau)):
0.0
Supply of the Jacobian by the user (no = 0, yes = 1):
0
Simple version of the program (no = 0, yes = 1):
0
Maximum number of function evaluations (0 = 200*(N+1)):
1000
Input of scaling factors (no = 0, yes = 1):
0
Initial step bound:
1
Number of steps between output of provisional results:
10
Maximum relative error in the functions:
0.0001
Is the Jacobian a band matrix (no = 0, yes = 1) ?
0
```

and yields the results

```
Output file for the PAN program NLIN:

Provisional root vector:
  1          6.80160000
  2          5.00000000
Provisional values of the functions at the root vector:
  1             0.18E+01
```

```
  2             0.95E+01

Provisional root vector:
  1           6.80160000
  2           5.00000000
Provisional values of the functions at the root vector:
  1             0.18E+01
  2             0.95E+01

-----  and finally -------------

Provisional root vector:
  1           6.80160000
  2           5.00000000
Provisional values of the functions at the root vector:
  1             0.18E+01
  2             0.95E+01

IFAIL =  5 in NAG routine C05NCF!
```

One sees that Powell's method here behaves like the modified Newton method. That is because the derivatives are only approximated by divided differences with $h = 0.01 = \sqrt{\text{`maximal relative error'}}$. All values for h smaller than 0.01 lead to the same result, while with $h = 0.1$ the root is calculated in machine precision. For this extreme case the method needs 71 steps. The interim results after 10, 20, ..., 70 steps are:

```
Root vector                Function values
6.801600   5.00000000         1.80160   9.5
5.000000   5.01702440         0.        5.8E-04
5.000000   5.00013842         0.        3.8E-8
5.000000   5.00000113         0.        2.5E-12
5.000000   5.00000001         0.        1.7E-15
5.000000   5.00000000         0.        1.1E-19
5.000000   5.00000000         0.        7.3E-24
```

Making the relative error tolerance even larger, $h = \sqrt{0.1}$, leads to the former error case again.

This example shows that even difficult situations may be overcome by using only the easy-to-use version of the program. On the other hand with that version the solution depends on the accuracy defined by the user, but we will not study this example further.

4.5 Nonlinear Equations Using IMSL

4.5.1 Solution of a Single Nonlinear Equation

NAG routine C05ADF corresponds to IMSL routine ZBREN using a similar algorithm. Additionally there is IMSL routine ZREAL, which searches for n roots of a function. IMSL does not contain a find-routine like C05AGF. ZBREN only knows one case of failure: 'No convergence after a maximum number of function evaluations'. IMSL contains no 'reverse communication' routines like NAG.

4.5.2 Systems of Nonlinear Equations

IMSL routines NEQNF and NEQNJ correspond to NAG routines C05NBF and C05PBF, respectively.

As with NAG routines, the Jacobian is calculated by divided differences in NEQNF. For NEQNJ the user must supply the Jacobian as a subroutine. IMSL routines use the Powell hybrid method like NAG routines, and error and parameter handling is very similar. More complex routines like C05NCF and C05PCF are not to be found in IMSL. There is no checking routine like NAG routine C05ZAF. The workspace used is similar in both libraries.

4.6 Exercises

4.6.1 Gamma Function

The Gamma function is defined as

$$\Gamma(x) := \int_0^\infty t^{x-1} e^{-t}\, dt, \quad x > 0.$$

Γ interpolates the values $n!$, $n \in \mathbb{N}$:

$$\Gamma(n) = (n-1)!$$

Formulate Newton's method for the determination of $\bar{x}$ in

$$\Gamma(\bar{x}) = p \ \text{ with } p \text{ given.}$$

Solve $\Gamma(\bar{x}) = 10$, $\bar{x} > 1$.

Approximate the integrals, which have to be evaluated, by the trapezoidal rule using step length $h = 2$ and replacing the interval $[0, \infty]$ by $[0, 16]$.

As a test problem take $\Gamma(\bar{x}) = 6$ which has the solution $\bar{x} = 4$.

4.6.2 Calculation of Effective Annual Interest Rates

A bank offers a loan of amount B with repayments of R at the end of each of n years such that after n years the loan and interest rate payments are paid back totally.

Let $f(p)$ be the remaining loan after n years for an interest rate p. This means that $f(\bar{p}) = 0$ for the correct interest rate $\bar{p}$ we are looking for.

1. Show that for $p > 0$ we have

$$\begin{aligned} f(p) &= (1+p)^n B - \left(\sum_{j=0}^{n-1} (1+p)^j \right) R \\ &= (1+p)^n B - \frac{(1+p)^n - 1}{p} R. \end{aligned}$$

2. Show that the annual interest rate p for the problem considered is uniquely determined if the total amount of repayments is larger than the loan:

$$n R > B \Longrightarrow \bar{p} \text{ is the only positive root of } f.$$

3. Define Newton's method for the problem $f(p) = 0$.
4. Let $B = £100,000$, $n = 20$ and $R = £10,000$. Use the program NLIN to determine $\bar{p}$. Try different starting values $p_0 \in [0, 1]$.

5

Matrix Eigenvalue Problems

Characteristic properties of natural systems can often be described by eigenvalues of certain differential equations, see Chapter 8. If the differential equation is discretized, a matrix eigenvalue problem has to be solved. This kind of problem will be considered in this chapter. We will restrict ourselves to real matrices. But the eigenvalues of a real matrix may be complex numbers. In considering numerical methods we will distinguish between the following problem classes:

- The standard and the general eigenvalue problem.
- Symmetric and nonsymmetric matrices with real or complex eigenvalues.
- Sparse matrices.
- Singular values of non-square matrices (cf. Chapter 1).
- Determination of one, several or all eigenvalues with or without eigenvectors.

This chapter is constructed slightly differently from the others. The first four sections are devoted to the description of the different problems and numerical methods for solving them. The fifth section describes the programs. The final section looks at some applications. It uses only one of the five programs written for this chapter, but the use of the others is similar and obvious, and the input files are almost identical apart from the matrix forms which are analogous to the forms used in Chapter 1.

5.1 The Standard Eigenvalue Problem

5.1.1 Introduction

Let $A \in \mathbb{R}^{n,n}$ be a matrix.

Determine complex numbers $\lambda_k \in \mathbb{C}$ and linearly independent vectors $x_k \in \mathbb{C}^n$, such that:

$$Ax_k = \lambda_k x_k, \quad k = 1, \ldots, n. \tag{5.1}$$

Generalizations and restrictions to this problem determine the classes of problems which we want to consider.

The eigenvalues are the roots of the *characteristic polynomial*

$$p(\lambda) := \det(A - \lambda I) \tag{5.2}$$

with $I \in \mathbb{R}^{n,n}$ the identity matrix. This property of the eigenvalues is only of theoretical interest; numerically it is used only in exceptional cases like very small or very sparse matrices, because the determination of the roots of the characteristic polynomial is numerically not stable and computationally complex.

Often the matrices are symmetric. This is a big advantage because then all eigenvalues are real and their numerical calculation is always well conditioned.

Theorem 5.1. *Let $A \in \mathbb{R}^{n,n}$ be symmetric. Then A has only real eigenvalues, and the eigenvectors are* orthogonal. *If $X \in \mathbb{R}^{n,n}$ is the matrix with eigenvectors x_i with $\|x_i\|_2 = 1$ as columns then X is an orthogonal matrix:*

$$X^T X = I. \tag{5.3}$$

The nonsymmetric case can be quite difficult. Therefore we will restrict our considerations to matrices similar to a diagonal matrix.

Definition 5.2.

1. *A is called* similar *to B, if there is a non-singular matrix $T \in \mathbb{C}^{n,n}$ with*
$$B := T^{-1}AT \tag{5.4}$$
and this transformation is called similarity *transformation.*
2. *A matrix $A \in \mathbb{R}^{n,n}$ is called* diagonalizable *if A is similar to a diagonal matrix Λ, i.e. if a nonsingular matrix $T \in \mathbb{C}^{n,n}$ exists such that*
$$\Lambda := T^{-1}AT, \quad \Lambda := diag(\lambda_1, \lambda_2, \ldots, \lambda_n), \quad \Lambda \in \mathbb{C}^{n,n}. \tag{5.5}$$

Lemma 5.3.

1. *If $T \in \mathbb{C}^{n,n}$ is a nonsingular matrix and if*
$$C := T^{-1}AT, \tag{5.6}$$
then A and C have the same eigenvalues. If $y \in \mathbb{C}^n$ is an eigenvector of C, then $x := Ty$ is an eigenvector of A:
$$Cy = \lambda y \Rightarrow Ax = \lambda x \quad with \quad x := Ty. \tag{5.7}$$

2. If $A \in \mathbb{R}^{n,n}$ is diagonalizable:

$$\Lambda := T^{-1}AT,$$

then the diagonal elements λ_i of Λ are the eigenvalues of A, and the columns t_i of T are the corresponding eigenvectors.

This lemma is used by the reduction methods which use matrix multiplications to obtain a simpler form like a tridiagonal or a Hessenberg matrix. The calculations can all be done in real arithmetic. Complex eigenvalues are represented by 2×2 diagonal blocks, complex eigenvectors by two real vectors which represent the real and the imaginary part respectively.

The steps of a typical method for *symmetric* matrices are described in the following algorithm:

(1) **Balancing**
Interchanging rows and corresponding columns of A
$\tilde{A} = P^TAP,$ P orthogonal: $P^{-1} = P^T$, see 5.1.2.

(2) **Reduction** of $\tilde{A}$ to tridiagonal form by multiplication with $n-2$ Householder matrices $U_1 \cdots U_{n-2} =: H$:
$B = H^T\tilde{A}H,$ H orthogonal: $H^{-1} = H^T$, see 5.1.3.

(3) **QR method**
This yields iteratively a diagonal matrix
$C = Q^TBQ,$ Q orthogonal: $Q^{-1} = Q^T$, see 5.1.4.

(4) **Eigenvalues and Eigenvectors**
Calculation according to Lemma 5.3 with $T = PHQ$.

We now describe the steps in more detail with special consideration to the alterations necessary for nonsymmetric matrices. For each of these steps we will give a short example.

5.1.2 Balancing

To achieve the best possible accuracy it is necessary to 'balance' the matrix if the elements vary considerably in size.

- **Symmetric matrix** A: Rows and corresponding columns are interchanged so that the diagonal elements are ordered according to size, beginning with the smallest value in a_{11}. This is a simple similarity transformation with a permutation matrix P for which $P^T = P^{-1}$.
- **Nonsymmetric matrix** A: The matrix A is balanced by a similarity transformation with a diagonal matrix D:

$$\tilde{A} = D^{-1}AD. \tag{5.8}$$

In $\tilde{A}$ the sums of the absolute values of the elements in corresponding rows and columns will have the same order of magnitude. To avoid rounding errors the elements of the diagonal matrices used for balancing should be restricted to be exact powers of the base employed, usually 2, 10, or 16.

Example 5.4. Let A be the symmetric matrix

$$A = \begin{pmatrix} 3 & -1 & -1 \\ -1 & 4 & -1 \\ -1 & -1 & 2 \end{pmatrix}.$$

Then the balanced one is

$$\tilde{A} = P^T A P = \begin{pmatrix} 2 & -1 & -1 \\ -1 & 3 & -1 \\ -1 & -1 & 4 \end{pmatrix} \quad \text{with} \quad P = \begin{pmatrix} 0 & 1 & 0 \\ 0 & 0 & 1 \\ 1 & 0 & 0 \end{pmatrix}.$$

■

5.1.3 Householder's Tridiagonalization

There are some similarity reduction methods which use a product of elementary matrices like Jacobi or Givens rotations to reduce A to a simpler form. We will consider here Householder's reduction to tridiagonal form for symmetric matrices, which is numerically very stable. It uses transformations with Householder matrices, see 1.5.2. Some of the other important methods are described in [99] or [44].

Householder's method can be applied to nonsymmetric matrices, too. Then it yields a reduction to Hessenberg form. A matrix B with elements b_{ij} has *upper Hessenberg form* if $b_{ij} = 0$ for $i > j + 1$:

$$\begin{pmatrix} \times & \times & \times & \times & \times & \times \\ \times & \times & \times & \times & \times & \times \\ 0 & \times & \times & \times & \times & \times \\ 0 & 0 & \times & \times & \times & \times \\ 0 & 0 & 0 & \times & \times & \times \\ 0 & 0 & 0 & 0 & \times & \times \end{pmatrix}.$$

The method consists of $n-2$ orthogonal similarity reductions:

$$\begin{aligned} B &= U_{n-2} \cdots U_1 A U_1 \cdots U_{n-2}, \quad \text{or} \\ B &= H^T A H, \quad \text{with} \\ H &= U_1 \cdots U_{n-2}, \end{aligned} \tag{5.9}$$

or in recursive form

$$\begin{aligned} A^{(1)} &:= A, \\ A^{(r+1)} &:= U_r A^{(r)} U_r, \\ B &:= A^{(n-1)} \end{aligned} \tag{5.10}$$

with the orthogonal Householder matrices ($U_r^2 = I$)

$$\begin{aligned} U_r &:= I - 2w_r w_r^T \quad \text{with} \\ \alpha &:= a_{r+1,r}^{(r)}, \\ \sigma &:= \sqrt{(a_{r+1,r}^{(r)})^2 + (a_{r+2,r}^{(r)})^2 + \cdots + (a_{n,r}^{(r)})^2}, \\ K_r &:= \sqrt{\frac{\sigma^2 + \operatorname{sign}(\alpha)\alpha\sigma}{2}}, \\ w_r &:= \frac{1}{2K_r}\left(0, \ldots, 0, \alpha + \operatorname{sign}(\alpha)\sigma, a_{r+2,r}^{(r)}, \ldots, a_{n,r}^{(r)}\right)^T, \\ \text{with} \quad \operatorname{sign}(\alpha) &:= 1, \quad \text{if } \alpha = 0. \end{aligned} \tag{5.11}$$

Computationally (5.10) is carried out as:

$$\begin{aligned} \text{Let} \quad v &:= A^{(i)} w_i \text{ and } \gamma := v^T w_i \\ \text{then} \quad A^{(i+1)} &:= A^{(i)} - 2w_i v^T - 2v w_i^T + 4\gamma w_i w_i^T. \end{aligned} \tag{5.12}$$

The sign rule for α is most important for the stability of the algorithm, because it prevents cancellation, which otherwise could occur in the kernel of the calculation. Obviously the rth step of Householder's method leaves the first r rows and columns of $A^{(r)}$ unchanged. It creates zeros in the rth column beyond the subdiagonal. After $n-2$ steps this leads to an upper Hessenberg form which for the symmetric case is the tridiagonal form. For the nonsymmetric case pivoting may be used first in each step.

The symmetric matrix A is now reduced to a tridiagonal matrix, for which the eigenvalue problem then has to be solved. For this problem the QR method is numerically stable.

Example 5.5. We continue with our example with the balanced matrix

$$\tilde{A} = \begin{pmatrix} 2 & -1 & -1 \\ -1 & 3 & -1 \\ -1 & -1 & 4 \end{pmatrix}.$$

Here only one step is necessary:

$$\begin{aligned} B &= U_1 \tilde{A} U_1 \\ w_1 &= (0, w_{12}, w_{13})^T \\ \alpha &= -1, \qquad \sigma = \sqrt{(-1)^2 + (-1)^2} = \sqrt{2} \end{aligned}$$

$$
\begin{aligned}
K_1 &= \sqrt{\frac{1}{2}(2+\sqrt{2})} = \sqrt{1+\frac{\sqrt{2}}{2}} \approx 1.3066 \\
w_1 &= \frac{1}{2K_1}(0,-1-\sqrt{2},-1)^T \approx (0,-0.9239,-0.3827)^T \\
v &:= \tilde{A}w_1 = \frac{1}{2K_1}(2+\sqrt{2},-2-3\sqrt{2},-3+\sqrt{2})^T \\
&\approx (1.3066,-2.389,-0.6069)^T \\
\gamma &= v^T w_1 = \frac{1}{4+2\sqrt{2}}(11+4\sqrt{2}) \approx 2.4393 \\
w_1 v^T &= \left(v w_1^T\right)^T = \frac{1}{4+2\sqrt{2}} \begin{pmatrix} 0 & 0 & 0 \\ -4-3\sqrt{2} & 8+5\sqrt{2} & 1+2\sqrt{2} \\ -2-\sqrt{2} & 2+3\sqrt{2} & 3-\sqrt{2} \end{pmatrix} \\
&\approx \begin{pmatrix} 0 & 0 & 0 \\ -1.2071 & 2.2071 & 0.5607 \\ -0.5 & 0.9142 & 0.2322 \end{pmatrix} \\
B &:= \tilde{A} - 2w_1 v^T - 2v w_1^T + 4\gamma w_1 w_1^T \\
&= \begin{pmatrix} 2 & \sqrt{2} & 0 \\ \sqrt{2} & 5/2 & -1/2 \\ 0 & -1/2 & 9/2 \end{pmatrix} \\
&\approx \begin{pmatrix} 2 & 1.4142 & 0 \\ 1.4142 & 2.5 & -0.5 \\ 0 & -0.5 & 4.5 \end{pmatrix}.
\end{aligned}
$$

■

5.1.4 The QR Method

A detailed mathematical description of the QR algorithm for tridiagonal or Hessenberg matrices may be found in [44]. We will briefly look at the steps of this method.

1. It can be assumed that all subdiagonal elements are not equal to zero, because, if this were the case, the eigenvalue problem can be divided into two smaller subproblems, which can be solved with less computational effort. For the nonsymmetric case a 'small' system of linear equations has also to be solved.

2. **QR decomposition**
For each matrix $B \in \mathbb{R}^{n,n}$ there exists a decomposition
$$B = QR \tag{5.13}$$
with an orthogonal matrix Q and an upper triangular matrix R.
3. For the cases considered the QR decomposition can be formed with $n-1$ orthogonal transformations. This reduces the computational effort considerably.
4. **QR transformation**
The QR decomposition is followed by an QR transformation:
$$B = QR \longrightarrow \tilde{B} = RQ. \tag{5.14}$$
This is a similarity transformation because with $R = Q^T B$ and $Q^T Q = I$ we have
$$\tilde{B} = RQ = Q^T BQ. \tag{5.15}$$
5. The QR transformation leaves the Hessenberg or tridiagonal form unchanged.
6. **QR algorithm**
QR decomposition and QR transformation are connected to an iterative algorithm:
$$\begin{aligned} B_1 &:= B, \\ B_k &= Q_k R_k, \quad B_{k+1} := R_k Q_k, \quad k = 1, 2, \dots \\ \text{or} \quad B_{k+1} &= Q_k^T B_k Q_k. \end{aligned} \tag{5.16}$$
This iteration converges for the nonsymmetric case to an upper block triangular matrix, which has single elements or 2×2 blocks on the diagonal, for example
$$\begin{pmatrix} [\times] & \times & \times & \times & \times & \times \\ 0 & \left[\begin{matrix} \times \\ \times \end{matrix}\right. & \left.\begin{matrix} \times \\ \times \end{matrix}\right] & \times & \times & \times \\ & & & \times & \times & \times \\ 0 & 0 & 0 & [\times] & \times & \times \\ 0 & 0 & 0 & 0 & [\times] & \times \\ 0 & 0 & 0 & 0 & 0 & [\times] \end{pmatrix}.$$
Then the eigenvalues can be taken from the diagonal or by solving 2×2 eigenvalue problems. From these 2×2 blocks one may obtain complex eigenvalues.
For the symmetric case the limit matrix is a diagonal matrix.
7. **QR algorithm with shift**
The convergence speed of the QR algorithm can be efficiently increased by a spectral shift:

$$\begin{aligned} B_1 &:= B, \\ B_k - \sigma_k I &= Q_k R_k, \\ B_{k+1} &:= R_k Q_k + \sigma_k I, \quad k = 1, 2, \ldots, \\ \text{where we have again:} \quad B_{k+1} &= Q_k^T B_k Q_k. \end{aligned} \tag{5.17}$$

For σ_k one may take the last diagonal element of B_k. For the case of complex σ_k we refer to [99] or [44]. The shift-idea was introduced by Francis, see [33].

8. **Determination of the eigenvectors**
The product of all orthogonal matrices Q_k used for the method is an orthogonal matrix itself:

$$Q := Q_1\, Q_2\, \cdots\, Q_m. \tag{5.18}$$

This matrix Q transforms B to an upper triangular block matrix for the nonsymmetric and to a diagonal matrix for the symmetric case. For the symmetric case the columns of Q are the eigenvectors of the tridiagonal matrix B. For the nonsymmetric case there must be solved the simple eigenvalue problem

$$Ry = \lambda y. \tag{5.19}$$

If y_k are the eigenvectors of R, then $x_k = Qy_k$ are the eigenvectors of B.
Because the Hessenberg or tridiagonal form has been developed with similarity transformations, the eigenvectors of B have to be back-transformed to those of A.
Calculating the transformation matrix Q is computationally complex. Therefore one often uses special methods for the determination of the eigenvectors, for example the power method found by von Mises; see Section 5.4.

Example 5.6. We will only calculate two steps without shift for the matrix B generated with the Householder transformation in the last example. These calculations can be carried out easily with MATLAB making use of the command `[Q,R] = qr(B)`.

$$\begin{aligned} B_1 = B &= \begin{pmatrix} 2 & \sqrt{2} & 0 \\ \sqrt{2} & 5/2 & -1/2 \\ 0 & -1/2 & 9/2 \end{pmatrix} \\ &\approx \begin{pmatrix} 2 & 1.4142 & 0 \\ 1.4142 & 2.5 & -0.5 \\ 0 & -0.5 & 4.5 \end{pmatrix} = Q_1 R_1 \end{aligned}$$

$$\text{with } Q_1 = \begin{pmatrix} -0.8165 & 0.5345 & -0.2182 \\ -0.5774 & -0.7559 & 0.3086 \\ 0 & 0.3780 & 0.9258 \end{pmatrix} \quad \text{and}$$

$$R_1 = \begin{pmatrix} -2.4495 & -2.5981 & 0.2887 \\ 0 & -1.3229 & 2.0788 \\ 0 & 0 & 4.0119 \end{pmatrix}$$

$$B_2 = R_1 Q_1 = \begin{pmatrix} 3.5000 & 0.7638 & 0 \\ 0.7638 & 1.7857 & 1.5164 \\ 0 & 1.5164 & 3.7143 \end{pmatrix} = Q_2 R_2$$

$$\text{with } Q_2 = \begin{pmatrix} -0.9770 & 0.1539 & 0.1475 \\ -0.2132 & -0.7053 & -0.6761 \\ 0 & -0.6920 & 0.7219 \end{pmatrix} \quad \text{and}$$

$$R_2 = \begin{pmatrix} -3.5824 & -1.1269 & -0.3233 \\ 0 & -2.1912 & -3.6398 \\ 0 & 0 & 1.6561 \end{pmatrix}$$

$$B_3 = R_2 Q_2 = \begin{pmatrix} 3.7403 & 0.4672 & 0 \\ 0.4672 & 4.0642 & -1.1460 \\ 0 & -1.1460 & 1.1955 \end{pmatrix}.$$

The diagonal elements of this matrix are very rough approximations to the eigenvalues of B, which in fact are

$$(4.6751, 3.5392, 0.7857).$$

Six more QR steps *without* shift lead to

$$B_9 = R_8 Q_8 = \begin{pmatrix} 4.5212 & 0.3888 & 0 \\ 0.3888 & 3.6931 & -0.0001 \\ 0 & -0.0001 & 0.7857 \end{pmatrix},$$

which shows the slow convergence of the original method without shift.

Four QR steps *with* shift yield

$$B_5 = R_4 Q_4 = \begin{pmatrix} 0.7858 & 0.0148 & 0 \\ 0.0148 & 3.5391 & -0.0000 \\ 0 & -0.0000 & 4.6751 \end{pmatrix},$$

which clearly shows the superiority of this method. ■

5.2 The General Eigenvalue Problem

5.2.1 Introduction

Let $A, B \in \mathbb{R}^{n,n}$ be two matrices.

Determine complex numbers $\lambda_k \in \mathbb{C}$ and complex vectors $x_k \in \mathbb{C}^n$ such that

$$Ax_k = \lambda_k B x_k \,, \quad k = 1, \dots n. \tag{5.20}$$

This problem is easily reduced to the standard eigenvalue problem, if B is nonsingular:

$$Ax = \lambda Bx \iff B^{-1}Ax = \lambda x. \tag{5.21}$$

But this method is not recommended, because it is numerically not stable if B is an ill-conditioned matrix. Furthermore the symmetry is normally lost on forming $B^{-1}A$, even if A and B are symmetric. A symmetric and stable transformation to the standard eigenvalue problem is possible, if both A and B are symmetric and one of them is positive-definite (spd); see Section 5.2.2.

If both matrices are symmetric, but only positive-semidefinite, then we obtain an algorithm with the proof of the following theorem, see [82].

Theorem 5.7. *If $A, B \in \mathbb{R}^{n,n}$ are two symmetric, positive-semidefinite matrices, then there is a nonsingular matrix $T \in \mathbb{R}^{n,n}$ such that both $T^{-1}AT$ and $T^{-1}BT$ are diagonal.*

For the general case without assumption on A or B the QZ algorithm can be used, see [71].

The description of the complete structure of the eigenspace of A relative to B including zero and infinity eigenvalues is possible with Kågstrøm's GUPTRI algorithm, see [59]. But this is far beyond the scope of this text.

5.2.2 The Symmetric Positive-Definite Case

For many applications at least one of the matrices A or B is symmetric and positive-definite. Then the general eigenvalue problem is easily reduced to the standard eigenvalue problem without losing numerical stability.

If A is symmetric positive-definite, then for $\lambda \neq 0$

$$Ax = \lambda Bx \iff Bx = \frac{1}{\lambda} Ax.$$

Therefore it is sufficient for the problem $Ax = \lambda Bx$ with A or B symmetric and positive-definite to consider the case 'B symmetric positive-definite' in detail. The following algorithm shows this case.

(1) **Cholesky decomposition of B**
Determine a lower triangular matrix L with
$$LL^T = B.$$

(2) **Inversion of L**
Determine the inverse L^{-1} of L, see below.

(3) **Transformation of A**
$$C := L^{-1} A (L^T)^{-1}.$$

(4) **Standard eigenvalue problem**
Determine eigenvalues λ_k and eigenvectors y_k of the matrix C:
$$Cy = \lambda y.$$

(5) **Back-transformation**
λ_k are the eigenvalues of the general eigenvalue problem,
$x_k := (L^T)^{-1} y_k$ its eigenvectors.

If

$$L = \begin{pmatrix} l_{11} & 0 & \cdots & 0 \\ l_{21} & \ddots & & \\ \vdots & & \ddots & \\ l_{n1} & \cdots & & l_{nn} \end{pmatrix}$$

and

$$L^{-1} = \begin{pmatrix} \gamma_{11} & 0 & \cdots & 0 \\ \gamma_{21} & \ddots & & \\ \vdots & & \ddots & \\ \gamma_{n1} & \cdots & & \gamma_{nn} \end{pmatrix},$$

then the calculation of the elements γ_{ij} is easily performed as

for $i = 1, \ldots, n$:
$$\gamma_{ii} := l_{ii}^{-1},$$
for $k = 1, \ldots, n-1$:
for $i = k+1, \ldots, n$:
$$\gamma_{ik} := -\gamma_{ii} \sum_{j=k}^{i-1} l_{ij} \gamma_{jk}.$$

Example 5.8. Consider the problem $Ax = \lambda Bx$ with

$$A = \begin{pmatrix} 192 & -56 & -80 \\ -56 & 75 & -34 \\ -80 & -34 & 165 \end{pmatrix}, \quad B = \begin{pmatrix} 64 & -8 & -8 \\ -8 & 17 & -3 \\ -8 & -3 & 66 \end{pmatrix}.$$

B is symmetric positive definite and the Cholesky decomposition $B = LL^T$ gives

$$L = \begin{pmatrix} 8 & 0 & 0 \\ -1 & 4 & 0 \\ -1 & -1 & 8 \end{pmatrix} \text{ and } L^{-1} = \begin{pmatrix} 0.1250 & 0.0312 & 0.0195 \\ 0 & 0.2500 & 0.0312 \\ 0 & 0 & 0.1250 \end{pmatrix},$$

which leads to $Cx = \lambda x$ with

$$C = L^{-1}A(L^T)^{-1} = \begin{pmatrix} 3 & -1 & -1 \\ -1 & 4 & -1 \\ -1 & -1 & 2 \end{pmatrix}.$$

This is the matrix A from Example 5.4. ■

5.3 Singular Value Decomposition

If the matrix A is non-square, say $A \in \mathbb{R}^{m,n}$ with $m \neq n$, then an eigenvalue problem with A is not defined. But it is possible to describe the characteristic structure of A by considering the eigenvalues of $A^T A$. From Chapter 1 we know that these are the squares of the singular values of A. Therefore there is a program in this chapter for the determination of singular values and singular vectors, too. The singular value decomposition

$$A = USV^T \tag{5.22}$$

with two orthogonal matrices $U \in \mathbb{R}^{m,m}$ and $V \in \mathbb{R}^{n,n}$ and a diagonal matrix $S \in \mathbb{R}^{m,n}$ was considered in Section 1.5.2. The left and right singular vectors are the column of U and V, respectively. They are the eigenvectors of AA^T and $A^T A$ to the eigenvalues σ_i^2, respectively:

$$A^T A = VSU^T USV^T = VS^2V^T \Longrightarrow (A^T A)V = S^2 V. \tag{5.23}$$

5.4 Sparse Matrices

Sparse matrices mainly occur in connection with the solution of partial differential equations. Solving eigenvalue problems in this connection means computing vibration modes, for example. Sparse matrices have as main characteristics diagonal dominance and symmetry. Therefore iterative algorithms are to be considered as most efficient for the solution of these problems.

One of these methods is the *power method*. We will describe this method for the simplest case only. Let $A \in \mathbb{R}^{n,n}$ be a symmetric matrix with n real eigenvalues of distinct modulus

$$|\lambda_1| > |\lambda_2| > \cdots > |\lambda_n|. \tag{5.24}$$

We firstly will describe the determination of the dominant eigenvalue λ_1 and its dominant eigenvector $x^{(1)} \in \mathbb{R}^n$:

(1) Select some start vector $z^{(0)} \in \mathbb{R}^n$ with $\|z^{(0)}\| = 1$
$i := 0$

(2) $u^{(i)} := Az^{(i)}$

$$z^{(i+1)} := \frac{u^{(i)}}{\|u^{(i)}\|}$$

(3) If
$\|z^{(i+1)} - z^{(i)}\| < \varepsilon \longrightarrow (5)$

(4) $i := i + 1$. Go to (2)

(5) Determine the index k, for which:
$|u_k^{(i)}| = \max_j |u_j^{(i)}|$

(6) $\tilde{\lambda}_1 := \|u^{(i)}\| \cdot \operatorname{sign}\left(\dfrac{z_k^{(i)}}{u_k^{(i)}}\right)$ and
$\tilde{x}^{(1)} := z^{(i+1)}$
$\tilde{\lambda}_1 \in \mathbb{R}$ and $\tilde{x}^{(1)} \in \mathbb{R}^n$ are approximations for λ_1 and $x^{(1)}$.

Theorem 5.9. *Assume that for $z^{(0)}$ we have:*

$$z^{(0)} = \sum_{j=1}^{n} c_j x^{(j)} \quad \textit{with} \quad c_1 \neq 0, \tag{5.25}$$

where $\{x^{(j)}\}$ is the orthogonal system of linearly independent eigenvectors. Then we have linear convergence of

$$\begin{aligned} \|u^{(k)}\| &\longrightarrow |\lambda_1| \quad \textit{with} \quad k \to \infty \\ \textit{and} \qquad z^{(k)} &\longrightarrow x^{(1)} \quad \textit{with} \quad k \to \infty \end{aligned} \tag{5.26}$$

with convergence rate $|\lambda_2|/|\lambda_1|$.

The largest computational part of this algorithm is matrix–vector multiplications in (2), for which the sparsity of the matrix is very useful. The elements of A have to be stored accordingly.

Example 5.10. Let

$$A = \begin{pmatrix} + & \times & \cdot & \cdot & \cdot & \times & \cdot & \cdot & \cdot & \cdot & \cdot & \cdot \\ \times & + & \times & \cdot & \cdot & \cdot & \times & \cdot & \cdot & \cdot & \cdot & \cdot \\ \cdot & \times & + & \times & \cdot & \cdot & \cdot & \times & \cdot & \cdot & \cdot & \cdot \\ \cdot & \cdot & \times & + & \times & \cdot & \cdot & \cdot & \times & \cdot & \cdot & \cdot \\ \cdot & \cdot & \cdot & \times & + & \cdot & \cdot & \cdot & \cdot & \times & \cdot & \cdot \\ \times & \cdot & \cdot & \cdot & \cdot & + & \times & \cdot & \cdot & \cdot & \times & \cdot \\ \cdot & \times & \cdot & \cdot & \cdot & \times & + & \times & \cdot & \cdot & \cdot & \times \\ \cdot & \cdot & \times & \cdot & \cdot & \cdot & \times & + & \times & \cdot & \cdot & \cdot \\ \cdot & \cdot & \cdot & \times & \cdot & \cdot & \cdot & \times & + & \times & \cdot & \cdot \\ \cdot & \cdot & \cdot & \cdot & \times & \cdot & \cdot & \cdot & \times & + & \cdot & \cdot \\ \cdot & \cdot & \cdot & \cdot & \cdot & \times & \cdot & \cdot & \cdot & \cdot & + & \times \\ \cdot & \cdot & \cdot & \cdot & \cdot & \cdot & \times & \cdot & \cdot & \cdot & \times & + \end{pmatrix}$$

with the following abbreviations: $+ \equiv 0.1111$, $\times \equiv -0.0278$, $\cdot \equiv 0$.
This matrix normally would have been stored in the form `(i, j, a(i,j))` for non-zero elements only, see Section 5.6. We are searching for the eigenvalue with the largest absolute value and the corresponding eigenvector. For this problem the eigenvector is expected to be oscillatory. Therefore we take as start vector

$$z^{(0)} := \frac{1}{\sqrt{12}}(-1, 1, -1, 1, -1, 1, -1, 1, -1, 1, -1, 1)^T,$$

It fulfills the norm condition

$$||z^{(0)}||_2 = 1,$$

and yields the following sequence of approximations. (We write for simplicity reasons $||\cdot||$ for $||\cdot||_2$.)

$$u^{(0)} := Az^{(0)} = \begin{pmatrix} -0.0481 \\ 0.0561 \\ -0.0561 \\ 0.0561 \\ -0.0481 \\ 0.0561 \\ -0.0642 \\ 0.0561 \\ -0.0561 \\ 0.0481 \\ -0.0481 \\ 0.0481 \end{pmatrix} \qquad z^{(1)} = \frac{u^{(0)}}{||u^{(0)}||} = \begin{pmatrix} -0.2587 \\ 0.3018 \\ -0.3018 \\ 0.3018 \\ -0.2587 \\ 0.3018 \\ -0.3449 \\ 0.3018 \\ -0.3018 \\ 0.2587 \\ -0.2587 \\ 0.2587 \end{pmatrix}$$

$$u^{(1)} := Az^{(1)} = \begin{pmatrix} -0.0455 \\ 0.0587 \\ -0.0587 \\ 0.0575 \\ -0.0443 \\ 0.0575 \\ -0.0707 \\ 0.0599 \\ -0.0575 \\ 0.0443 \\ -0.0443 \\ 0.0455 \end{pmatrix} \qquad z^{(2)} = \frac{u^{(1)}}{\|u^{(1)}\|} = \begin{pmatrix} -0.2418 \\ 0.3119 \\ -0.3119 \\ 0.3055 \\ -0.2354 \\ 0.3055 \\ -0.3755 \\ 0.3182 \\ -0.3055 \\ 0.2354 \\ -0.2354 \\ 0.2418 \end{pmatrix}$$

$$\ldots \qquad\qquad \ldots$$

If we calculate the approximation $\tilde{\lambda}_1$ (Step (6) in the algorithm above) for each step, we get the following sequence of approximations for λ_1:

Step No.	1	2	3	4	5	6
$\tilde{\lambda}_1$	0.1860	0.1882	0.1892	0.1897	0.1900	0.1901
	7	8	9	10	11	12
	0.1902	0.1903	0.1903	0.1903	0.1904	0.1904

After twelve steps we get the eigenvalue approximation $\tilde{\lambda}_1 = 0.1904$, which is correct to the four digits shown. Next we show the eigenvector approximation $\tilde{x}^{(1)}$ as well as the correct eigenvector $x^{(1)}$:

$$\tilde{x}^{(1)} = \begin{pmatrix} -0.2235 \\ 0.3482 \\ -0.3510 \\ 0.2858 \\ -0.1632 \\ 0.2916 \\ -0.4277 \\ 0.3734 \\ -0.2919 \\ 0.1647 \\ -0.1790 \\ 0.2133 \end{pmatrix}, \quad x^{(1)} = \begin{pmatrix} -0.2290 \\ 0.3569 \\ -0.3527 \\ 0.2742 \\ -0.1489 \\ 0.2961 \\ -0.4361 \\ 0.3749 \\ -0.2804 \\ 0.1505 \\ -0.1795 \\ 0.2158 \end{pmatrix}.$$

The convergence would have been even slower if we would have taken another start vector as $z^{(0)} := (1,1,1,1,1,1,1,1,1,1,1,1)^T/\sqrt{12}$. ■

If we are interested in the m eigenvalues $\lambda_1, \lambda_2, \ldots, \lambda_m$ of A, then we may take an algorithm of Bauer, called simultaneous iteration, see [86]. With m starting vectors m eigenvalues and eigenvectors are iteratively approximated.

(1) **Starting vectors**: Select

$$U_0 = \begin{pmatrix} \vdots & \vdots & & \vdots \\ u_1^{(0)} & u_2^{(0)} & \cdots & u_m^{(0)} \\ \vdots & \vdots & & \vdots \end{pmatrix} \in \mathbb{R}^{n,m}$$

arbitrarily, but linearly independent.
$k := 0$.

(2) **Orthogonalization**:
Determine an upper triangular matrix
$R_k = \{r_{ij}^{(k)}\} \in \mathbb{R}^{m,m}$ such that:
$Z_k = U_k R_k$ and $Z_k^T Z_k = I_m$
(for example Schmidt's orthogonalization).

(3) **Matrix multiplication**:
$U_{k+1} := AZ_k$.

(4) **Convergence test**:
Is $\sum_{i \neq j} |r_{ij}^{(k)}| < \varepsilon$?
If not, then $k := k + 1$ and go to (2).

(5) **Results:**
eigenvector approximations: $\tilde{V} := Z_k$.
eigenvalue approximations: $\tilde{D} := \mathrm{diag}(r_{11}, r_{22}, \ldots, r_{mm})$.

For this method we have the following convergence properties:

$$\lim_{k \to \infty} Z_k = V,$$

$$\lim_{k \to \infty} R_k = D \quad \text{with} \tag{5.27}$$

$$D = \mathrm{diag}(\lambda_1, \ldots, \lambda_m) \quad \text{and} \tag{5.28}$$

$$AV = VD. \tag{5.29}$$

For accuracy and convergence reasons in practice we choose the number of simultaneously iterated vectors to be larger than the number of eigenvalues to be determined. For $m = n$ the method is identical with the QR method. For further details we refer the reader to [86].

5.5 Programs Solving Different Eigenvalue Problems

We have written five programs for this chapter, which call more than ten NAG routines. Figure 5.1 shows a decision tree for easier selection of these programs.

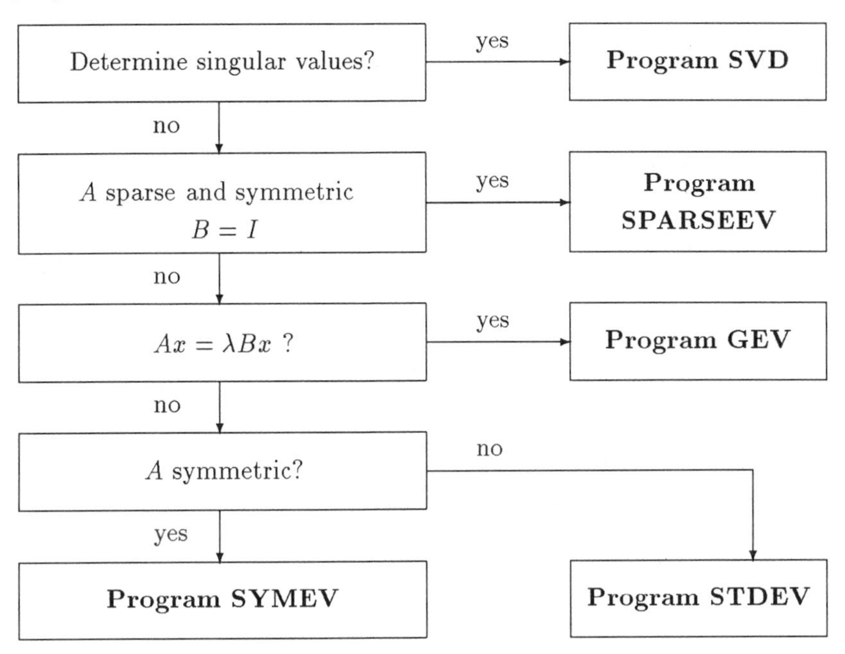

Figure 5.1. Decision tree for eigenvalue problems $Ax = \lambda Bx$

The NAG library often contains for one problem *Black Box* as well as *General Purpose routines*. We selected the General Purpose routines for the programs STDEV and SYMEV. This allows the experienced user to control the different components of the algorithm individually. For the other programs we selected the Black Box routines, for which a single call is sufficient for solving the problem.

5.5.1 STDEV: Standard Eigenvalue Problem

STDEV, see B.5.1, solves the standard eigenvalue problem

$$Ax = \lambda x$$

for nonsymmetric matrices $A \in \mathbb{R}^{n,n}$. The parameters used by the NAG routines called are to be found in Table 5.1. The program is organized in five steps:

(1)	Balancing: Permutation or scaling
(2)	Reduction of A to upper Hessenberg form
(3)	Accumulation of the Householder transformations from (2)
(4)	Solution of the transformed eigenvalue problem
(5)	Back-transformation of balancing (1)

These steps correspond to the following NAG routine calls:

(1)	F01ATF (N,IB,A,NMAX,K,L,D)
(2)	F01AKF (N,K,L,A,NMAX,IND)
(3)	F01APF (N,K,L,IND,A,NMAX,V,NMAX)
(4)	F02AQF (N,K,L,EPS,A,NMAX,V,NMAX,EWR,EWI,IND,IFAIL)
(5)	F01AUF (N,K,L,N,D,V,NMAX)

N	integer	Order of matrix A.
IB	integer	Basis of floating-point numbers.
A	array	Matrix: A(NMAX,NMAX).
NMAX	integer	Maximum order.
K,L	integer	Results of balancing (F01ATF): It is A(I,J) = 0, if I > J *and* (J=1,..., K–1 or I=L+1,...,N)
D	array	Information about permutations and scaling from F01ATF: D(NMAX).
IND	int.array	F01...: Information about row and column interchanges. F02AQF: Number of iterations per eigenvalue: IND(NMAX).
V	array	Matrix of accumulated transformations or matrix with eigenvectors: V(NMAX,NMAX).
EPS	real	Machine precision (X02AAF).
EWR	array	Real part of the eigenvalues: EWR(NMAX).
EWI	array	Imaginary part: EWI(NMAX).
IFAIL	integer	Error parameter. Set IFAIL=–1 before calling the routines.
	In F02AQF the following error message is possible: IFAIL=1 : Number of accumulated iterations > 30 N.	

Table 5.1. Parameters of program STDEV

After being returned by F01APF the matrix V contains the accumulated Householder matrices, which is used by F02AQF. If we want to solve an isolated eigenvalue problem with a Hessenberg matrix with F02AQF, then V must hold the identity matrix. After being returned by F02AQF the ith column of V contains the eigenvector corresponding to the ith eigenvalue of the original matrix A, if the eigenvalue is real. The eigenvector does not fulfill $||x_i|| = 1$. If the ith and $(i+1)$th eigenvalue are complex, then the ith and $(i+1)$th column of V contain the real and imaginary part, respectively, of the eigenvector corresponding to the eigenvalue with positive imaginary part.

5.5.2 SYMEV: Symmetric Eigenvalue Problem

The program SYMEV, see Section B.5.2, solves the eigenvalue problems

$$Ax = \lambda x$$

for a symmetric matrix $A \in \mathbb{R}^{n,n}$.

N	integer	Order of matrix A.
A	array	Matrix: A(NMAX,NMAX).
NMAX	integer	Maximum order.
TOL	real	TOL shall contain: (Smallest positive number)/(machine precision). (X02AKF( )/X02AJF( )).
D	array	Diagonal of tridiagonal matrix: D(NMAX).
E	array	Subdiagonal of tridiagonal matrix: E(NMAX), stored in E(2) to E(N).
U	array	Matrix of accumulated transformations or matrix with eigenvectors: U(NMAX,NMAX).
EPS	real	Machine precision (X02AAF).
IFAIL	integer	Error parameter. Set IFAIL=–1 before calling the routines.
	In F02AMF the following error message is possible: IFAIL=1 : Number of accumulated iterations > 30 N.	

Table 5.2. Parameters of program SYMEV

The program is organized in four steps:

(1)	Balancing of A with Permutations.
(2)	Reduction of A to tridiagonal form.
(3)	Solution of the tridiagonal eigenvalue problem.
(4)	Back-transformation of (1).

We programmed steps (1) and (4) ourselves; the other steps correspond to the following NAG routine calls:

(2)	F01AJF (N,TOL,A,NMAX,D,E,U,NMAX)
(3)	F02AMF (N,EPS,D,E,U,NMAX,IFAIL)

The parameters of all these routines used in SYMEV are described in Table 5.2. The subdiagonal elements stored in E are destroyed in F02AMF.

After being returned by F01AJF the matrix U contains the accumulated Householder matrices, which are used by F02AMF. If we want to solve an isolated eigenvalue problem with a tridiagonal matrix with F02AMF, then U must hold the identity matrix. After being returned by F02AMF the ith column of U contains the eigenvector corresponding to the ith eigenvalue of the original matrix A or of A with respect to B for the general eigenvalue problem. The eigenvectors do not fulfill $||x_i|| = 1$.

5.5.3 GEV: General Eigenvalue Problem

The program GEV, see B.5.3, solves the general eigenvalue problem

$$Ax = \lambda Bx, \quad A, B \in \mathbb{R}^{n,n}.$$

If both matrices are symmetric and B is positive definite the program uses the routines of SYMEV, adding two steps to transform and back-transform the general to the standard eigenvalue problem:

(0)	Transformation to the standard eigenvalue problem with Cholesky's method applied to B. F01AEF (N,A,NMAX,B,NMAX,DL,IFAIL)
(5)	Back-transformation of (1). F01AFF (N,1,N,B,NMAX,DL,U,NMAX)

If A and B are symmetric but only A is positive-definite, then one may solve the problem in inverse form

$$Bx = \mu Ax \tag{5.30}$$

with $\mu = \lambda^{-1}$ for $\lambda \neq 0$.

If one of the matrices is not symmetric or if both are not positive-definite, GEV calls the Black Box routine

F02BJF (N,A,NMAX,B,NMAX,EPS,ALFR,ALFI,BETA,
WANTQ,Q,NMAX,ITER,IFAIL),

which uses the QZ algorithm of Moler and Stewart.

N	int.	Order of matrices A, B.
A	array	Matrix: A(NMAX,NMAX).
NMAX	int.	Maximum order.
B	array	Matrix: B(NMAX,NMAX). F02BJF: B is destroyed. F02AEF: Upper triangular part of B is preserved, L is stored in the lower part without its diagonal.
DL	array	F02AEF: Diagonal of L: DL(NMAX).
EPS	real	Tolerance: $a_{ij} := 0$, if $a_{ij} < \|\|A\|\| \cdot$EPS.
ALFR	array	Real part of α: ALFR(NMAX).
ALFI	array	Imaginary part of α: ALFI(NMAX).
BETA	array	Vector β: BETA (NMAX).
WANTQ	log.	Eigenvectors are calculated, if WANTQ=.TRUE.
Q	array	Matrix of eigenvectors : Q(NMAX,NMAX).
ITER	array	Number of iterations per eigenvalue: ITER(NMAX).
IFAIL	int.	Error parameter. Set IFAIL=−1 before calling the routines.
	In F02BJF the following error message is possible: IFAIL>0 : Number of accumulated iterations > 30 N. In F02AEF the following error message is possible: IFAIL=1 : B numerically not positive-definite.	

Table 5.3. Parameters of program GEV

According to this algorithm NAG routine F02BJF does not yield the eigenvalues λ_j directly, but rather gives numbers ALFR(j), ALFI(j) and BETA(j), which are connected to the eigenvalues by

$$\lambda_j = \frac{\alpha_j}{\beta_j} = \frac{\mathrm{ALFR}(j) + i \cdot \mathrm{ALFI}(j)}{\mathrm{BETA}(j)}, \quad i = \sqrt{-1}. \tag{5.31}$$

If BETA(j) is zero, then $\lambda_j = \pm\infty$. Conjugate complex eigenvalues have successive numbers, i.e.

$$\lambda_j = \frac{\alpha_j}{\beta_j} = \overline{\lambda_{j+1}} = \overline{\left(\frac{\alpha_{j+1}}{\beta_{j+1}}\right)},$$

which does not mean that α_j and α_{j+1} are conjugate complex numbers.

If the logical parameter WANTQ is .FALSE., no eigenvectors are calculated. For WANTQ=.TRUE. the eigenvector for a real eigenvalue λ_j is to be found in the jth column of Q. For complex eigenvalues $(\lambda_j, \lambda_{j+1})$ the eigenvector corresponding to the eigenvalue with positive imaginary part is stored in the columns j (real part) and $j+1$ (imaginary part), respectively.

5.5.4 SPARSEEV: Sparse Eigenvalue Problems

Program SPARSEEV, see B.5.4, solves different sparse eigenvalue problems using NAG routine F02FJF. This routine has very wide possibilities. It finds the m eigenvalues of largest absolute value and the corresponding eigenvectors for the eigenvalue problem $Cx = \lambda x$, where C is a real matrix of order n such that $BC = C^T B$ for a given positive-definite matrix B. For example, if A is symmetric, then putting $C = A$ and $B = I$ yields the standard symmetric problem $Ax = \lambda x$; putting $C = B^{-1}A$ yields the generalized symmetric-definite problem $Ax = \lambda Bx$. But the use of F02FJF requires a sophisticated user, because three subroutines in prescribed form have to be programmed to specify the case to be solved.

SPARSEEV does that for the user, but it only can solve standard eigenvalue problems with a symmetric matrix A. Four cases can be selected by key numbers:

1	Calculation of m eigenvalues of largest absolute value
2	Calculation of m eigenvalues of smallest absolute value
3	Calculation of m eigenvalues furthest from σ
4	Calculation of m eigenvalues closest to σ

The only work left to the user is the description of the matrix. Most often sparse matrices are used in connection with the solution of partial differential equations. For example when solving two-dimensional problems an area has to be covered by a grid. For each point of the grid one row of the matrix is generated. The program constructing the grid will normally create the corresponding sparse matrix as well. We will see an example for a grid and the corresponding matrix in Section 5.6.

5.5.5 SVD: Singular Value Decomposition

In Section 1.5.2 we considered the singular value decomposition in some detail. The program SVD, see Section B.5.5, computes the singular values as well as the left and right singular vectors of an arbitrary matrix $A \in \mathbb{R}^{m,n}$. It uses NAG routine F02WEF.

5.6 Applications

5.6.1 Vibrations of a Diaphragm

We consider a domain in the plane $\Omega \in \mathbb{R}^2$, which is covered by a thin elastic diaphragm. It is fixed at the boundaries $\partial\Omega$ of the domain. We will approximately calculate the two smallest frequencies of this diaphragm. Therefore we firstly have to discretize the respective partial differential equation

$$\begin{aligned} \frac{\partial^2 u}{\partial x^2} + \frac{\partial^2 u}{\partial y^2} &= \lambda u \quad \text{for} \quad (x,y) \in \Omega, \\ u(x,y) &= 0 \quad \text{for} \quad (x,y) \in \partial\Omega. \end{aligned} \tag{5.32}$$

As an example we take the domain of Example 1.12:

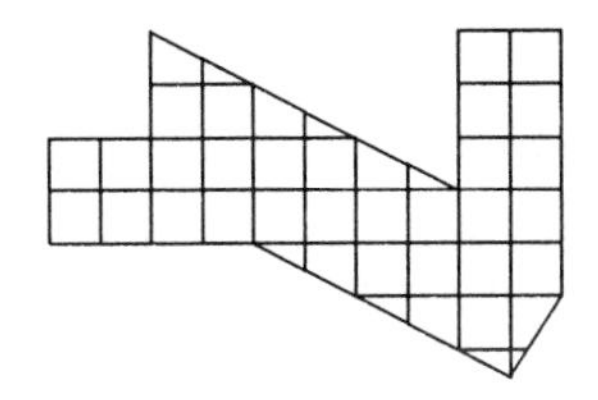

We do not describe the discretization. It leads to the eigenvalue problem $Ax = \lambda x$ with A as in Section 1.12. Neglecting the factor $1/h^2$ it consists of elements with value '4' on the main diagonal and '-1' for all other non-vanishing elements ($\times$ in 1.12). We considered three grids with $h = 1$, $h = 1/2$ and $h = 1/4$. ($h = 1$ leads to the grid shown.) The input can be done with help of the relevant template within our PAN system. But this will be tedious and error-prone, especially for $h < 1$. Normally a grid generation program will design the grid and the resulting matrix, or a program which has the design of the grid as input will give the matrix as output.

The results of SPARSEEV for the three values of h considered are shown in the following table:

h	Order n	Number of matrix elements $\neq 0$	λ_1	λ_2
1	23	51	1.147	1.399
1/2	115	308	1.270	1.513
1/4	509	1451	1.319	1.551

The accuracy of the eigenvalues is not satisfactory. Further grid refinements would be necessary for better results. In Figure 5.1 we see the discrete eigenfunctions for the two eigenvalues. The domain has been rotated by 90 degrees merely for better representation. In addition the highest frequency eigenfunction is shown just for fun.

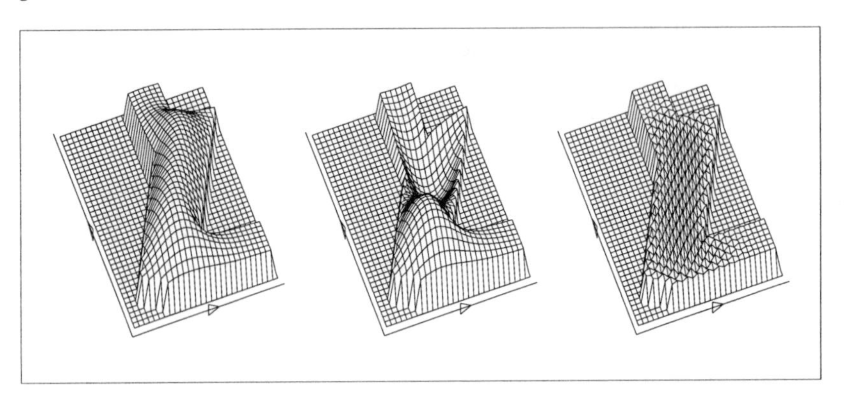

Figure 5.2. Eigenfunction of a diaphragm

5.6.2 Axially Loaded Beam

In Example 8.1 the deflection of an axially loaded beam is considered. For certain material constants and loads we have to solve the differential equation (8.5). For calculating the modes – called elastic curve within this connection – we have to replace the right-hand side in (8.5) by zero and to introduce the eigenvalue λ as a parameter to get instead:

$$\begin{array}{ll}\text{Differential equation:} & -y''(x) - \lambda(1+x^2)\,y(x) = 0,\\ \text{Boundary conditions:} & y(-1) = y(1) = 0.\end{array} \tag{5.33}$$

The solution of this eigenvalue problem yields the frequencies and eigenfunctions of the beam.

This problem is the same as (8.15) with $q(x) = -(1+x^2)$. Discretizing it as in (8.16) we get the standard matrix eigenvalue problem (8.17). Here we only take seven points in the interval $[-1, 1]$

$$\{x_i\} = (-0.75, -0.5, -0.25, 0, 0.25, 0.5, 0.75)$$

to get the matrix (rounded to three digits)

$$A = \begin{pmatrix} 1.28 & -0.64 & 0 & 0 & 0 & 0 & 0\\ -0.8 & 1.6 & -0.8 & 0 & 0 & 0 & 0\\ 0 & -0.941 & 1.88 & -0.941 & 0 & 0 & 0\\ 0 & 0 & -1 & 2 & -1 & 0 & 0\\ 0 & 0 & 0 & -0.941 & 1.88 & -0.941 & 0\\ 0 & 0 & 0 & 0 & -0.8 & 1.6 & -0.8\\ 0 & 0 & 0 & 0 & 0 & -0.64 & 1.28 \end{pmatrix}.$$

For this special example we could have used the symmetry of the problem to reduce its size. Now we have to solve an eigenvalue problem with a non-symmetric tridiagonal matrix. This can be done with the special NAG routine F02AQF, but for demonstration purposes we prefer to use the program STDEV and the template for that program to generate the following input and output files.

Input and Output Files for STDEV

```
Input file for the PAN program STDEV:
Matrix type (0 = Input, 1 = Random, 2 = Hilbert):
0
Row and column dimension of Matrix A:
7 7
Matrix A:
 1.28    -0.64    0       0       0      0      0
-0.8      1.6    -0.8     0       0      0      0
 0       -0.941   1.88   -0.941   0      0      0
 0        0      -1       2      -1      0      0
 0        0       0      -0.941   1.88  -0.94   0
 0        0       0       0      -0.8    1.6   -0.8
 0        0       0       0       0     -0.64   1.28

Output file for the PAN program STDEV:
Matrix A:
 1.2800 -0.6400  0.0000  0.0000  0.0000  0.0000  0.0000
-0.8000  1.6000 -0.8000  0.0000  0.0000  0.0000  0.0000
 0.0000 -0.9410  1.8800 -0.9410  0.0000  0.0000  0.0000
 0.0000  0.0000 -1.0000  2.0000 -1.0000  0.0000  0.0000
 0.0000  0.0000  0.0000 -0.9410  1.8800 -0.9410  0.0000
 0.0000  0.0000  0.0000  0.0000 -0.8000  1.6000 -0.8000
 0.0000  0.0000  0.0000  0.0000  0.0000 -0.6400  1.2800

  1. eigenvalue:          3.54057428
Corresponding eigenvector:
  1                       0.06482870
  2                      -0.22398451
  3                       0.49062312
  4                      -0.63681371
  5                       0.49043571
  6                      -0.22889704
  7                       0.06480393
 ...                      ...
```

The seven eigenvalues and eigenvectors shown are not sorted. Sorting then gives the following results:

i	λ_i	$x^{(i)}$						
1	0.13	0.201	0.360	0.458	0.491	0.458	0.360	0.201
2	0.45	−0.384	−0.496	−0.327	0	0.327	0.496	0.384
3	0.95	−0.482	−0.251	0.277	0.526	0.277	−0.251	−0.482
4	1.53	0.483	−0.188	−0.500	0	0.500	0.188	−0.483
5	2.14	0.370	−0.497	−0.035	0.506	−0.035	−0.497	0.370
6	2.78	−0.200	0.468	−0.491	0	0.491	−0.468	0.200
7	3.54	−0.065	0.229	−0.491	0.637	−0.491	0.229	−0.065

Obviously the physical characteristics of the problem are mirrored by the discrete solution: the change of sign corresponds to the roots of the continuous eigenfunctions, the number of which increases from zero for the smallest eigenvalue to $n-1$ for the largest one.

We will treat a non-symmetric example of the same problem by replacing the function

$$q(x) = -(1+x^2) \quad \text{by} \quad q(x) = -e^x\,(1+x^2),$$

which can be interpreted as representing a beam with a thicker left end. The new matrix is easily calculated in the same way as the example above and we get the following results:

i	λ_i	$x^{(i)}$						
1	0.12	−0.151	−0.289	−0.402	−0.476	−0.494	−0.433	−0.268
2	0.41	−0.351	−0.595	−0.655	−0.491	−0.126	0.310	0.483
3	0.84	0.423	0.584	0.372	-9.915	−0.486	−0.315	0.402
4	1.41	0.493	0.475	-4.972	−0.516	−0.257	0.495	−0.187
5	2.19	0.556	0.212	−0.484	−0.303	0.542	−0.233	4.443
6	3.27	0.582	−0.239	−0.467	0.569	−0.253	5.492	-6.234
7	4.57	0.492	−0.676	0.500	−0.214	5.237	-7.129	5.433

Again we observe the typical characteristics; in addition to those mentioned above we see in this case that the roots of the eigenfunctions are nearer the right interval boundary point, i.e. the thin end of the beam.

This beam is to be seen in Figure 5.3 together with its first three vibration modes.

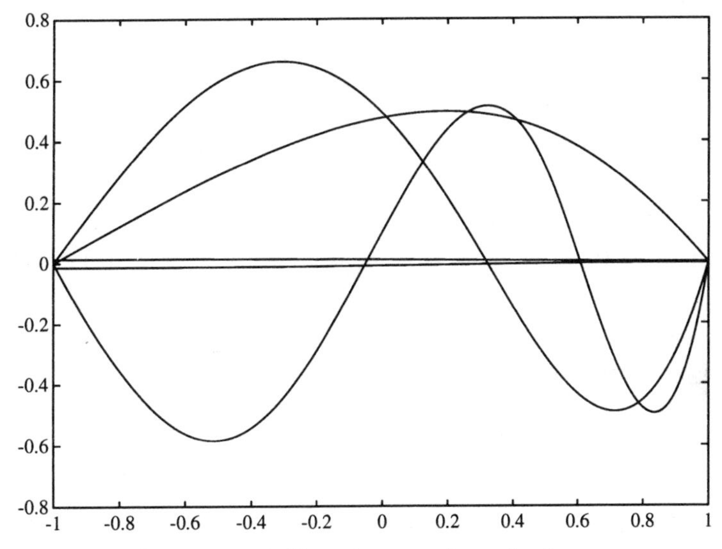

Figure 5.3. Vibration modes of a beam

5.7 Eigenvalue Problems with IMSL

There are two main differences between IMSL and NAG:

- IMSL can calculate the performance index, which holds information about the accuracy of a eigenvalue/eigenvector pair (λ, x) according to the equation $Ax = \lambda x$.
- IMSL holds no routines for sparse eigenvalue problems.

5.7.1 The Standard Eigenvalue Problem

The Non-Symmetric Standard Eigenvalue Problem

Both libraries use the same algorithm (balancing, reduction to Hessenbergform, QR algorithm).
For non-symmetric real matrices we find the following routines in IMSL:

- EVLRG for determining all eigenvalues, corresponding to F02AFF in the NAG library.
- EVCRG for determining all eigenvalues and eigenvectors, corresponding to F02AGF.
- EVLRH for determining all eigenvalues of Hessenberg matrices, corresponding to F02APF.
- EVCRH for determining all eigenvalues and eigenvectors of Hessenberg matrices, corresponding to F02AQF.

- EPIRG for determining the performance index.

IMSL has no corresponding routines to the following NAG routines:

- F01ATF (Balancing),
- F01AKF (Reduction to Hessenberg form),
- F01APF (Accumulation of Householder matrices),
- F01AUF (Back transformation of balancing),
- F02BKF (Calculation of certain eigenvectors of Hessenberg matrices)

With IMSL it is not possible to calculate fewer than n eigenvectors. Contrary to IMSL, NAG routines destroy the matrix, but need no working space. Usage and error handling are similar for both libraries.

The Symmetric Standard Eigenvalue Problem

For real symmetric matrices IMSL holds the following routines:

- EVLSF for determining all eigenvalues, corresponding to F02AAF in the NAG library.
- EVCSF for determining all eigenvalues and eigenvectors, corresponding to F02ABF.
- EVASF for determining smallest or largest eigenvalues, and EVESF for determining the corresponding eigenvectors. There are no equivalent NAG routines.
- EVBSF for determining all eigenvalues in an interval, and EVFSF for determining the corresponding eigenvectors. There are no single equivalent NAG routines; one rather has to perform these calculations step-wise with NAG routines.
- EPISF for determining the performance index.

Contrary to IMSL, the NAG library accepts linear array input for the matrix in the symmetric case

$$a_{11}\ ;\ a_{21}, a_{22}\ ;\ \ldots\ ;\ a_{n1}, a_{n2} \ldots a_{nn},$$

which saves about 50% storage. This is possible with NAG routine F01AYF (reduction to a tridiagonal matrix), for example. The calculation of eigenvalues and corresponding eigenvectors within a specified interval for a tridiagonal matrix is possible with NAG routine F02BEF. NAG routine F02BBF calculates some eigenvalues without eigenvectors for tridiagonal matrices not within a interval, but according to numbers of ordered eigenvalues.

The usage of NAG routines is sometimes more complicated than with IMSL. For example, if one wishes to calculate all eigenvalues within a given interval with the corresponding eigenvectors, one has to call three routines: First F01AYF/F01AGF, then F02BEF and finally F01AHF. With IMSL one routine call is sufficient for the same problem.

Both libraries use the same algorithms (reduction to a tridiagonal matrix Householder transformations, the QL algorithm for all eigenvalues of the tridiagonal matrix, bisection and inverse iteration for eigenvalues of tridiagonal matrices within a certain interval). Usage and error handling are similar for both libraries.

5.7.2 The General Eigenvalue Problem $Ax = \lambda Bx$

A and B Symmetric, B Positive Definite

In IMSL one finds three routines:

- GVLSP determines all eigenvalues, corresponding to F02ADF in the NAG library.
- GVCSP determines all eigenvalues and eigenvectors, corresponding to F02AEF.
- GPISP determines the performance index.

IMSL and NAG use the same method for this problem. First a Cholesky decomposition of B is used to lead to a standard eigenvalue problem. Then this is solved. Usage and error handling are similar for both libraries.

A and B Non-Symmetric

IMSL routines GVCRG and GVLRG calculate all eigenvalues and eigenvectors or only the eigenvalues. They correspond to NAG routine F02BJF. GPIRG determines the performance index.

Both NAG and IMSL use the QZ algorithm from Moler and Stewart, [71]. For treating the case $\lambda = \infty$, both libraries determine the eigenvalues λ_j as α_j / β_j with $\alpha_j \in \mathbb{C}$, $\beta_j \in \mathbb{R}$. While F02BJF has the error message 'too many iterations are required ...', GVCRG and GVLRG have no error messages at all. The IMSL routines work without a user-defined error tolerance.

5.7.3 Singular Value Decomposition

NAG routine F02WEF for calculating the singular value decomposition corresponds to IMSL routine LSVRR. Error handling of F02WEF and LSVRR is similar. Both libraries use the Golub–Reinsch method.

5.8 Exercises

5.8.1 Singular Value Decomposition (SVD) of A

Let $A \in \mathbb{R}^{m,n}$ be given and $A = USV^T$ be a known SVD of A.

1. Determine the SVD of A^T.
2. Assuming that A is quadratic and non-singular, determine the SVD of A^{-1}.
3. Show: If $N := \{v_{k_1}, \ldots, v_{k_l}\}$ is the set of right singular vectors corresponding to all l zero singular values $s_{k_i} = 0$, $i = 1, \ldots, l$, then N is an orthogonal basis of the null space (kernel) of A.
4. Show: If $M := \{u_{k_1}, \ldots, u_{k_r}\}$ is the set of left singular vectors corresponding to all r non-zero singular values $s_{k_i} > 0$, $i = 1, \ldots, r$, then M is an orthogonal basis of the image space (range) of A.
5. Show that rank(A) is equal to the number of positive singular values.
6. Let A be quadratic ($m = n$). Show that A is non-singular if and only if all singular values are positive.
7. Let A be quadratic. Show that then the modulus of the determinant of A is
$$|\det(A)| = \sum_{i=1}^{n} s_i.$$
8. Show that
$$||A||_2 = s_1$$
and, if A is quadratic and non-singular, that
$$||A^{-1}||_2 = s_n^{-1}.$$

5.8.2 Eigenvalue/Eigenvector Approximations

1. Show that the following theorem holds.

Theorem 5.11. *Let $A \in \mathbb{R}^{n,n}$ be symmetric, $\lambda \in \mathbb{R}$ and $x \in \mathbb{R}^n$ be given with $||x||_2 = 1$. Define a residual vector*
$$\eta := Ax - \lambda x.$$
Let $\sigma(A) := \{\lambda_1, \lambda_2, \ldots, \lambda_n\}$ be the spectrum of A. Then we have
$$\min_{\lambda_i \in \sigma(A)} |\lambda - \lambda_i| \leq ||\eta||_2.$$

2. Use Theorem 5.11 to find estimates $|\lambda - \lambda_i|$ for all three eigenvalues of the matrix

$$A = \begin{pmatrix} 3 & 1 & 2 \\ 1 & 3.5 & 2 \\ 2 & 2 & 4 \end{pmatrix},$$

for which only an approximation $\tilde{T}$ of matrix T (according to the second part of Lemma 5.3) is given:

$$\tilde{T} = \begin{pmatrix} -0.58 & -0.62 & 0.48 \\ -0.36 & 0.77 & 0.55 \\ 0.73 & -0.15 & 0.68 \end{pmatrix}.$$

(The columns of $\tilde{T}$ are approximations of eigenvectors x_1, x_2, x_3.)

5.8.3 The Sound of a Drum

1. According to the procedure given in Example 1.13 devise the matrix A for the region shown below.

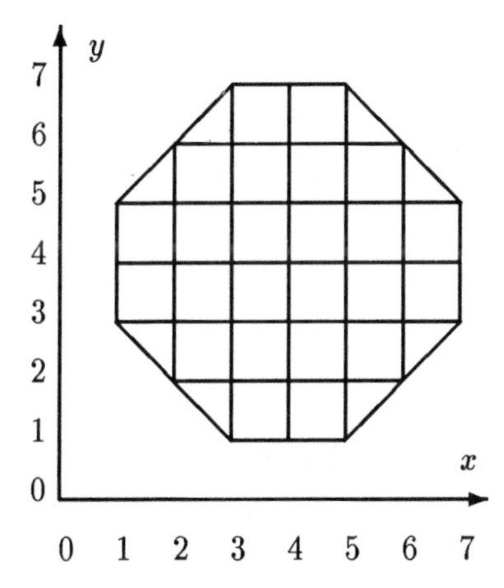

2. The first frequency of a drum of that shape can be approximated by calculating the smallest eigenvalue of A. Do that, but make use of all possible symmetries. That changes A, naturally. You should end up with a matrix of order $n = 5$.
3. Write a program which prepares an input file for one of the PAN eigenvalue programs using different grid spacing $h_i := 2^{-i}$, $i = 0, 1, 2, \ldots$. Approximate the first frequency λ_1/h_i^2 of the drum for different h_i such that the different values give an impression of the accuracy of the values for different grid spacings.

6
Numerical Integration

In many applications the evaluation of definite integrals – also called *quadrature* – is part of a mathematical problem. Often quadrature cannot be done analytically, or it is too complicated within a program with a large number of modules running. In such cases numerical integration is necessary.

We will consider only definite integrals approximated numerically by sums:

$$I := \int_a^b f(x)\, dx \approx \tilde{I} := \sum_{i=1}^{n} w_i f(x_i). \tag{6.1}$$

The abscissae x_i and weights w_i determine the integration method.

Indefinite integrals can be interpreted as solutions of differential equations and can therefore be calculated with the methods of Chapter 7.

We do not give formulae for the integration of sets of data points. These can be interpolated or approximated with the methods of Chapter 3 and can then be integrated analytically (see Section 3.3 for an example).

We will describe widely used methods such as Newton–Cotes methods, Romberg's method using extrapolation, and Gaussian integration. Then we will deal with algorithms, which are more complicated but also more effective, and which are therefore used in most software libraries. They are interlacing rules based on Gaussian quadrature, but have the additional facilities of adaptive or whole-interval automatic integration. For comparing different quadrature rules one counts the number of function evaluations necessary. Programs and examples are given in Section 6.2.

6.1 Quadrature Methods

6.1.1 Newton–Cotes Rules

The idea behind these rules is to interpolate the function f to be integrated and then to integrate the interpolating polynomial p analytically. The data points necessary for the interpolation are calculated from f at $m + 1$ equidistant points. As with interpolation, this method is stable only for small values of m. On the other hand the break-points that appear when several interpolations are linked from interval to interval create no problems for the integration approach. Therefore we give the rules for only $m = 1$ and $m = 2$:

$$\text{Trapezoidal rule:} \quad \int_a^b f(x)\,dx \approx \frac{b-a}{2}(f(a)+f(b)). \tag{6.2}$$

$$\text{Simpson's rule:} \quad \int_a^b f(x)\,dx \approx \frac{b-a}{6}\left(f(a)+4f\left(\frac{a+b}{2}\right)+f(b)\right). \tag{6.3}$$

To decrease the error caused by these rough approximations one can link several such rules together. Let there be $n+1$ equidistant points

$$x_j = a + jh, \quad j = 0, 1, \ldots, n, \quad \text{with} \quad h = \frac{b-a}{n}. \tag{6.4}$$

Linking n trapezoidal rules or $n/2$ Simpson's rules (with n even) together we arrive at the following composite formulae:

$$\begin{aligned} \tilde{I} = T(h) &= \frac{h}{2}\left(f(x_0) + 2f(x_1) + \cdots + 2f(x_{n-1}) + f(x_n)\right), \\ \tilde{I} = S(h) &= \frac{h}{3}(f(x_0) + 4f(x_1) + 2f(x_2) + \cdots + \\ &\quad +2f(x_{n-2}) + 4f(x_{n-1}) + f(x_n)). \end{aligned} \tag{6.5}$$

If one knows upper bounds for the 2nd or 4th derivative of f, one can estimate the error made by applying these rules to f:

$$\begin{aligned} |I - T(h)| &\leq \frac{|b-a|}{12} h^2 \max_{x\in[a,b]} |f''(x)|, \\ |I - S(h)| &\leq \frac{|b-a|}{180} h^4 \max_{x\in[a,b]} |f^{(4)}(x)|. \end{aligned} \tag{6.6}$$

Example 6.1.

$$I = \int_0^{\pi/2} \frac{5.0}{e^{\pi}-2} \exp(2x)\cos(x)\,dx = 1.0. \tag{6.7}$$

Figure 6.1 shows the function f and the area covered by the trapezoidal rule for $n = 4$. The results for both rules can be found in the following table:

Rule	h	$\tilde{I}$	$I - \tilde{I}$	Using (6.6)
Trapezoidal	$\pi/8$	0.926	0.074	0.12
Simpson	$\pi/8$	0.9925	0.0075	0.018

■

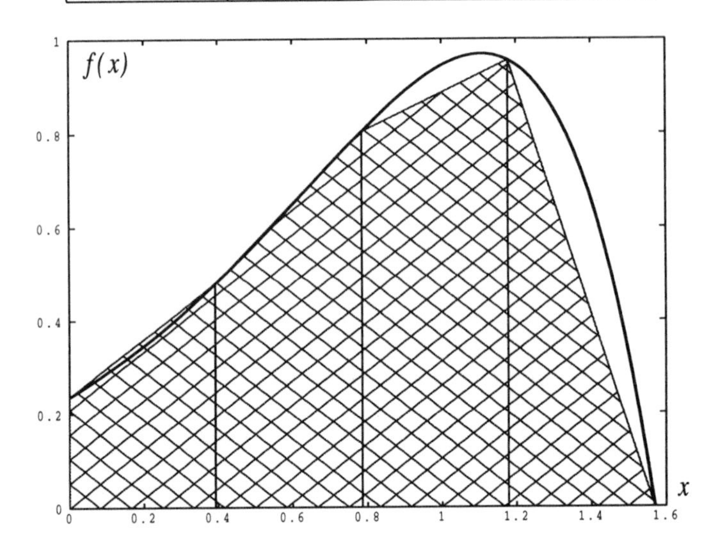

Figure 6.1. The trapezoidal rule for Example 6.1 ($n = 4$)

6.1.2 Romberg's Rule

Romberg applied Richardson's method of 'extrapolation to $h = 0$' to the error series of the trapezoidal rule to achieve a miraculous decrease of the errors of both the trapezoidal rule and Simpson's rule. He made use of the fact that the power series for the error developed for the variable h depends on h^2-terms rather than on h-terms:

$$T(h) - I = c_1 h^2 + c_2 h^4 + \cdots. \tag{6.8}$$

The coefficients c_i in this series are independent of h. Therefore one can combine two values $T(H)$ and $T(h)$ with $H \neq h$ to cause the first term of the series to disappear. For example if $h = 2H$ then

$$\tilde{I} := \frac{1}{3}\left(4T(H) - T(h)\right) \tag{6.9}$$

and the accuracy is increased to

$$I - \tilde{I} = O(h^4). \tag{6.10}$$

This method can be continued systematically, and one obtains high accuracy simply with a few more function evaluations. We will give the Romberg algorithm for step lengths which are always divided by two:

Romberg integration

Choice of step lengths

for $k = 0, 1, \ldots, m$

$$h_k := \frac{b-a}{2^k}$$

Recursive evaluation of $m+1$ trapezoidal rules

$T_{0,0} := T(h_0)$

for $k = 1, \ldots, m$

$$T_{k,0} := T(h_k) = \tfrac{1}{2} T_{k-1,0} + h_k(f(a+h_k) + f(a+3h_k) + \cdots + f(b-3h_k) + f(b-h_k))$$

Extrapolation to $h^2 = 0$

for $j = 1, 2, \ldots, m$

for $k = j, j+1, \ldots, m$:

$$T_{k,j} := \frac{4^j T_{k,j-1} - T_{k-1,j-1}}{4^j - 1}$$

The values $T_{k,j}$ can be presented in a triangle scheme. The value $T_{m,m}$ computed last is the most accurate one. Its error can be estimated by

$$|T_{m,m} - I| \leq \frac{(b-a)^{2m+3}\, B_{m+1}}{2^{m(m+1)}\,(2m+2)!} \max_{x\in[a,b]} |f^{(2m+2)}(x)| \tag{6.11}$$

with the Bernoulli numbers $(B_1, B_2, \ldots) = (1/6, 1/30, 1/42, 1/30, \ldots)$.

We apply this algorithm to Example 6.1 to obtain

k	h_k	$T_{k,0}$	$T_{k,1}$	$T_{k,2}$	$T_{k,3}$
0	$\pi/2$	0.18			
1	$\pi/4$	0.72	0.9044		
2	$\pi/8$	0.93	0.9925	0.998386	
3	$\pi/16$	0.98	0.9995	0.999974	0.999999

For the value $T_{3,3}$ the error estimate (6.11) yields the upper bound

$$|I - T_{3,3}| \leq 3 \times 10^{-5}. \tag{6.12}$$

Romberg's rule can be more efficient for other sequences of step lengths (see [92]).

6.1.3 Gaussian Quadrature

A Gaussian approach leads to rules which are usually even more accurate than Romberg's method: for the rule

$$\int_a^b f(x)\,dx \approx \sum_{i=1}^{n} w_i f(x_i) \tag{6.13}$$

determine weights w_i *and* abscissae x_i such that the rule will integrate polynomials of highest possible degree exactly.

A Newton–Cotes rule with n points integrates polynomials of maximum degree $n-1$ exactly; Gauss rules with the same number of points integrate polynomials of maximum degree as high as $2n-1$ exactly. Below we will give a small example to clarify the rule design principle. Then we will give the algorithm.

The high accuracy of Gauss rules brings with it two practical disadvantages:

- The weights and abscissae depend on the interval of integration.
- There are different weights and abscissae for each n.

The first point is easily covered by a linear transformation as in the following example. We will come back to the second point later.

Example 6.2. Consider the case $n = 2$:

$$\int_{-1}^{1} f(x)\,dx \approx \sum_{i=1}^{2} w_i f(x_i). \tag{6.14}$$

Determine $x_1,\ x_2,\ w_1,\ w_2$ such that for each polynomial p of maximum degree 3

$$\int_{-1}^{1} p(x)\,dx = \int_{-1}^{1} (a_0 + a_1 x + a_2 x^2 + a_3 x^3)\,dx = w_1 p(x_1) + w_2 p(x_2). \tag{6.15}$$

Integration and comparison of coefficients gives

$$\begin{aligned} w_1 &= 1 & w_2 &= 1 \\ x_1 &= \frac{-1}{\sqrt{3}} & x_2 &= \frac{1}{\sqrt{3}}\,. \end{aligned}$$

This means that

$$\int_{-1}^{1} f(x)\,dx \approx f\left(\frac{-1}{\sqrt{3}}\right) + f\left(\frac{1}{\sqrt{3}}\right). \tag{6.16}$$

This rule can easily be transformed to a general interval of integration:

$$\int_a^b f(x)\,dx \approx \frac{b-a}{2}\left(f(u_1)+f(u_2)\right) \tag{6.17}$$

with

$$u_i = \frac{a+b}{2} + \frac{b-a}{2}x_i, \quad i = 1,2. \tag{6.18}$$

■

In general we will get the following algorihmical steps for a Gauss rule for an interval $[a,b]$.

Gauss–Legendre quadrature

for $\int_a^b f(u)\,du$ with given abscissae and weights $x_i,\ w_i,\ i = 1,\cdots,n$ (for the original interval $[-1,1]$).

Transformation to $[a,b]$:

for $i = 1,\ldots,n$

$$u_i := \frac{a+b}{2} + \frac{b-a}{2}x_i$$

Calculation of transformed integral:

$$\hat{I} := \sum_{i=1}^{n} w_i f(u_i)$$

Back-transformation to $[-1,1]$:

$$\tilde{I} := \hat{I}\,\frac{b-a}{2}$$

The second disadvantage mentioned may cause real practical problems. For each n one has to store $2n$ numbers each with a sufficiently high number of digits. Below we will see how one can develop interlacing rules, which in a sequence of rules with increasing n use abscissae of the preceding rule. This decreases the amount of values to be stored, together with the fact that only specific values for n are used in the relevant methods.

Introducing a weight function ω into the approach (6.13)

$$\int_a^b f(x)\,\omega(x)\,dx = \sum_{i=1}^{n} w_i f(x_i) \tag{6.19}$$

makes Gaussian integration a powerful instrument for handling difficult quadratures, for example integrals over infinite intervals or improper integrals with integrands which have certain singularities. The determination of weights w_i and abscissae x_i now depends on the weight function as well. It is carried out using orthogonal polynomials (see [92]).

This is why Gaussian integration rules are normally referred to by a conjugate name, the second part of which defines the system of orthogonal polynomials. Of these the simplest form (6.13) (interval $[-1,1]$ and weight function $\omega(x) \equiv 1$) is called Gauss–Legendre integration. The most important rules are to be found in the following table:

Polynomial system	Weight function $\omega(x)$	Interval
Legendre: P_n	1	$[-1,1]$
Chebyshev of the first kind: T_n	$(1-x^2)^{-1/2}$	$[-1,1]$
Chebyshev of the second kind: U_n	$(1-x^2)^{1/2}$	$[-1,1]$
Laguerre: L_n	e^{-x}	$[0,\infty)$
Hermite: H_n	e^{-x^2}	$(-\infty,\infty)$

6.1.4 Interlacing Gauss Rules (Kronrod, Patterson)

The distinct nature of the weights and abscissae for different values of n constitutes one (perhaps the only) disadvantage of Gauss rules. Kronrod and Patterson found different ways of overcoming this disadvantage for specific rules and values of n. They both found sequences of *interlacing* rules, which for increasing n use abscissae from the preceding rule. Such Gauss rules are called *optimal*.

First we will mention Gauss–Kronrod quadrature. For specific values of n Kronrod developed an optimal $(2n+1)$-points rule by adding $n+1$ points:

$$\tilde{I}_1 := \sum_{i=1}^{n} w_i f(x_i) \quad \rightarrow \quad \tilde{I}_2 := \sum_{j=1}^{2n+1} v_j f(y_j) \tag{6.20}$$
$$\text{with } \{x_1, x_2, \ldots, x_n\} \subset \{y_1, y_2, \ldots, y_{2n+1}\}.$$

The combined rule integrates polynomials up to degree $3n+1$ exactly (instead of $4n+1$ for an independent $(2n+1)$-points rule). Theoretical details as well as weights and abscissae for pairs of rules with $n = 7$, 10, 15, 20, 25 and 30 can be found in [78]. The original paper of Kronrod, [64], is based on theoretical investigations by Szegö.

Patterson, [76], developed a sequence of interlaced Gauss–Legendre rules with (1, 3, 7, 15, 31, 63, 127, 255, 511) points, which integrate polynomials up to optimal degrees (1, 5, 11, 23, 47, 95, 191, 383, 767) exactly.

If one wants to use interlaced rules for infinite intervals of integrations, one can use the following transformations:

$$\int_a^{\pm\infty} f(x)\,dx = \pm \int_0^1 f\left(a \pm \frac{1-t}{t}\right) t^{-2}\,dt \tag{6.21}$$

for finite a or

$$\begin{aligned}\int_{-\infty}^{\infty} f(x)\,dx &= \int_0^{\infty} (f(x) + f(-x))\,dx \\ &= \int_0^1 \left(f\left(\frac{1-t}{t}\right) + f\left(\frac{t-1}{t}\right)\right) t^{-2}\,dt. \end{aligned}\tag{6.22}$$

6.1.5 Automatic Whole-Interval and Adaptive Rules

It is important that software library routines for integration use methods which control the accuracy of the integration automatically with an accuracy requirement parameter. This is normally achieved by computing a sequence of approximate integrals and error estimates

$$(I_k, \varepsilon_k), \quad k = 1, 2, \ldots\,. \tag{6.23}$$

The calculation will terminate when the estimate ε_k for a particular value of k becomes smaller than a given tolerance τ.

One possibility for such sequences is to take the interlacing rules of Patterson. These rules are all of the same type; they are always applied to the whole interval, but with increasing numbers of points. Therefore they are called *whole-interval* automatic quadrature rules.

If the integrand is strongly varying, or if it has discontinuities within the integration interval, then these whole-interval rules may not be successful or may not yield sufficient accuracy. In these instances *adaptive* automatic quadrature rules should be chosen. This means that the interval $[a, b]$ is divided into subintervals, which the same rule with the same number of points is applied to. This process is repeated recursively until for each subinterval an error estimate will be smaller than a given tolerance (divided by the relative subinterval width). The idea of adaptiveness is independent of a specific rule.

One cannot use the mathematical error estimates given above for automatic methods, because they need the evaluation of derivatives. Instead, one can apply two rules of the same type with different numbers of points, for example a pair of Kronrod rules, and take their difference $|\tilde{I}_1 - \tilde{I}_2|$ as an estimate for the error made. This automatic error control can fail, but there are only a few reported cases (see [67]).

In Section 6.2.1 we will compare the two different automatic rule principles with two extreme examples.

6.1.6 Multi-Dimensional Integration

For multi-dimensional integration the first idea may be to use *product rules.* For a simple two-dimensional integral

$$I := \int_a^b \int_c^d f(x,y)\, dy\, dx \tag{6.24}$$

we can take the one-dimensional abscissae and weights (x_i, w_i) for the interval $[a,b]$ and (y_i, v_i) for the interval $[c,d]$ to obtain

$$I \approx \sum_{i=1}^{n} w_i \int_c^d f(x_i, y)\, dy \approx \sum_{i=1}^{n}\sum_{j=1}^{n} w_i v_j\, f(x_i, y_j). \tag{6.25}$$

This method is easily generalized to variable integration limits:

$$\int_a^b \int_{\phi_1(y)}^{\phi_2(y)} f(x,y)\, dx\, dy. \tag{6.26}$$

The efficiency of the product rules decreases rapidly with increasing dimension. Therefore more efficient rules have been developed for particular applications. Good examples of an important application are Gauss rules for special geometric forms like triangles or tetrahedra, to be applied with finite element methods.

As well as the two methods mentioned, Monte Carlo methods can be used. They normally yield only a very limited accuracy, but can be combined with number theoretic methods.

Finally there are transformation methods (Sag–Szekeres), which allow the application of simple product integration methods like the trapezoidal rule.

6.2 Programs and Examples

The NAG Fortran Library contains more than 20 routines for numerical integration, about half of which are for multi-dimensional quadrature. This large number is reached because of the many special cases mentioned above and because there are automatic routines and user-controlled routines using the same method. The software package QUADPACK contains only routines for one-dimensional integration, but there are more than 30 such routines (see [78]); some of them are implemented as NAG routines as well. IMSL has about 20 integration routines, two of which are for multi-dimensional integration (see Section 6.3).

Most of the library routines use Gauss rules of different kinds (i.e. different weight functions and intervals). Additionally one will find routines using the methods of Clenshaw and Curtis, which are especially useful for singular integrands.

After having determined the integral fulfilling the tolerance specifications, some routines apply an additional extrapolation method to improve the accuracy of the method.

We have written four programs for numerical quadrature. They do not cover all of the special cases which are covered by the NAG routines, but they should be sufficient to evaluate most of the 'everyday' integrals with high accuracy. The decision tree in Figure 6.2 briefly shows the possibilities of these four programs:

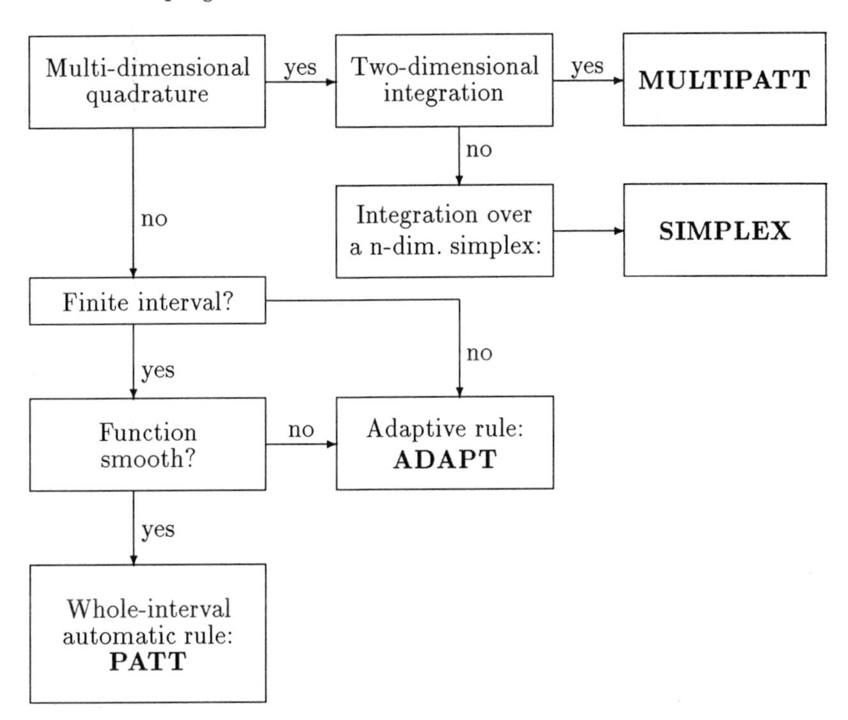

Figure 6.2. Decision tree for numerical integration

6.2.1 Automatic One-Dimensional Quadrature

The program ADAPT, see B.6.1, is a frame program for NAG routines D01AJF and D01AMF for integration over finite, semi-infinite or infinite intervals using Gauss–Kronrod rule pairs with (10, 21) and (7, 15) points, respectively.

The program PATT, see B.6.2, is a frame program for NAG routine D01ARF for whole-interval automatic quadrature using Patterson's method.

Both programs are well documented by their comments and the descriptive example input files, which can be created within the PAN system from the corresponding template programs. We will apply the program ADAPT to two extreme examples and use the example with finite integration interval to compare ADAPT with PATT.

Example 6.3. We define two functions, which both vary strongly and have a discontinuity:

$$f_1(x) = \begin{cases} \sin(30x)\exp(3x) & \text{if } x < 13\pi/60 \\ 5\exp(-(x-13\pi/60)) & \text{if } x \geq 13\pi/60 \end{cases}$$

$$f_2(x) = \begin{cases} \sin(30x)\exp(3x) & \text{if } x < 13\pi/60 \\ 5\exp(-(x-13\pi/60)^2) & \text{if } x \geq 13\pi/60. \end{cases}$$

For these functions we will determine the integrals

$$I_1 = \int_0^3 f_1(x)\,dx = 4.56673516941143,$$

$$I_2 = \int_0^\infty f_2(x)\,dx = 4.48957118586546.$$

For both methods the accuracy is controlled absolutely as well as relatively by the difference of two formulae using different numbers of points. For the Kronrod rules these are the two fixed numbers given above. The Patterson quadrature is carried out for a growing number of points anyway, and after each additional evaluation the difference between the new integral and that for the preceding number of points is compared with the given tolerance.

For the semi-infinite interval D01AMF carries out transformation (6.21). Patterson rules (D01ARF) can only be applied to finite intervals.

As relative and absolute tolerance we take $\tau = 1.0 \times 10^{-9}$. The Kronrod rule results are given in the following table:

Integral	Kronrod result	Error	Error estimate
I_1	4.5667351695951	1.84×10^{-10}	1.32×10^{-9}
I_2	4.4895711603026	2.55×10^{-12}	8.22×10^{-12}

Function f_1 and the subintervals for I_1 created by ADAPT are given in Figure 6.3.

The program ADAPT needs 29 points to reach the required accuracy, 21 of which lie within interval $[0.67, 0.7]$ around the discontinuity $\bar{x} = 13\pi/60 \approx 0.6806784$. The results are very satisfactory, the error estimates being only slightly larger than the errors themselves. We found examples which lead to error estimates smaller than the error. In all these cases we obtained a warning from the NAG routine used but the results were always acceptable.

Now for comparison we will evaluate the first integral I_1 with PATT, using whole-interval automatic quadrature. This is obviously not favourable for a function with a discontinuity. And indeed even the maximum of 511 points is not sufficient to obtain an accuracy of 1.0×10^{-3}. If we go down to a relative and absolute tolerance of 1.0×10^{-2} we obtain an integral value $I_P = 4.556027$ and an error estimate of 0.02421 for the absolute error, which underestimates the real error by a factor of 2.

If one is obliged to use a whole-interval quadrature for a function like f_1, there is a way out of disaster. One can divide the whole interval at the discontinuity oneself and then apply Patterson's method to both intervals. If we do that for the intervals $[0, 13\pi/60]$ and $[13\pi/60, 3]$, we obtain a total value for the integral which is even more accurate than the one obtained with Gauss–Kronrod rules.

Integral	Patterson rule	Error	Error estimate
I_1	4.5667351694115	1×10^{-13}	6×10^{-12}

■

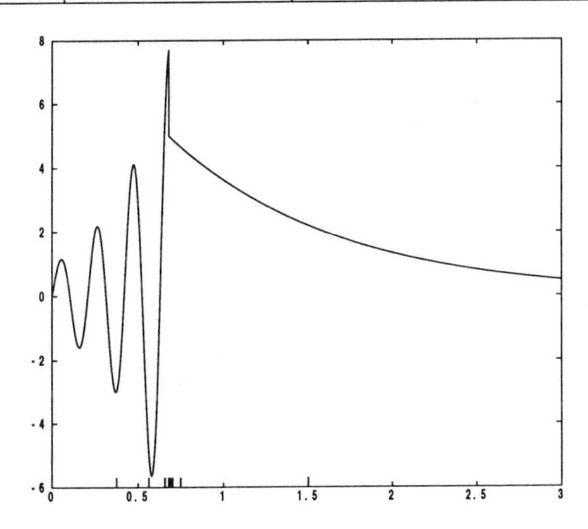

Figure 6.3. Interval subdivision for I_1

This example makes clear that even extreme functions can be integrated successfully, if the user does not simply apply a routine blindly, but looks for a convenient one or does some additional work. The program PATT has a built-in possibility that allows the interval to be subdivided by the user (see the input file below). D01ARF uses this possibility to evaluate several integrals for sub-intervals using a Legendre expansion calculated for the total interval, which means that no additional accuracy is obtained. But the program PATT allows a subdivision of the interval into several subintervals, summing up independent calls to D01ARF. It should be mentioned again that the routines always give a warning in cases when they cannot be guaranteed to yield reliable results.

Input Files for ADAPT and PATT

Now we will give the input files for I_1 which lead to the results shown above, beginning with the Fortran function

```
      double precision function f(x)
      double precision x,pi,s,x01aaf
      integer count
      common count
c
      pi = x01aaf(pi)
      s=1.3d1*pi/6.0d1
      if (x.lt. s) then
        f=sin(30.0d0*x)*exp(3.0d0*x)
      else
        f=5.0d0*exp(-(x-s))
      endif
      count = count + 1
      return
      end
```

```
Input file for the PAN program ADAPT:
Absolute accuracy epsilon:
1.0E-9
Relative accuracy delta:
1.0E-9
Interval (0=[a,b], 1=[a,inf), 2=(-inf,b], 3=(-inf,inf)):
0
a and b:
0 3
```

```
Input file for the PAN program PATT:
Absolute accuracy epsilon:
1.0E-9
Relative accuracy delta:
1.0E-9
Left limit:
0.0
```

```
Right limit:
3.0
Several subintervals (no = 0, yes = 1):
1
Independent integration (no = 0, yes = 1):
1
Dimension of vector of inner points:
1
Vector of inner points:
0.680678408277788535
```

6.2.2 Two-Dimensional Patterson Quadrature

The program MULTIPATT, see B.6.3, computes the double-integral (6.26). The integrand f as well as the lower and upper limit functions Φ_1 and Φ_2 must be defined in FUNCTION subprograms.

As an example we will calculate the volume of the two meeting shallow-water waves, which we defined in Section 3.7.2. We have to integrate the function F. In Section 3.7.2 we interpolated this function with two-dimensional B-splines. But we did not find a NAG routine which can integrate a two-dimensional B-spline representation. So we evaluated

$$\int_{-8}^{7} \int_{-8}^{7} F(x, y)\, dx\, dy$$

with MULTIPATT setting $\Phi_1(y) = -8$ and $\Phi_2(y) = 7$. We have to write the corresponding three Fortran functions F, PHI1 and PHI2 for D01DAF, then compile them and link them with the compiled program MULTIPATT. With an absolute tolerance of 0.01 the result is

```
Value of the integral:      81.669784961233
```

If we decrease the tolerance to 0.00001, we obtain a slightly different result

```
Value of the integral:      81.676300875934
```

with a much increased effort. If we further increase the tolerance up to 1.0×10^{-14}, the last value changes only its last digit. The tolerance must be as small as 1.0×10^{-15} to obtain an error message (IFAIL=170) indicating that the inner integral does not converge.

The small input file for this example as well as the necessary functions can be created by the corresponding function template programs of PAN. We will give one input file here.

Input File for MULTIPATT

```
Input file for the PAN program MULTIPATT:
Absolute accuracy epsilon:
```

```
0.01
Left limit:
-8.0
Right limit:
7.0
```

6.2.3 Multi-Dimensional Quadrature over a Simplex

The program SIMPLEX, see B.6.4, will serve mainly as an example for integration methods which are applied in larger connections, primarily with finite element methods for the solution of partial differential equations. For finite element methods two- and three-dimensional domains are divided in a large number of small domains of the same geometric structure, often triangles for two dimensions or tetrahedra for three dimensions, i.e. in general simplices. In each of these small domains a simple function must be integrated, often a linear or quadratic polynomial. This could easily be done analytically, but that would be less efficient than a numerical quadrature, which is fast and sufficiently accurate. We will test the program SIMPLEX by integrating an extreme function. Let f be

$$f(x, y, z) := \sqrt{x\, y\, (z+1)}.$$

The first derivative of this function, which is shown for $z = 0$ in Figure 6.4, has a singularity for all values (x, y, z) with $x = 0$ or $y = 0$. This makes it harder to reach sufficient accuracy for the integral.

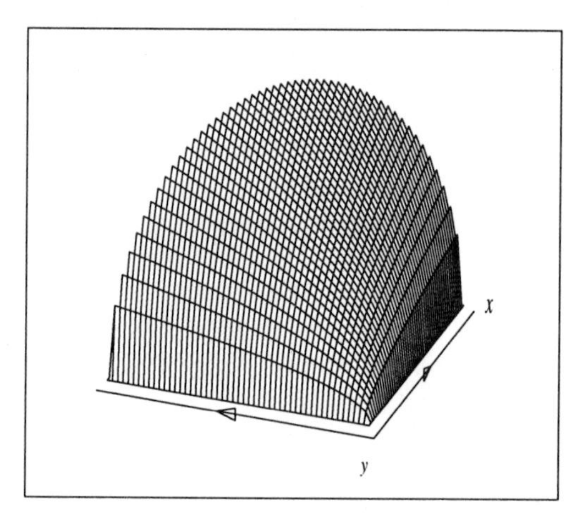

Figure 6.4. The root function $\sqrt{xy(z+1)}$ for $z = 0$

The function has to be written as a Fortran function, compiled and then linked with the program. It can also be defined by the function template program of PAN. Then the input file has to be created with an editor or by using the corresponding template program of PAN. The program then computes a sequence of approximations for the integral, from order 1 up to the maximum order according to NAG routine D01PAF. The true value of the integral is $I = 3.3048735\ldots$.

Input File for SIMPLEX

```
Input file for the PAN program SIMPLEX:
Maximum order of approximation:
6
Dimension of the integral (>1):
3
Row and column dimension of matrix of simplex coordinates:
4  3
Matrix of simplex coordinates:
 0   0  0
10   0  0
 0  10  0
 0   0  0.1
```

This input file leads to the following results, where we have included the true absolute error in brackets:

```
Output file for the PAN program SIMPLEX:

Order  Value of the integral  Estimate of and absolute error
  1         4.21842849              4.200          (0.913)
  2         3.53256198              0.690          (0.228)
  3         3.41134272              0.120          (0.106)
  4         3.36629344              0.045          (0.061)
  5         3.34459033              0.022          (0.040)
  6         3.33251718              0.012          (0.028)
```

One sees that for high orders the estimate for the error may become smaller than the error itself.

6.3 Numerical Integration Using IMSL

IMSL routines QDAG and QDAGI for one-dimensional adaptive integration correspond to NAG routines D01AJF and D01AMF, respectively, but in QDAG the user can choose different Gauss–Kronrod rule pairs from (7, 15) points to (30, 61) points; D01AJF always takes the pair (10, 21). QDAGI uses (10, 21) points, D02AMF (7, 15). Parameters and error warnings are similar for these routines.

NAG routine D01ARF for whole-interval automatic quadrature corresponds to IMSL routine QDNG. QDNG cannot evaluate indefinite integrals. For the evaluation of definite integrals the routines are very similar. D01ARF uses rules with a maximum of 511 points, while QDNG use only rules with a maximum of 87 points.

The evaluation of indefinite integrals is only possible in IMSL with Gaussian quadrature in GQRUL or GQRCF. The corresponding NAG routines are D01BBF and D01BCF. The computation of two-dimensional integrals

$$I = \int_a^b \int_{\phi_1(y)}^{\phi_2(y)} f(x,y)\, dx\, dy$$

can be done with NAG routine D01DAF and IMSL routine TWODQ. D01DAF uses Patterson rules, TWODQ adaptive Gauss–Kronrod rules. Therefore the IMSL routine will be more convenient for discontinuous integrands or for integrals with discontinuities in the limit functions ϕ_k.

For n-dimensional integration IMSL has the routine QAND, which evaluates the integral

$$I = \int_{a_1}^{b_1} \cdots \int_{a_n}^{b_n} f(x_1, \ldots, x_n)\, dx_n \ldots\, dx_1$$

with a Gauss rule. The corresponding NAG routine is D01FBF, but NAG also has several other routines for multi-dimensional integration.

6.4 Exercises

6.4.1 Gauss Quadrature

Determine the point x_1 and weight w_1 such that

$$\int_{-1}^{1} p(x)\, dx = w_1\, p(x_1)$$

for each polynomial of degree 1.

6.4.2 Gauss–Kronrod Quadrature

Determine points y_1, y_2, y_3 and weights v_1, v_2, v_3 such that one y_i is equal to x_1 from the Gauss rule for $n = 1$, and that

$$\int_{-1}^{1} p(x)\, dx = \sum_{i=1}^{3} v_i\, p(y_i)$$

for each polynomial of degree 4 or less.

Hint: Try it with symmetric y_i and use the basic polynomials.

7

Initial Value Problems for Ordinary Differential Equations

Ordinary differential equations (ODEs) are equations for functions of one independent variable. Each equation can depend on the unknown function and its derivatives up to a specified order.

Often differential equations are model equations for application problems in science, nature and engineering. The changes of a system with time are described by them; for example they could describe a chemical reaction. The solution normally is uniquely defined if one knows the initial state ('initial conditions' (IC)) or values at the two boundary points of a specified interval ('boundary conditions' (BC)). These two possibilities are known as initial value and boundary value problems, respectively. Mathematically they are of quite different nature as far as the theory as well as the numerical methods are concerned. This chapter considers initial value problems; the next chapter will consider boundary value problems.

Initial value problems cover a wide range of problems with different methods of solution. Therefore one may find software packages and libraries especially devoted to the solution of these problems. The most important methods are to be found in the NAG Fortran or IMSL libraries.

7.1 Introduction

In this section we will learn something about the theoretical basis and the standard numerical methods. We will introduce the important terms *consistency*, *stability* and *convergence*. The construction and characteristic differences of standard methods will be shown with simple example methods applied to a test equation.

7.1.1 The Problem

Let $f \in C([a,b] \times \mathbb{R}^n, \mathbb{R}^n)$ be a given function and $y_0 \in \mathbb{R}^n$ a given vector. A function $y \in C^1([a,b], \mathbb{R}^n)$ has to be determined for which:

$$\begin{array}{llll} \text{ODE:} & y'(x) & = & f(x, y(x)) \quad \forall x \in [a, b], \\ \text{IC:} & y\,(a) & = & y_0 . \end{array} \tag{7.1}$$

$y' = f(x, y)$ is called a first-order system of differential equations for a vector valued function y, depending on a single variable x. Writing system (7.1) out in full, we get[†]:

$$\begin{aligned} y_1'(x) &= f_1(x, y_1(x), y_2(x), \cdots, y_n(x)), \\ y_2'(x) &= f_2(x, y_1(x), y_2(x), \cdots, y_n(x)), \\ \cdots \quad &\cdots \quad \cdots \\ y_{n-1}'(x) &= f_{n-1}(x, y_1(x), y_2(x), \cdots, y_n(x)), \\ y_n'(x) &= f_n(x, y_1(x), y_2(x), \cdots, y_n(x)), \\ y\,(a) &= y_0 = (y_{10}, y_{20}, \cdots, y_{n0})^T . \end{aligned}$$

This system of differential equations is called *explicit*, because the highest derivative (here the first) is isolated on the left-hand side. We will not consider *implicit* differential equations if they cannot be transformed into explicit ones easily.

Lemma 7.1. *Every explicit system of differential equations of order m can be transformed into an equivalent one of order 1.*

Proof: We only show (the rest is simple combinatorial analysis):

A single differential equation of order n can always be transformed into a system of order 1 with n equations:

Let a single differential equation

$$z^{(n)} = g(x, z, z', \cdots, z^{(n-1)}) \tag{7.2}$$

be given with initial values

$$z(a) = z_0, \quad z'(a) = z_0', \quad \cdots \quad , \quad z^{(n-1)}(a) = z_0^{(n-1)}.$$

Define a vector valued function

$$\begin{aligned} y(x) &= (y_1(x), \ y_2(x), \ \ldots \ , \ y_n(x))^T \\ &:= \left(z(x), \ z'(x), \ \ldots \ , \ z^{(n-1)}(x)\right)^T . \end{aligned} \tag{7.3}$$

Then obviously

[†] y-values with indices are functions here, but single values below, therefore we always write them with their variable.

$$
\begin{aligned}
y_1'(x) &:= y_2(x), \\
y_2'(x) &:= y_3(x), \\
&\vdots \\
y_{n-1}'(x) &:= y_n(x), \\
y_n'(x) &:= g(x, z, z', \ldots, z^{(n-1)}) = g(x, y_1(x),\ y_2(x),\ \ldots,\ y_n(x))
\end{aligned}
\tag{7.4}
$$

and

$$
y(a) = y_0 = \left(z_0,\ z_0',\ \ldots,\ z_0^{(n-1)}\right)^T. \tag{7.5}
$$

Equation (7.4) with (7.5) is a system of type (7.1). Its vector valued solution $y(x)$ yields the solution $z(x) = y_1(x)$ of (7.2). ∎

We will not consider the questions of existence and uniqueness of solutions here. But we will name two important conditions for existence of a solution as well as for numerical error estimates:

Assumption 7.2.

1. *The function* $f : \mathbb{R}^{n+1} \to \mathbb{R}^n$ *is defined and continuous in the strip region*

$$
S := \{(x, y) \mid x \in [a, b],\, y \in \mathbb{R}^n\} \quad \textit{with} \quad -\infty < a < b < \infty. \tag{7.6}
$$

2. *Lipschitz condition: There exists a constant* $L \in \mathbb{R}$, *such that*

$$
\|f(x, y) - f(x, \tilde{y})\| \le L\|y - \tilde{y}\| \quad \forall (x, y), (x, \tilde{y}) \in S. \tag{7.7}
$$

There are systems of practical importance which do not fulfill these assumptions and nevertheless can have a numerically stable solution, but normally they need additional care both theoretically and as far as the numerical method is concerned.

7.1.2 A Model Problem

The following equation (here for $n = 1$ only) often serves as a model or test equation for numerical methods:

$$
\begin{aligned}
y' &= \lambda y, \\
y(0) &= 1.
\end{aligned}
\tag{7.8}
$$

This equation is linear with the solution $y(x) = e^{\lambda x}$. Therefore it cannot mirror the complex nonlinear structure of many differential equations, but varying the parameter λ can show how stable a numerical method is, as we soon will see.

7.1.3 Three Example Methods

We are looking for approximations of solution values $y(x_j)$ at equidistant grid points

$$x_j := a + jh, \quad j = 0, 1, \cdots, N \quad \text{with } h := \frac{b-a}{N}. \tag{7.9}$$

Numerical methods can be constructed using the following Lemma.

Lemma 7.3. *The system (7.1) is equivalent to the following system of integral equations:*

$$y(x) = y_0 + \int_a^x f(\xi, y(\xi))\, d\xi. \tag{7.10}$$

Another possibility is using the truncated Taylor series:

$$\begin{aligned} y(x_{j+1}) &= y(x_j) + hy'(x_j) + \frac{h^2}{2!}y''(x_j) + \frac{h^3}{3!}y'''(x_j) + \cdots \\ &= y(x_j) + hf(x_j, y(x_j)) + \cdots . \end{aligned} \tag{7.11}$$

With $y_j \approx y(x_j)$ and $f_j := f(x_j, y_j)$, this leads to a first simple method:

Euler's explicit method

$$y_{j+1} = y_j + hf_j, \quad j = 0, 1, \cdots, N-1$$

The same method could have been obtained from the integral equation (7.10) by substituting the integral by the rectangular formula:

$$\int_{x_j}^{x_{j+1}} f(\xi, y(\xi))\, d\xi \approx hf(x_j, y_j). \tag{7.12}$$

Euler's method is explicit in the form given. But if we take the function value for the rectangular formula at the right interval boundary point, then we get Euler's implicit method instead, because then the unknown solution value is contained in the function value $f(x_{j+1}, y_{j+1})$ on the right-hand side:

Euler's implicit method

$$y_{j+1} = y_j + hf_{j+1}, \quad j = 0, 1, \cdots, N-1$$

For this method one has to solve a system of (in general nonlinear) equations for each solution step. Therefore one must have important reasons when opting for an implicit method.

Both Euler methods are *one-step methods*, which means, that one calculates y_{j+1} using only the approximate value y_j at the point x_j.

Let us now develop a simple *multistep method*. If we integrate (7.10) over two intervals of length h and take the mid-point value for the rectangular formula

$$\int_{x_j}^{x_{j+2}} f(\xi, y(\xi))\, d\xi \approx 2hf(x_{j+1}, y_{j+1}), \tag{7.13}$$

we get

Mid-point rule

$$y_{j+2} = y_j + 2hf_{j+1}, \quad j = 0, 1, \cdots, N-2$$

This is an explicit two-step method, because we need two preceding values y_j and y_{j+1} for calculating y_{j+2}. This is not possible in the first step, therefore we need a single-step method for calculating y_1 from y_0. This method is then called the *starting method*.

7.1.4 Application to the Model Equation

The solution of the model equation $y(x) = e^{\lambda x}$ increases monotonically for $\lambda > 0$ and decreases monotonically for $\lambda < 0$. We are interested to see if the methods can reproduce at least this growth characteristic.

Euler's explicit method:

$$\begin{aligned}
y_0 &= 1 \\
y_1 &= 1 + \lambda h \\
y_2 &= y_1 + h\lambda(1 + \lambda h) = (1 + \lambda h)^2 \\
&\vdots \\
y_j &= (1 + \lambda h)^j .
\end{aligned}$$

Euler's implicit method:

$$\begin{aligned}
y_0 &= 1 \\
y_1 &= \frac{1}{1 - \lambda h} \\
y_2 &= y_1 + h\lambda y_2 = \frac{1}{(1 - \lambda h)^2} \\
&\vdots \\
y_j &= \frac{1}{(1 - \lambda h)^j} .
\end{aligned}$$

Mid-point rule:

$$\begin{aligned} y_0 &= 1 \\ y_1 &= e^{\lambda h} \quad \text{(true solution)} \\ y_2 &= 1 + 2h\lambda e^{\lambda h} \\ &\vdots \\ y_j &= e^{\lambda j h}\left(1 - \frac{\lambda^3 jh}{6}h^2 + \frac{\lambda^3}{12}h^3\right) \\ &\quad \underbrace{+(-1)^j e^{-\lambda j h}\frac{\lambda^3}{12}h^3}_{\text{(see 3 below)}} + O(h^4). \end{aligned}$$

For all three methods the values y_j converge to the true solution value $y(\bar{x}) = e^{\lambda\bar{x}}$, if

$$h \to 0, \quad j \to \infty \text{ such, that } \bar{x} = jh. \tag{7.14}$$

But for fixed $h > 0$ and $\lambda \ll 0$ there are big differences.

1. Euler's explicit method:
   If $\lambda < 0$ is so small that $|1+\lambda h| > 1$, then the values y_j will grow and oscillate heavily; the numerical solution becomes *unstable* and totally wrong.
2. Euler's implicit method:
   Independently of $\lambda < 0$ the solution remains *stable*; it decreases monotonically.
3. Mid-point rule (explicit):
   The solution is a better approximation of the true solution $y(x) = e^{\lambda x}$ than those calculated with the two Euler methods, but it contains an oscillating part (marked by the brace) which grows for $\lambda < 0$ and finally superposes the other solution parts†.

The sentence 'Implicit methods are more stable than explicit ones' is true in general, see Section 7.1.7. In Section 7.2.4 we will see a worked out nonlinear example comparing different one-step methods.

7.1.5 Discretization Error and Consistency

The error $y_j - y(x_j)$ of a method consists of an error part which has been made at the actual step by the method chosen, and of a part which has been carried forward from errors of former steps.

---

†The formula for y_j is to be found in [92] for example.

For analyzing a method one first locally looks at its behaviour for one step assuming that all values introduced in this step are correct. This can only lead to *local* statements.

Definition 7.4.

1. *A one-step method is defined by its increment function* Φ*:*

$$y_{j+1} = y_j + h\Phi(x_j, y_j). \tag{7.15}$$

2. *The local discretization error of a one-step method is defined as*

$$d_{j+1} := \frac{1}{h}(y(x_{j+1}) - y(x_j)) - \Phi(x_j, y(x_j)). \tag{7.16}$$

3. *A one-step method is called consistent with (7.1) if we have asymptotically:*

$$\max_{j=1,\cdots N} ||d_j|| \longrightarrow 0 \quad \textit{with} \quad h \to 0. \tag{7.17}$$

This limit means that $N \to \infty$*, because* $Nh \equiv b - a$*.*

4. *A one-step method has order* p *if for* $K > 0$

$$\max_{j=1,\cdots N} ||d_j|| \leq Kh^p = O(h^p) \quad \textit{with} \quad h \to 0. \tag{7.18}$$

This 'order' is a 'consistency order' for the moment, but we will see that it is always the same as the global 'convergence order'. The discretization error is defined with true function values only; it is defined as the difference between the growth of the true solution and that of the numerical method in one step.

Both Euler methods are consistent and of order $p = 1$, as is easily seen, if one looks at the construction of these methods.

7.1.6 Convergence and Stability

Consistency and order are important features for the precision a method can yield. But one has to consider the accumulated error rather than the local one. The consistency order fails to indicate the order of the accumulated discretization error, but this gap can be filled.

Definition 7.5.

1. *The accumulated discretization error is given as*

$$g_j = y(x_j) - y_j. \tag{7.19}$$

2. *A method for solving (7.1) is called convergent if*

$$\max_{j=1,\cdots,N} ||g_j|| \longrightarrow 0 \quad \textit{with} \quad h \to 0. \tag{7.20}$$

3. A one-step method has convergence order p if with a constant $K > 0$:

$$\max_{j=1,\cdots N} ||g_j|| \leq Kh^p = O(h^p) \quad with \quad h \to 0. \tag{7.21}$$

If a one-step method is convergent, then it is consistent as well with

$$||g_j|| = O(h^p).$$

This means that the convergence order is always equal to the consistency order. Therefore the term 'order' is sufficient.

There are consistent methods which are not convergent. Such methods are called *asymptotically unstable.* This type of stability is the link between consistency and convergence. There are many more stability concepts. We only will mention two typical ones.

- A one-step method is called *asymptotically stable* (which means stable with $h \to 0$), if small perturbations of the method lead only to small perturbations of the approximation values.
- For the numerical treatment of ordinary differential equations another class of stability is very important, too. It is the stability for $h > 0$. We could see this with the examples in Section 7.1.4. Within this class one can find many different stability definitions in the literature; see for example [50]. The most important one is *absolute stability*.

A sufficient and important assumption for asymptotic stability is the Lipschitz condition, which follows.

Lemma 7.6. *A one-step method with increment function Φ is asymptotically stable if Φ fulfills a Lipschitz condition: A number $L > 0$ exists such that*

$$||\Phi(x_j, y_j) - \Phi(x_j, \tilde{y}_j)|| \leq L||y_j - \tilde{y}_j|| \quad \forall (x_j, y_j), (x_j, \tilde{y}_j) \in S. \tag{7.22}$$

We now come to the central theorem for the theory of numerical methods for ODEs.

Theorem 7.7. *A consistent one-step method of order p, which is asymptotically stable, is convergent with order p:*

Consistency and Stability = **Convergence**

The Lipschitz condition for Φ makes an error estimation possible.

Theorem 7.8. *Let a one-step method be consistent with order p, and let it fulfill the Lipschitz condition (7.22). Define the so-called Lipschitz function E_L as*

$$E_L(x) := \begin{cases} \dfrac{e^{L(x-a)}}{L} & \text{if } L > 0 \\ x & \text{if } L = 0. \end{cases}$$

Then

$$\|y_j - y(x_j)\| \leq h^p K E_L(x_j - a). \tag{7.23}$$

This error estimate grows exponentially with x; therefore it most often overestimates the true error.

7.1.7 Stiff Differential Equations

A system of ordinary differential equations is called *stiff* if the solution contains decreasing components with extremely different growth parameters. Stiff systems have become more and more important during recent decades for some applications. We will give an example from the most important group of applications.

Example 7.9. *Stiff systems with reaction kinetics*

Reaction kinetics is an important subject in chemistry.

Let a two-step reaction for materials y_1, y_2, y_3 be given:

$$y_1 \xrightarrow{k_1} y_2 \xrightarrow{k_2} y_3$$

with the initial concentrations

$$y_1(0) = 1, \quad y_2(0) = 1, \quad y_3(0) = 1.$$

For $k_1 = 1$ and $k_2 = 101$ this leads to a linear ODE system with constant coefficients:

$$\begin{aligned} y_1' &= -y_1, \\ y_2' &= y_1 - 101y_2, \\ y_3' &= 101y_2. \end{aligned}$$

Its solution can be easily calculated analytically:

$$\begin{aligned} y_1(x) &= e^{-x}, \\ y_2(x) &= 0.01e^{-x} + 0.99e^{-101x}, \\ y_3(x) &= 3 - 1.01e^{-x} - 0.99e^{-101x}. \end{aligned}$$

One sees that a slowly decreasing solution component is combined with a rapidly decreasing one. The coefficient matrix

$$\begin{pmatrix} -1 & 0 & 0 \\ 1 & -101 & 0 \\ 0 & 101 & 0 \end{pmatrix}$$

of the system has the eigenvalue $\lambda_3 = 0$ and the two quite different negative ones

$$\lambda_1 = -1 \quad \text{and} \quad \lambda_2 = -101.$$

■

The numerical solution of such systems with explicit methods mostly yields totally wrong results and may be difficult even with implicit methods. Therefore special methods have been developed for those systems. One class of them will be considered in Section 7.4.

If the system is nonlinear its behaviour can vary between 'stiff' and 'non-stiff', because the 'growth parameters' now depend on x. The growth parameters are the eigenvalues λ_i, $i = 1, \cdots, n$, of the Jacobian matrix

$$f_y(x,y) := \begin{pmatrix} \frac{\partial f_1}{\partial y_1}(x,y) & \cdots & \frac{\partial f_1}{\partial y_n}(x,y) \\ \vdots & & \vdots \\ \frac{\partial f_n}{\partial y_1}(x,y) & \cdots & \frac{\partial f_n}{\partial y_n}(x,y) \end{pmatrix}. \tag{7.24}$$

Now a system is called stiff at a point x if f_y only has non-positive eigenvalues at x which differ greatly:

1. $\text{Re}\, \lambda_i \leq 0, \;\; i = 1, \cdots, n$
2. For $\lambda_i < 0: \;\; \max_i |\text{Re}\, \lambda_i| \gg \min_i |\text{Re}\, \lambda_i|$

As a quantity for measuring stiffness we can define a stiffness index S:

$$S := \frac{\max_i |\text{Re}\, \lambda_i|}{\min_i |Re\, \lambda_i|} \quad \text{for } \lambda_i < 0. \tag{7.25}$$

Often we know that the problem to be solved leads to a stiff system. Then we will look for special methods anyway. But it is possible, too, to determine the stiffness with a stiffness test before we solve the problem numerically. Libraries like NAG give the possibility of such a test and the program STIFFTEST uses the test of NAG routine D02BDF.

7.2 One-Step Methods

7.2.1 Runge–Kutta Methods

The most important one-step methods are the *Runge–Kutta methods*. They can be written as

$$\begin{aligned} y_{j+1} &= y_j + h\Phi(x_j, y_j) \quad \text{with} \\ \Phi(x_j, y_j) &= \sum_{l=1}^{m} \gamma_l k_l \quad \text{with} \\ k_l &= f(x_j + \alpha_l h, y_j + h\sum_{s=1}^{m} \beta_{ls} k_s), \quad l = 1, \cdots, m. \end{aligned} \tag{7.26}$$

These method are *explicit* if $\beta_{ls} = 0$ for $s \geq l$. They are called *semi-implicit* if $\beta_{ls} = 0$ for $s > l$, and *implicit* otherwise. m is called the stage of the Runge–Kutta method, and $m^2 + 2m$ parameters can be used to control the order and stability characteristics of the method. For the construction of a reasonable Runge–Kutta method one has to fulfill the following conditions and targets.

1. A Runge–Kutta method is consistent if

$$\gamma_1 + \gamma_2 + \cdots + \gamma_m = 1. \tag{7.27}$$

2. Each k_l shall be an h^2 approximation of $y'(x_j + \alpha_l h)$:

$$\alpha_l = \sum_{s=1}^{m} \beta_{ls}. \tag{7.28}$$

3. The order shall be as large as possible.

For $m > 1$ infinitely many methods fulfilling these conditions exist. For $m = 1$, (7.27) yields $\gamma_1 = 1$ and hence

$$y_{j+1} = y_j + hf(x_j + \alpha_1 h, y_j + h\beta_{11}k_1).$$

The method is explicit only if $\beta_{11} = 0$, which includes $\alpha_1 = 0$ according to (7.28). And this is the explicit Euler method.

With $\alpha_1 = \beta_{11} = 1$ we have

$$\begin{aligned} y_{j+1} &= y_j + hk_1 \\ &= y_j + hf(x_j + h, y_j + hk_1) \\ &= y_j + hf(x_j + h, y_{j+1}). \end{aligned}$$

This is the implicit Euler method.

The parameters of a Runge–Kutta method can be given in a scheme:

$$\begin{array}{c|cccc} \alpha_1 & \beta_{11} & \beta_{12} & \cdots & \beta_{1m} \\ \vdots & \vdots & \vdots & \vdots & \vdots \\ \alpha_m & \beta_{m1} & \beta_{m2} & \cdots & \beta_{mm} \\ \hline & \gamma_1 & \gamma_2 & \cdots & \gamma_m \end{array}$$

For explicit methods this scheme is triangular.

7.2.2 Classical Runge–Kutta Method

The oldest and best known Runge–Kutta method is of stage and order four, and was developed by Runge and Kutta about 90 years ago.

$$y_{j+1} = y_j + \frac{h}{6}(k_1 + 2k_2 + 2k_3 + k_4) \tag{7.29}$$

$$\text{with} \quad \begin{aligned} k_1 &:= f(x_j, y_j), \\ k_2 &:= f\left(x_j + \tfrac{1}{2}h, y_j + \tfrac{1}{2}hk_1\right), \\ k_3 &:= f\left(x_j + \tfrac{1}{2}h, y_j + \tfrac{1}{2}hk_2\right), \\ k_4 &:= f\left(x_j + h, y_j + hk_3\right). \end{aligned}$$

It is explicit with the following scheme:

$$\begin{array}{c|cccc} 0 & & & & \\ \frac{1}{2} & \frac{1}{2} & & & \\ \frac{1}{2} & 0 & \frac{1}{2} & & \\ 1 & 0 & 0 & 1 & \\ \hline & \frac{1}{6} & \frac{2}{6} & \frac{2}{6} & \frac{1}{6} \end{array}$$

There are many more fourth-order methods, but it can be shown that no consistent and stable method of stage four or five exists, which has a higher order than four. One criterion for developing new methods is the importance of adjusting the steplength h during the calculation.

7.2.3 Step-Control Policy, Merson's Method

For solving nonlinear differential equations effectively it is necessary to use a variable step length. If f is varying rapidly the step length should be small. If the step length were small in sub-intervals of smooth f the rounding error would become unnecessarily large. Therefore one uses a step-control policy. For this purpose one would like to have a realistic estimate of the accumulated error. But this is not possible, because all easy-to-calculate estimates like (7.23) for the accumulated error normally overestimate the true error. Therefore one uses local error control, for which several possibilities exist.

One may solve the system in parallel with two methods of different orders or of the same order but different stages. Then the difference of the two results at each point x_j is a good estimate for the error made. Using two methods need not double the number of operations, if one uses coupled formulae. Runge–Kutta methods can be coupled using the surplus degrees of freedom for constructing two methods which have most of the coefficients in common. One such pair of formulae is England's method, which consists of a four-stage method of order four and a six-stage method of order five, see [32]. A slightly different possibility is used by Merson who coupled a four-stage and a five-stage method, both of order four. The additional fifth function evaluation is used for a local error estimate:

$$
\begin{array}{c|cccccc}
0 & & & & & k_1^{[4]} = k_1^{[5]} \\
\frac{1}{3} & \frac{1}{3} & & & & k_2^{[4]} = k_2^{[5]} \\
\frac{1}{3} & \frac{1}{6} & \frac{1}{6} & & & k_3^{[4]} = k_3^{[5]} \\
\frac{1}{2} & \frac{1}{8} & 0 & \frac{3}{8} & & k_4^{[4]} = k_4^{[5]} \\
\cdots & \cdots & \cdots & \cdots & \cdots & \cdots \\
1 & \frac{1}{2} & 0 & -\frac{3}{2} & 2 & \quad k_5^{[5]} \\
\hline
y_{j+1}^{[4]} & \frac{1}{2} & 0 & -\frac{3}{2} & 2 & \gamma_l^{[4]} \\
\hline
y_{j+1}^{[5]} & \frac{1}{6} & 0 & 0 & \frac{2}{3} & \frac{1}{6} \quad \gamma_l^{[5]}
\end{array}
$$

Merson's combined four/five-stage methods of order four

With the scheme shown two methods are defined by the two vectors of coefficients $\{\gamma_l^{[4]}\}$ and $\{\gamma_l^{[5]}\}$. The error estimate then reads as follows:

$$y(x_{j+1}) - y_{j+1}^{[5]} \approx \frac{1}{5}(y_{j+1}^{[5]} - y_{j+1}^{[4]}) = \frac{h}{30}(-2k_1 + 9k_3 - 8k_4 + k_5). \tag{7.30}$$

This can be shown to be a correct estimate for linear differential equations and is taken as an approximate estimate for nonlinear problems, which is not always justified.

A further important possibility for an error estimate is *extrapolating the step length to* $h = 0$. One calculates solutions at x_j for two different step lengths h and qh $(0 < q < 1)$ in parallel and extrapolates the two solution values to $h = 0$. For a method of order p we get

$$
\begin{array}{rrcl}
h: & y_j & = & y(x_j) + h^p\, e_j \;+\; O(h^{p+1}), \\
qh: & \bar{y}_j & = & y(x_j) + (qh)^p\, e_j \;+\; O((qh)^{p+1})
\end{array}
$$

and therefore approximately

$$h^p\, e_j \approx \frac{\bar{y}_j - y_j}{q^p - 1} =: T. \tag{7.31}$$

This technique is the basis for a numerical method on its own, which we will not discuss further; see [92].

After having estimated the error for a step there are different step-control policies, one of which we will show in the following diagram.

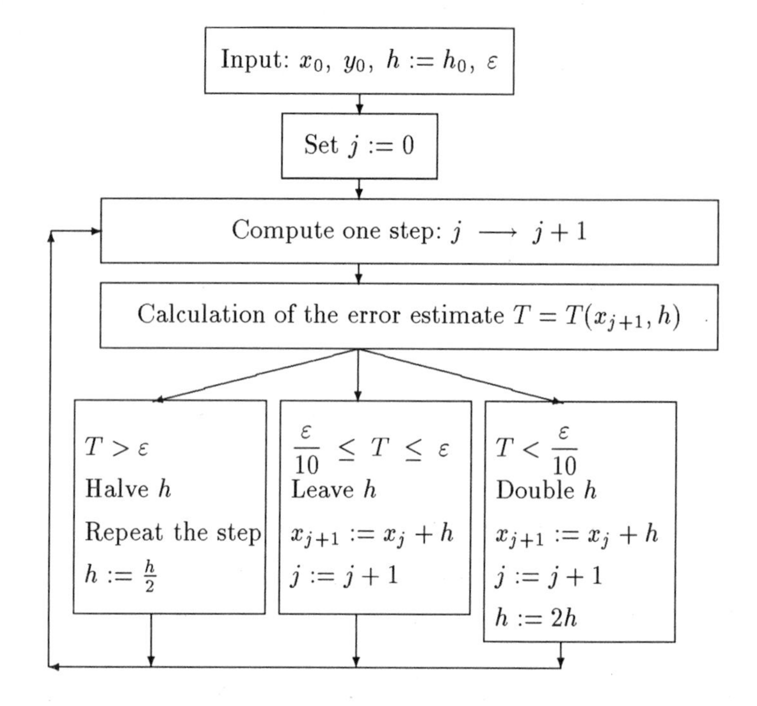

Figure 7.1. Step-control policy

7.2.4 A Nonlinear Example

We will now apply some of the methods we have got to know so far to a nonlinear example, which will be worked out in some detail to compare the different one-step methods. The differential equation

$$\begin{array}{lrcl} & y' & = & -200xy^2, \\ \text{with} & y(0) & = & 1 \\ \text{or with} & y(-3) & = & \dfrac{1}{901} \end{array} \tag{7.32}$$

possesses the solution

$$y(x) = \frac{1}{1+100x^2}, \tag{7.33}$$

which is to be seen in Figure 7.2.

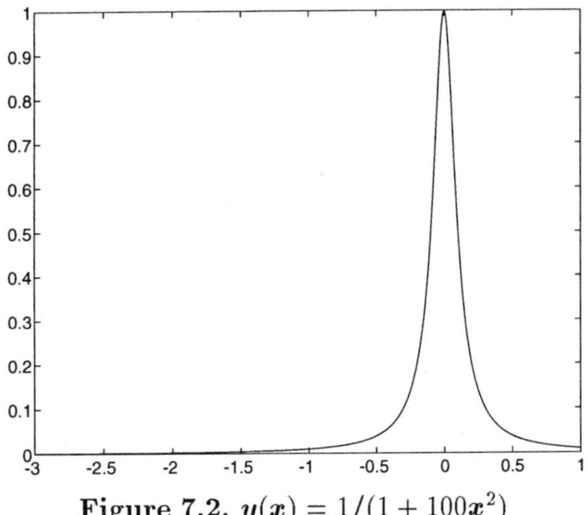

Figure 7.2. $y(x) = 1/(1 + 100x^2)$

For the interval $[0, 1]$ we see a medium-fast decreasing solution and expect to have stability problems with explicit methods of low order and large step length. For the interval $[-3, 1]$ the solution has very different growth parameters, which should mean that it is important to apply a reasonable step-control policy.

We first will apply Euler's explicit and implicit methods, classical and Merson's Runge–Kutta methods computing 10 steps with fixed step length $h = 0.1$ from $x_0 = 0$ to $x_{10} = 1$:

Euler's explicit method

$$\begin{aligned} y_{n+1} &= y_n + hf(x_n, y_n) \\ &= y_n - h(200\, x_n\, y_n^2) \end{aligned}$$

which with $h = 0.1$ yields

$$\begin{aligned} y_{n+1} &= y_n - 20x_n y_n^2 \\ \Longrightarrow & \\ y_1 &= y_0 - 20x_0 y_0^2 \\ &= 1 - 0 = 1 \\ y_2 &= 1 - 20 \cdot 0.1 = -1 \\ y_3 &= -1 - 20 \cdot 0.2\,(-1)^2 = -5 \\ y_4 &= -5 - 20 \cdot 0.3\,(-5)^2 = -155 \\ & \dots \end{aligned}$$

This sequence of solution values shows that for Euler's explicit method the step length $h = 0.1$ is too large. A step length $h = 0.05$ would have given a decreasing sequence of values from $y_0 = 1$ to $y_{20} = 0.0088$, which is wrong by about 10%. This means that $h = 0.1$ is outside the stability limits for finite $h > 0$, but $h = 0.05$ is not.

Euler's implicit method

$$\begin{aligned} y_{n+1} &= y_n + h\, f(x_{n+1}, y_{n+1}) \\ &= y_n - h\,(200\, x_{n+1}\, y_{n+1}^2) \end{aligned}$$

which with $h = 0.1$ yields

$$y_{n+1} = y_n - 20\, x_{n+1}\, y_{n+1}^2$$

This is the nonlinear equation which has to be solved in each solution step. Because it is quadratic here, it can be solved analytically:

$$y_{n+1} = \frac{1}{40 x_{n+1}} \left(-1 \pm \sqrt{1 + 80 x_{n+1} y_n} \right) .$$

For a reasonable solution one has to take the +-sign, which would have been the result of any reasonable iterative solution of the nonlinear equation, too. This yields the following values:

$$\begin{aligned} y_1 &= \frac{1}{4}\left(-1 + \sqrt{1+8}\right) = 0.5 \\ y_2 &= 0.25 \\ &\ldots \end{aligned}$$

Classical Runge–Kutta method

We only show the first step with $x_0 = 0$, $y_0 = 1$, $h = 0.1$ in detail.

$$\begin{aligned} k_1 &= f(0,1) = -200 \cdot 0 \cdot 1^2 = 0 \\ k_2 &= f(0.05, 1 + 0.05 k_1) = f(0.05, 1) = -10 \\ k_3 &= f(0.05, 1 + 0.05 k_2) = f(0.05, 0.5) = -2.5 \\ k_3 &= f(0.1, 1 + 0.1 k_3) = f(0.1, 0.75) = -11.25 \end{aligned}$$

This yields

$$\begin{aligned} y_1 &= y_0 + \frac{h}{6}(k_1 + 2k_2 + 2k_3 + k_4) \\ &= 1 + \frac{0.1}{6}(0 - 20 - 5 - 11.25) = 0.395833 \\ y_2 &= 0.18259 \\ &\ldots \end{aligned}$$

Merson's five-stage method

This is calculated equivalently to the classical Runge–Kutta method and will not be given in detail here. We will show the results of these calculations with fixed step length $h = 0.1$ from $x_0 = 0$ to $x_N = 1$ in the following table and in Figure 7.3.

x	Classical Runge–Kutta Dotted	True Solution Solid	Merson Merson Dash-dot	Implicit Euler Dashed
0.0000	1.0000	1.0000	1.0000	1.0000
0.1000	0.3958	0.5000	0.4866	0.5000
0.2000	0.1826	0.2000	0.1934	0.2500
0.3000	0.0958	0.1000	0.0983	0.1371
0.4000	0.0575	0.0588	0.0582	0.0826
0.5000	0.0379	0.0385	0.0382	0.0537
0.6000	0.0268	0.0270	0.0269	0.0372
0.7000	0.0199	0.0200	0.0199	0.0270
0.8000	0.0153	0.0154	0.0153	0.0203
0.9000	0.0121	0.0122	0.0122	0.0158
1.0000	0.0099	0.0099	0.0099	0.0126

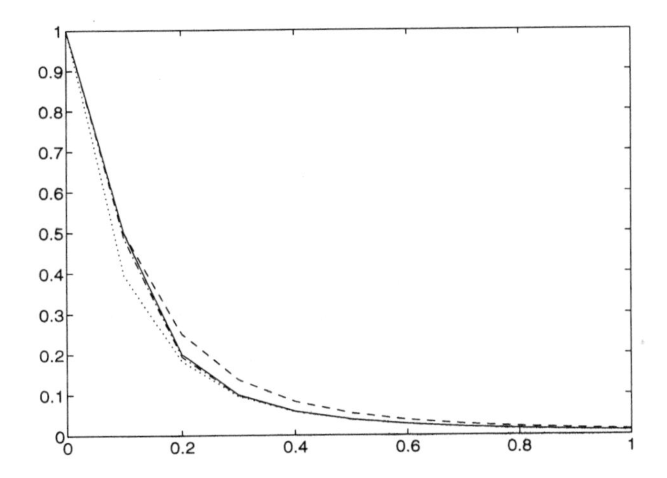

Figure 7.3. True solution versus different one-step methods

Obviously Merson's five-stage Runge–Kutta method of order four is superior to the other methods.

Fourth order methods with step-control policy

Now we will integrate the differential equation from $x_0 = -3$ to $x_N = 1$ and compare the amount of work necessary for a given tolerance $\varepsilon = 0.00001$. We measure this amount as usual by counting the function evaluations of f. We controlled the step size by a formula of Ceschino–Kuntzmann for the classical Runge–Kutta method and by (7.30) for Merson's method.

	Classical Runge–Kutta	Merson
Min. step length	0.0015625	0.0125
Max. step length	0.2	0.2
No. of function evaluations	1044	340

With both methods we obtain about the same error, which means, that again Merson's method is the far more economical than the classical Runge–Kutta method with Ceschino–Kuntzmann's step-control. If we had taken the smallest step length as a fixed one for getting the minimal error, we would have needed 10240 function evaluations for the classical Runge–Kutta method or 1600 function evaluations for Merson's method, which shows the necessity of a step-control policy.

7.3 Multistep Methods

We considered the Runge–Kutta methods as an important class within the large field of one-step methods. Similarly now we will consider the Adams predictor–corrector methods as an important class of multistep methods. Backward differentiation formulae, which are special multistep methods, are considered in the next section.

A linear m-step method uses values from m preceding steps for calculating a new approximate value.

Let there be given

the coefficients	$a_0,\ a_1,\ \cdots,\ a_m$ with $a_m := 1$
and	$b_0,\ b_1,\ \cdots,\ b_m,$
and the solution values	$y_j,\ y_{j+1},\ \cdots, y_{j+m-1}.$

Calculate

the new solution value y_{j+m} with

$$a_0\ y_j + a_1\ y_{j+1} + \cdots a_m\ y_{j+m} = h \cdot (b_0\ f_j + b_1\ f_{j+1} + \cdots b_m\ f_{j+m})\,. \quad (7.34)$$

This method is

explicit	if $b_m = 0$,
implicit	if $b_m \neq 0$.

As with implicit one-step methods one has in general to solve a system of nonlinear equations for each step of an implicit multistep method, while for an explicit method, (7.34) can be resolved for y_{j+m} ($a_m = 1,\ b_m = 0$):

$$y_{j+m} = -\sum_{k=0}^{m-1} a_k\ y_{j+k}\ +\ h \sum_{k=0}^{m-1} b_k\ f_{j+k}\,. \quad (7.35)$$

7.3.1 Adams Methods

Special methods can be constructed using numerical integration. The formula

$$y(x_{j+m}) = y(x_l) + \int_{x_l}^{x_{j+m}} f(x, y(x))\, dx \tag{7.36}$$

opens up different possibilities. First one may take different values for l, normally $l = j+m-1$ or $l = j+m-2$. Then one replaces the function f formally by an interpolating polynomial P:

$$y_{j+m} = y_l + \int_{x_l}^{x_{j+m}} P(x)\, dx. \tag{7.37}$$

For an explicit method m points with known values are used for interpolation; for an implicit method the unknown value f_{j+m} is also used for formal polynomial interpolation:

$$P(x_{j+k}) = f(x_{j+k}, y_{j+k}), \quad k = 0, 1, \cdots, r, \tag{7.38}$$

with $r = m-1$ for explicit and $r = m$ for implicit methods. Then this polynomial is integrated (formally) over the last or the two last sub-intervals using (7.37). This yields the coefficients a_k and b_k. Figure 7.4 illustrates this construction for $l = j+m-1$:

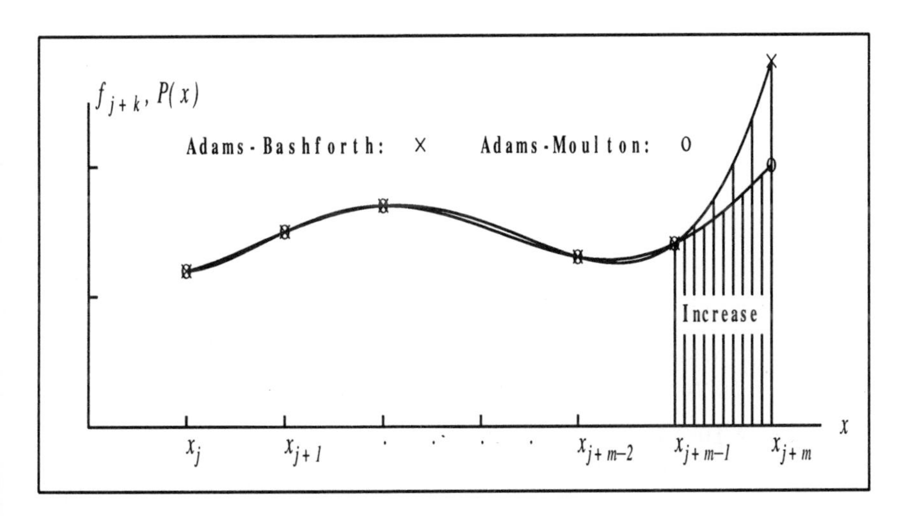

Figure 7.4. Construction of Adams multistep methods

The following table shows the four classes of methods constructed in this manner:

Method	r	l	Type
Adams–Bashforth	$m-1$	$j+m-1$	explicit
Nyström	$m-1$	$j+m-2$	explicit
Adams–Moulton	m	$j+m-1$	implicit
Milne–Simpson	m	$j+m-2$	implicit

For $m = 2$ we get the following special methods:

Adams–Bashforth: $$y_{j+2} = y_{j+1} + \frac{h}{2}(3f_{j+1} - f_j). \tag{7.39}$$

Nyström: $$y_{j+2} = y_j + 2hf_{j+1}. \tag{7.40}$$

Adams–Moulton: $$y_{j+2} = y_{j+1} + \frac{h}{12}(5f_{j+2} + 8f_{j+1} - f_j). \tag{7.41}$$

Milne–Simpson: $$y_{j+2} = y_j + \frac{h}{3}(f_{j+2} + 4f_{j+1} + f_j). \tag{7.42}$$

For $m = 2$ Nyström's method is identical with the mid-point rule. The numerical integration used with Milne–Simpson's method is Simpson's rule, see Chapter 6.

7.3.2 Consistency of Linear Multistep Methods

For linear multistep methods the discretization error is given as

$$d_{j+m} = \frac{1}{h}\sum_{k=0}^{m} a_k y(x_{j+k}) - \sum_{k=0}^{m} b_k f(x_{j+k}, y(x_{j+k})). \tag{7.43}$$

With the coefficients of the method one defines two polynomials:

$$\text{First characteristic polynomial:} \quad \rho(\zeta) = \sum_{k=0}^{m} a_k \zeta^k.$$

$$\text{Second characteristic polynomial:} \quad \sigma(\zeta) = \sum_{k=0}^{m} b_k \zeta^k.$$

With these polynomials consistency can be easily examined.

Theorem 7.10. *A linear multistep method is consistent if the starting method is consistent and if*

$$\rho(1) = 0 \quad \textit{and} \quad \rho'(1) = \sigma(1). \tag{7.44}$$

Additional conditions have to be fulfilled for the asymptotic stability of multistep methods. We will not give them in detail, but mention again that implicit methods are more stable in general than explicit ones. Therefore in practice combinations of both have been shown to be quite efficient. These combinations are called predictor–corrector methods. They use an implicit method without solving a system of nonlinear equations.

7.3.3 Predictor–Corrector Methods

Let **P** be an explicit multistep method, called the predictor, with coefficients a_k^* and b_k^*:

$$\mathbf{P}: \quad y_{j+m}^{[0]} = -\sum_{k=0}^{m-1} a_k^* \, y_{j+k} \; + \; h \sum_{k=0}^{m-1} b_k^* \, f_{j+k}. \tag{7.45}$$

The computed value is written $y_{j+m}^{[0]}$, because it is only a first approximation to the final value for y_{j+m}. Now the right-hand side f is evaluated with $y_{j+m}^{[0]}$:

$$\mathbf{E}_0: \quad f_{j+m} = f(x_{j+m}, y_{j+m}^{[0]}). \tag{7.46}$$

Let **C** be an implicit multistep method, called the corrector, with coefficients a_k and b_k:

$$\mathbf{C}: \quad y_{j+m}^{[1]} = h b_m f_{j+m} - \sum_{k=0}^{m-1} a_k \, y_{j+k} \; + \; h \sum_{k=0}^{m-1} b_k \, f_{j+k}. \tag{7.47}$$

Then

$$\mathbf{P} \; \mathbf{E}_0 \; \mathbf{C}$$

is one step of a simple predictor–corrector method with $y_{j+m} := y_{j+m}^{[1]}$. No system of nonlinear equations has to be solved; or, it is approximately solved with the predictor method to get an initial value and then with the corrector as one step of an iterative method.

This method can obviously be varied. One reasonable extension is to evaluate the function f with the corrector value again to get a better value for the next step:

$$\mathbf{P} \; \mathbf{E}_0 \; \mathbf{C} \; \mathbf{E}_1.$$

Evaluation and corrector can be repeated iteratively:

$$\mathbf{P} \; (\mathbf{E} \; \mathbf{C})^r \; \mathbf{E}$$

with a fixed number r. Predictor–corrector methods are quite popular because they combine the simplicity of explicit methods with (almost) the stability of implicit methods.

7.3.4 Starting Methods

Multistep methods use m preceding values y_l, $l = j, \ldots, j+m-1$, for obtaining y_{j+m}. Therefore we need a starting method for computing the values $y_1, \ldots, y_{m-1}$ in addition to the initial value y_0. As a starting method we may take a one-step method, which should have at least the order of the multistep method used subsequently.

This easiest possibility can be replaced by using several multistep methods of one family, varying stage m and as a consequence varying order p, and starting with $m = p = 1$. Then it is important to start with a small step length because of the small order at the beginning of the calculations, and to use a step-control policy. Another trick for getting more precision with the starting method is the use of negative step lengths. As an example we will look at a combination of Adams methods of growing order:

$$\begin{aligned} P: &\quad y_{-1}^{[0]} := y_0 - hf_0 \quad y_1^{[0]} := y_0 + hf_0, \\ C: &\quad y_{\pm 1}^{[l]} = y_0 \pm \frac{h}{2}(f_0 + f_{\pm 1}^{[l-1]}), \quad l = 1,2,3, \\ &\quad \text{with } f_{\pm 1}^{[l]} := f(x_0, y_{\pm 1}^{[l]}). \end{aligned} \tag{7.48}$$

For the third corrector approximation one has

$$y_{\pm 1}^{[3]} - y(x_{\pm 1}) = O(h^3). \tag{7.49}$$

which is, together with a small starting step length, sufficiently precise for a multistep method of order four.

7.3.5 Adams Methods with Variable Step Length and Order

Adams methods can be formulated with variable step length and variable order. Step length and order are controlled by local error estimates. This leads to a complex method with variable coefficients and will not be given here; see [50] for further details. In the next section we will consider the same policy for another class of methods for which this procedure is much easier.

7.4 Backward Differentiation Formulae (BDF)

7.4.1 Introduction

Adams predictor–corrector methods are not suitable for stiff systems because the explicit method would yield bad approximations which could not be improved satisfactorily by the corrector. The iteration $(\mathbf{E}\ \mathbf{C})^r$ may not converge at all.

We now will derive a class of stable implicit multistep methods which can be applied to stiff systems with success: the *backward differentiation formulae*. Most linear multistep methods are based on numerical integration. The underlying idea of BDF is totally different as they are based on numerical differentiation of a given function.

Let x_k be equidistant points with step size h and let q be defined as the polynomial interpolating the points

$$(x_k, y_k), \qquad k = j, j+1, \cdots, j+m. \tag{7.50}$$

The values y_k for $k = j, j+1, \cdots, j+m-1$, are known as usual; y_{j+m} is to be determined. The interpolation polynomial q is defined with backward differences of y_{j+m}:

$$q(x) = q(x_{j+m-1} + sh) = \sum_{k=0}^{m} (-1)^k \binom{-s+1}{k} \nabla^k y_{j+m}, \quad s \in [0,1]. \tag{7.51}$$

The unknown value y_{j+m} is now determined by inserting q into the differential equation

$$\left.\frac{\partial q}{\partial x}\right|_{x_{j+m}} = f(x_{j+m}, y_{j+m}) \tag{7.52}$$

and solving the resulting equation or system. This procedure results in the implicit m-step method

$$\sum_{k=0}^{m} \delta_k \nabla^k y_{j+m} = h f_{j+m} \tag{7.53}$$

with coefficients

$$\delta_k = (-1)^k \frac{d}{ds} \left.\binom{-s+1}{k}\right|_{s=1}, \tag{7.54}$$

which are easily calculated by direct differentiation of

$$(-1)^k \binom{-s+1}{k} = \frac{1}{k!}(s-1)s(s+1)\cdots(s+k-2),$$

which yields

$$\delta_0 = 0 \qquad \delta_k = \frac{1}{k} \qquad \text{for} \quad k = 1, \cdots, m. \tag{7.55}$$

7.4.2 Consistency of BDF

The implicit BDF are defined by

$$\sum_{k=1}^{m} \frac{1}{k} \nabla^k y_{j+m} = h f(x_{j+m}, y_{j+m}). \tag{7.56}$$

They are consistent with order $p = m$. By resolving (7.56) for $m = 1, \ldots, 6$ one obtains the following formulae:

$$m = 1: \quad y_{j+1} = y_j + hf_{j+1} \qquad \text{(implicit Euler)}.$$

$$m = 2: \quad y_{j+2} = \frac{4}{3}y_{j+1} - \frac{1}{3}y_j + \frac{2}{3}hf_{j+2}.$$

$$m = 3: \quad y_{j+3} = \frac{18}{11}y_{j+2} - \frac{9}{11}y_{j+1} + \frac{2}{11}y_j + \frac{6}{11}hf_{j+3}.$$

$$m = 4: \quad y_{j+4} = \frac{48}{25}y_{j+3} - \frac{36}{25}y_{j+2} + \frac{16}{25}y_{j+1} - \frac{3}{25}y_j + \frac{12}{25}hf_{j+4}.$$

$$m = 5: \quad y_{j+5} = \frac{300}{137}y_{j+4} - \frac{300}{137}y_{j+3} + \frac{200}{137}y_{j+2} - \frac{75}{137}y_{j+1} + \frac{12}{137}y_j + \frac{60}{137}hf_{j+5}.$$

$$m = 6: \quad y_{j+6} = \frac{360}{147}y_{j+5} - \frac{450}{147}y_{j+4} + \frac{400}{147}y_{j+3} - \frac{225}{147}y_{j+2} + \frac{72}{147}y_{j+1} - \frac{10}{147}y_j + \frac{60}{147}hf_{j+6}.$$

BDF with $m > 6$ are not used because of missing stability.

7.4.3 BDF with Variable Step Length

If we introduce variable step lengths into BDF we cannot use backward differences like before. Instead we can use recursive divided differences

$$\begin{aligned} y[x_j] &:= y_j \\ y[x_j, \cdots, x_{j-k}] &:= \frac{y[x_j, \cdots, x_{j-k+1}] - y[x_{j-1}, \cdots, x_{j-k}]}{x_j - x_{j-k}} \end{aligned} \tag{7.57}$$

for defining Newton's interpolating polynomial for the points (x_k, y_k), with $k = j, j+1, \cdots, j+m$:

$$q(x) = \sum_{k=0}^{m} \left[\prod_{i=0}^{k-1} (x - x_{j+m-i}) \right] y[x_{j+m}, x_{j+m-1}, \cdots, x_{j+m-k}]. \tag{7.58}$$

Insertion into the differential equation

$$\left. \frac{\partial q}{\partial x} \right|_{x_{j+m}} = f(x_{j+m}, y_{j+m}) \tag{7.59}$$

then yields the BDF with variable step length

$$\sum_{k=1}^{m}\left[\prod_{i=1}^{k-1}(x_{j+m}-x_{j+m-i})\right] y[x_{j+m},x_{j+m-1},\cdots,x_{j+m-k}] = f_{j+m}. \tag{7.60}$$

With this formula the coefficients $\gamma_k = \gamma_k(j+m)$ of the BDF

$$\sum_{k=0}^{m}\gamma_k y_{j+k} = h_{j+m-1} f_{j+m} \quad (h_k := x_{k+1}-x_k) \tag{7.61}$$

can be calculated easily (for each step!), more easily than for Adams methods with variable step length.

The resulting nonlinear system per step is normally solved with an iterative method. Starting values for y_{j+m} can be obtained by interpolating the points (x_{j+k-1}, y_{j+k-1}) for $k = 0,\cdots,m$ with a polynomial $\tilde{p}$ and setting

$$y_{j+m}^{[0]} := \tilde{p}(x_{j+m}). \tag{7.62}$$

As the *starting method* for an m-stage BDF one takes BDF of stage $1,2,\ldots,(m-1)$. This has to be done with a step-control policy, see 7.4.4, or even better with variable step length and order, see Section 7.4.5.

7.4.4 Step-Control Policy

The discretization error for BDF is given as

$$d_{j+m} = C_{m+1}\, h_{j+m-1}^{m}\, y^{(m+1)}(x_{j+m}) \tag{7.63}$$

with known error coefficients C_{m+1}. It can be estimated by

$$h_{j+m-1}\, d_{j+m} \approx T := \frac{y_{j+m} - y_{j+m}^{[0]}}{x_{j+m}-x_j}. \tag{7.64}$$

Now the step length can be controlled using this estimate T. But it is to be preferred to control both the step length and the order, because this is more efficient and easily possible with BDF.

7.4.5 Variable Step Length and Order

Most libraries like NAG use variable step length and order with BDF. The order can be controlled as follows. In (7.63) derivatives $y^{(m)}, y^{(m+1)}, y^{(m+2)}$ are replaced by

$$\begin{aligned}
y^{(m)} &\approx y[x_{j+m},\cdots,x_j], \qquad (7.65)\\
y^{(m+1)} &\approx \nabla y[(x_{j+m})] := \frac{y[x_{j+m},\cdots,x_j]-y[x_{j+m-1},\cdots,x_{j-1}]}{h_{j+m-1}},\\
y^{(m+2)} &\approx \nabla^2 y[(x_{j+m})] := \frac{\nabla y[(x_{j+m})]-\nabla y[(x_{j+m-1})]}{h_{j+m-1}}.
\end{aligned}$$

Then the following numbers are calculated for orders $m-1, m, m+1$:

$$\begin{aligned} q_{m-1} &:= \frac{1}{1.3}\left[\frac{\varepsilon}{C_m y[x_{j+m},\cdots,x_j]}\right]^{\frac{1}{m-1}}, \\ q_m &:= \frac{1}{1.2}\left[\frac{\varepsilon}{C_{m+1}\nabla y[(x_{j+m})]}\right]^{\frac{1}{m}}, \\ q_{m+1} &:= \frac{1}{1.4}\left[\frac{\varepsilon}{C_{m+2}\nabla^2 y[(x_{j+m})]}\right]^{\frac{1}{m+1}}, \end{aligned} \tag{7.66}$$

where ε is the given tolerance. Now the order that yields the largest value for q is used. A few additional strategies must be introduced, such as not changing the order when repeating a step. A starting method similar to (7.48) with Adams methods is possible. This combined BDF with variable step length and order is very efficient and stable and well suited for weakly stiff systems.

7.5 Programs

There are four programs for this chapter. Corresponding to the three classes of methods described we have three programs at our disposal. In RUKU, see Section B.7.1, the combined Runge–Kutta–Merson methods of stage four and five and order four are implemented using NAG routine D02PAF. NAG routines D02QGF and D02QWF are used by program ADAMS, see Section B.7.2, to solve initial value problems with a variable-order, variable step length Adams method. As a third method the backward differentiation formulae with variable order and variable step length are realized in the program BDF, see Section B.7.3, to solve mildly stiff systems. It uses NAG routine D02QDF. Additionally there is a program STIFFTEST, see Section B.7.4, which performs a stiffness test using NAG routine D02BDF.

7.6 Applications

One can find a lot of interesting applications for initial value problems. Most often these applications model continuous time-dependent processes. Some examples are the development of temperature in nuclear fuel rods, reaction kinetics from chemistry, see 7.1.7, population and competition models from economics, physiological indicator models like doping tests, epidemics, species behaviour, pest control or symbiosis from medicine and life science, or models for an arms race or anti-aircraft defence.

Some of these problems – like the first one mentioned – must be controlled; the differential equation is 'fed' by an input function, which itself depends on the solution. Thus we arrive at the subject of 'optimal control', which is a combination of differential equations and optimization, and is beyond the scope of this text. A classical result for optimal control is the 'bang-bang principle' for controlling rockets, which you should not use with your car on public roads, because it means that you have to drive at full speed or putting the brakes full on, to save petrol.

Many of the examples mentioned are to be found in [16], [19], [54], [57], [92], [95].

Lotka–Volterra's Competition Model

The Lotka–Volterra equations are autonomous, which means that the right-hand side does not depend on the independent variable explicitly. The nonlinear right-hand side is quadratic. Lotka–Volterra equations describe predator–prey models in biology as well as competition models from economics, see [54] or [19], from which we have taken an example for numerical exploration.

Two populations of size P_1 and P_2, respectively, are living from the same resource R. Let the initial sizes of these populations be P_{10} and P_{20} and the development be described by the differential equations

$$\begin{aligned} P_1' &= 0.004\, P_1\, (50 - P_1 - 0.75\, P_2), \\ P_2' &= 0.001\, P_2\, (100 - P_2 - 3\, P_1). \end{aligned}$$

On the right-hand side the terms $\alpha_i P_i - \beta_i P_i^2$ stand for the growth of the respective population without competition, while the mixed terms $-\gamma_i P_i P_j$ stand for the rivalry of the two populations, here $-0.003\, P_1\, P_2$ in both equations. First of all we are interested in stationary points, for which the size of the populations are balanced, which means that the solutions of the differential equation become constant.

For $P_2 = 0$ and $P_1 \neq 0$ we have $0 = P_1' = 0.004\, P_1\, (50 - P_1)$, which means that the first population grows or shrinks to the limit value $P_1 = 50$ and then remains constant.

For $P_1 = 0$ and $P_2 \neq 0$ we have $0 = P_2' = 0.001\, P_2\, (100 - P_2)$, which means that the second population grows or shrinks to the limit value $P_2 = 100$ and then remains constant.

The point $(P_1, P_2) = (0, 0)$ obviously is a stationary one, too. Another stationary point is found if we set the bracketed terms on the right-hand sides to zero:

$$\begin{aligned} P_1 + 0.75\, P_2 &= 50, \\ 3\, P_1 + P_2 &= 100. \end{aligned}$$

This gives the point (20,40). Now it is important to know if these stationary points are *locally stable*, which means attracting a solution. Whether a stationary point is attracting (stable) or rejecting (unstable) can be tested by checking the eigenvalues of the Jacobian Df. If their real parts are negative, the stationary point is attracting. For the problem in question we have

$$Df = \begin{pmatrix} 0.2 - 0.008P_1 - 0.003P_2 & -0.003P_1 \\ -0.003P_2 & 0.1 - 0.003P_1 - 0.002P_2 \end{pmatrix}.$$

For checking the stability for the stationary points we have to solve four 2×2 eigenvalue problems, which in fact leads to quadratic equations. The results are:

Point	λ_1	λ_2	locally stable?
(0, 0)	0.2	0.1	No
(0,100)	-0.1	-0.1	Yes
(20, 40)	0.0272	-0.1472	No
(50, 0)	-0.2	-0.05	Yes

The biological interpretation is as follows.
Neither both populations will die nor both will survive, because both points $(0,0)$ and $(20,40)$ are unstable. Only the initial values P_{10} and P_{20} determine the final state.

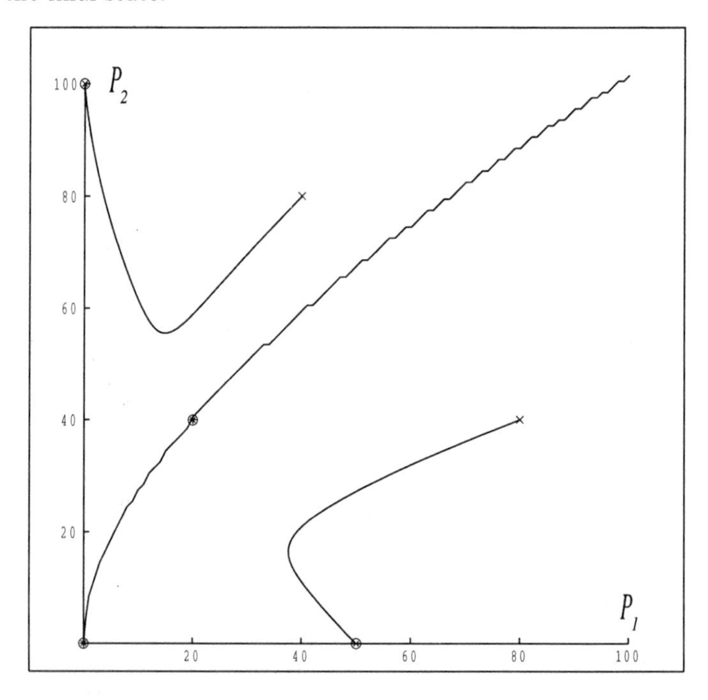

Figure 7.5. Solutions of the competition model

In Figure 7.5 the stationary points are marked by small circles. Two orbital curves for solutions with initial values (40,80) and (80,40) are to be seen, which end at the locally stable stationary points (0,100) and (50,0), respectively. In addition the curve is shown which separates the two sets of initial value points, for which the solution converges to the respective stable stationary point.

The solutions are easily obtained with any of the three programs RUKU, ADAMS or BDF. For RUKU we give input and output files. The input files for the other two programs are similar; the differences can be seen in the program heads which are contained in the electronic version of this text only.

Input and Output Files for RUKU for Graphical Output with Standard Control Parameters

```
Input file for the PAN program RUKU:
Starting point:
0.0
Dimension of starting vector:
2
Starting vector:
40.0 80.0
Graphical output (no = 0, X Window = 1, PostScript = 2):
1
Output of solution curves (no = 0, yes = 1):
1
Left margin:
0.0
Right margin:
150.0
Lower margin:
0.0
Upper margin:
100.0
Number of solution curves:
2
Output of an orbital curve (no = 0, yes = 1):
1
Left margin:
0.0
Right margin:
100.0
Lower margin:
```

```
0.0
Upper margin:
100.0
Component in x direction:
1
Component in y direction:
2
Input of control parameters (no = 0, yes = 1):
0
Tolerance for the local error test:
0.0
Equidistant solution points (no = 0, yes = 1):
1
Number of solution points:
9
Final solution point:
135.0

Output file for the PAN program RUKU:

Solution at     15.00000000:  1          14.65634005
                              2          55.55696914

Solution at     30.00000000:  1          10.57107211
                              2          60.14051468

Solution at     45.00000000:  1           6.92229087
                              2          69.17948322

Solution at     60.00000000:  1           3.54618757
                              2          80.19401173

Solution at     75.00000000:  1           1.33119110
                              2          90.00504336

Solution at     90.00000000:  1           0.38232590
                              2          96.02998318

Solution at    105.00000000:  1           0.09401358
                              2          98.67859852

Solution at    120.00000000:  1           0.02165430
                              2          99.60581365
```

```
Solution at    135.00000000:    1         0.00487731
                                2        99.88990329
```

7.7 Initial Value Problems with IMSL

IMSL has three routines for initial value problems: IVPRK, IVPAG and IVPBS. IVPRK uses the Runge–Kutta–Verner method, which combines two Runge–Kutta formulae of orders five and six. It is similar to NAG routine D02PAF. IVPAG uses Gear's method for stiff differential equations or Adams–Moulton method of variable order at the discretion of the user. The maximum orders are five for Gear and twelve for Adams–Moulton, respectively. IVPAG needs the Jacobian, which can be programmed by the user, numerically approximated or approximated by a diagonal matrix. IVPAG can treat implicit differential equations of the form $Ay' = f(x,y)$, where A can be the identity matrix, a constant matrix or a matrix depending on x. The type of the Jacobian must be identical with the type of A with the following possibilities:

- full
- banded
- full, symmetric positive-definite
- banded, symmetric positive-definite.

The NAG library has routines for solving implicit differential equations of the form $A(x,y)y' = f(x,y)$, too. IVPBS uses the Bulirsch–Stoer method of rational extrapolation. It fulfills high-precision demands and varies the order up to seven.

7.8 Exercises: Autonomous Initial Value Problems

Consider the following family of autonomous initial value problems

$$\begin{aligned} y' &= f(y) \quad \forall x \in [a,b], \\ y\,(a) &= y_0. \end{aligned}$$

with the additional assumptions:

1. $f \in C^\infty(\mathbb{R} \to \mathbb{R})$
2. There is a number $c > 0$ with $|f(y(x))| \geq c \;\; \forall x$.

For this class we will find a high-precision method approximating the solution values $y(x_i) \approx y_i$ at $N+1$ points $x_0 < x_1 < \cdots < x_N$.

1. Show: For $\bar{y} = y(x_{i+1})$ we have

$$
\begin{aligned}
F_{y(x_i)}(\bar{y}) &= 0 \quad \text{if we define} \\
F_\alpha(\beta) &:= \int_\alpha^\beta \frac{1}{f(y)}\,dy - (x_{i+1} - x_i).
\end{aligned}
\tag{7.67}
$$

2. Replace the integral in (7.67) by the Gauss–Legendre quadrature formula for n points and call the new function $\tilde{F}_\alpha(\beta)$ approximating $F_\alpha(\beta)$.
   The initial value method is now defined as

Let y_0 be given
For $i = 0, 1, \ldots, N-1$
Calculate y_{i+1} by solving $\tilde{F}_{y_i}(y_{i+1}) = 0$

3. Calculate $\tilde{F}'_{y_i}$ and formulate Newton's method for solving $\tilde{F}_{y_i}(y_{i+1}) = 0$. As a starting value take $y_{i+1}^{(0)} := y_i$.
4. What is the maximum convergence order one can achieve with this method and how many iteration steps are necessary to achieve it?
5. Solve

$$
\begin{aligned}
y'(x) &= 2 - \frac{1}{y^2(x)} \quad \forall x \in [0,1], \\
y(0) &= 1,
\end{aligned}
$$

   with $N = 2$, $t_0 = 0$, $t_1 = 1/2$, $t_2 = 1$ and $n = 2$. Take three Newton iteration steps per integration step.

8

Boundary Value and Eigenvalue Problems for Ordinary Differential Equations

The mathematical theory for (two-point) boundary and eigenvalue problems for ordinary differential equations (ODEs) is not as standardized as that for initial value problems. But that does not mean that these problems do not play an important role for a lot of engineering and other applications. There are some problems of practical importance, where the mathematician cannot determine the question of existence or uniqueness of solutions, but they are regularly solved approximately with numerical methods.

We will describe a finite difference method for a special class of problems and briefly the two main initial value methods called shooting and multiple or parallel shooting for a more general class of boundary value problems, but again restricted to a single differential equation of order 2.

Two-point boundary value problems may be interpreted as one-dimensional partial differential equations. Therefore we can apply the relevant methods like variational methods to them. In this sense Chapter 9 may give additional ideas for the numerical solution of boundary value problems for ODEs.

8.1 Introduction

With two-point boundary value problems one has to solve ordinary differential equations according to conditions at two interval-end points, the boundary conditions (BC). There may be one ODE of higher than first order or a system of more than one ODE. Most often the ODE is explicit. Then an equivalent transformation to a system of first order is possible (see Lemma 7.1).

Eigenvalue problems for differential equations belong to the same class of problems, but in addition the right-hand side f depends on a parameter λ called the eigenvalue. Then the complete solution in general consists of an infinite system of eigenvalues and eigenfunctions $(\lambda_i, y_i(x)), i = 1, \dots, \infty$, but one most often seeks only one or a few of them.

During this chapter we restrict ourselves to a single ODE of second order with separated boundary conditions and on finite intervals (a, b).

8.1.1 Boundary Value Problem

Let there be given a continuous function $f \in C([a,b] \times \mathbb{R} \times \mathbb{R}, \mathbb{R})$ and two real numbers α and β. A function $y \in C^2([a,b], \mathbb{R})$ has to be determined, for which:

$$\begin{array}{ll} \text{ODE:} & y'' = f(x, y, y'), \\ \text{BC:} & y(a) = \alpha \quad y(b) = \beta. \end{array} \tag{8.1}$$

8.1.2 Eigenvalue Problem

Let there be given a continuous function $f \in C([a,b] \times \mathbb{R} \times \mathbb{R}, \mathbb{R})$. Complex numbers $\lambda \in \mathbb{C}$ called *eigenvalues* and functions $y \in C^2([a,b], \mathbb{R})$ called *eigenfunctions* have to be determined, for which:

$$\begin{array}{ll} \text{ODE:} & y'' = \lambda\, f(x, y, y'), \\ \text{BC:} & y(a) = 0 \quad y(b) = 0. \end{array} \tag{8.2}$$

More complex boundary and eigenvalue problems may arise with

- Higher order systems
- Implicit differential equations
- General boundary conditions, for example for problem (8.1) we may have:

$$r_j(y(a), y'(a), y(b), y'(b)) = 0, \quad j = 1, 2 \tag{8.3}$$

  with certain functions r_j, which have to fulfill regularity conditions
- Nonlinear dependence on parameter λ

We will not consider the assumptions for existence and uniqueness in general, but we will get to know a class of problems for which these questions are more easily to be answered.

8.1.3 Sturm–Liouville Boundary Value Problem

(Charles Sturm (1803–1855), Joseph Liouville (1809–1882).)

$$\begin{array}{ll} \text{ODE:} & Lu := -(p(x)y'(x))' + q(x)y(x) = g(x) \text{ for } x \in (a,b) \\ \text{BC:} & B_1 := a_1\, y(a) + a_2\, y'(a) = \alpha, \\ & B_2 := b_1\, y(b) + b_2\, y'(b) = \beta. \end{array} \tag{8.4}$$

This problem is *self-adjoint*. It has a unique twice differentiable solution y if the corresponding homogeneous problem $Lu = 0, B_1 = 0, B_2 = 0$ possesses only the trivial solution $y = 0$.

8.1.4 Sturm–Liouville Eigenvalue Problem

$$\begin{array}{lll}\text{ODE:} & -(p(x)y'(x))' + q(x)y(x) = \lambda g(x)y(x) \text{ for } x \in (a,b) & \\ \text{BC:} & a_1\, y(a) + a_2\, y'(a) = 0, & (8.5) \\ & b_1\, y(b) + b_2\, y'(b) = 0. & \end{array}$$

If the assumptions as above are fulfilled and also $p(x) > 0$, $q(x) \geq 0$, $g(x) > 0$, $a_1, b_1, b_2 \geq 0$ and $a_2 \leq 0$ with at least two of these values not equal to zero, then all eigenvalues λ_i are real, distinct and positive. If they are ordered like

$$0 < \lambda_1 < \lambda_2 < \cdots < \lambda_k \to \infty$$

then for each λ_k the corresponding eigenfunction y_k oscillates with $k-1$ zeros in (a,b).

8.1.5 Deflection of a Beam

Example 8.1. Let a homogeneous, elastic beam of length 2 be movably supported at both ends. If its modulus of elasticity E is constant then its bending resistance is $EJ(x)$, $-1 \leq x \leq 1$, with moment of inertia J. $J(x)$ is proportional to $D^4(x)$, where D is the diameter of the beam. If the beam is loaded axially with a force P and transversely with a force $h(x)$, then the ODE used to find the bending moment M is

$$\begin{array}{ll}\text{ODE:} & M''(x) + \dfrac{P}{EJ(x)}M(x) = -h(x). \\ \text{BC:} & M(-1) = M(1) = 0.\end{array}$$

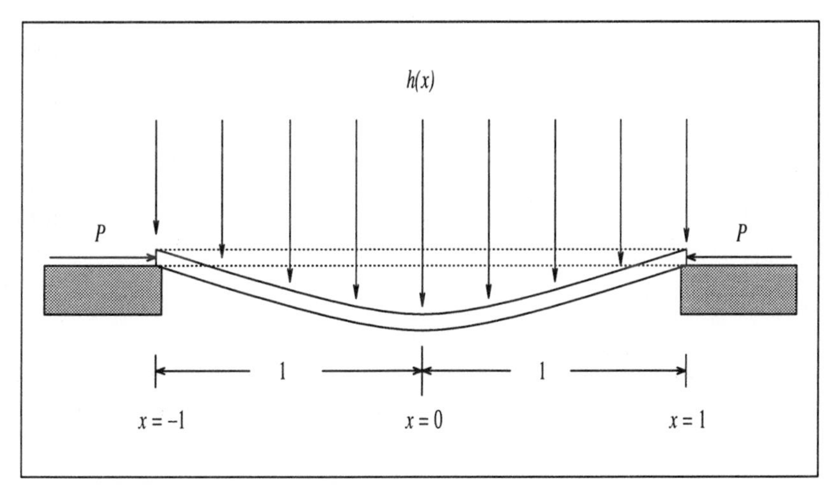

Figure 8.1. Deflection of a beam

We will simplify the problem. Let $h(x) \equiv h$, $J(x) := \dfrac{J_0}{1+x^2}$, $P := EJ_0$.

This means that the beam is thinner near the end-points than in the middle and all forces are constant. Let the variables be transformed by $y = -\frac{M}{h}$ then we have to solve the Sturm–Liouville problem:

$$\begin{array}{ll} \text{ODE:} & -y''(x) - (1+x^2)y(x) = 1, \\ \text{BC:} & y(-1) = y(1) = 0. \end{array} \tag{8.6}$$

■

8.2 Finite Difference Methods

Let the interval $[a, b]$ be divided into $n+1$ subintervals $[x_i, x_{i+1}]$ of equal lengths with

$$\begin{aligned} x_i &:= a + ih, \quad i = 0, 1, \ldots, n+1, \quad \text{with} \\ h &:= \frac{b-a}{n+1}. \end{aligned} \tag{8.7}$$

Now the differential equation is replaced by an algebraic equation for each inner point x_i of this grid with the solution values $y(x_i)$ as unknowns. For this purpose the derivative values are approximated by finite (divided) differences. This results in a system of n equations for the n unknowns $y(x_i)$, which is linear if the differential equation is linear. Given boundary values are introduced and transferred to the right-hand side. We will soon see an example, but first we will look at the most important finite differences for the first four derivatives.

$$\begin{aligned} y'(x_i) &= \frac{y(x_{i+1}) - y(x_{i-1})}{2h} + O(h^2), \\ y''(x_i) &= \frac{y(x_{i+1}) - 2y(x_i) + y(x_{i-1})}{h^2} + O(h^2), \\ y'''(x_i) &= \frac{y(x_{i+2}) - 2y(x_{i+1}) + 2y(x_{i-1}) - y(x_{i-2})}{2h^3} + O(h^2), \\ y^{(4)}(x_i) &= \frac{y(x_{i+2}) - 4y(x_{i+1}) + 6y(x_i) - 4y(x_{i-1}) + y(x_{i-2})}{h^4} + O(h^2). \end{aligned} \tag{8.8}$$

These finite differences all approximate the respective derivative with a precision $O(h^2)$. There are finite differences of higher accuracy; see for example [16].

Example 8.2.

$$\begin{aligned} -y''(x) + q(x)y(x) &= g(x) \\ y(a) = \alpha \qquad & y(b) = \beta. \end{aligned} \tag{8.9}$$

Let $q_i := q(x_i)$ and $g_i := g(x_i)$; then we get the following linear equations for $y_i \approx y(x_i)$:

$$\begin{aligned} y_0 &= \alpha, \\ \frac{-y_{i+1} + 2y_i - y_{i-1}}{h^2} + q_i y_i &= g_i, \quad i = 1, \dots, n, \\ y_{n+1} &= \beta. \end{aligned} \tag{8.10}$$

If we multiply all these equations by h^2 and transfer the known boundary values to the right-hand side, we get the linear system

$$Ay = k \quad \text{with} \tag{8.11}$$

$$A = \begin{pmatrix} 2+q_1h^2 & -1 & 0 & \cdots & 0 \\ -1 & 2+q_2h^2 & -1 & 0 & \\ 0 & \ddots & \ddots & \ddots & \ddots \\ \vdots & \ddots & \ddots & \ddots & -1 \\ 0 & \cdots & 0 & -1 & 2+q_nh^2 \end{pmatrix},$$

$$\begin{aligned} y &= (y_1, y_2, \dots, y_n)^T, \\ k &= (h^2 g_1 + \alpha, h^2 g_2, \dots, h^2 g_{n-1}, h^2 g_n + \beta)^T. \end{aligned}$$

This is a symmetric tridiagonal system. If $q_i \geq 0$, then it is positive-definite and can be solved with a Cholesky band solver with complexity $O(n)$. ■

Theorem 8.3. Error estimate
If a solution y to (8.9) exists, which is four times continuously differentiable with

$$|y^{(4)}| \leq M \quad \forall x \in [a, b],$$

and if $q(x) \geq 0$, then

$$|y(x_i) - y_i| \leq \frac{Mh^2}{24}(x_i - a)(b - x_i). \tag{8.12}$$

The proof is to be found for example in [92].

Remark 8.4.

1. *In general it is recommended to solve the differential equation in parallel with two different step lengths h and qh and then to extrapolate to the limit $h = 0$. For $q = 1/2$ we get*

$$\frac{1}{3}\left(4y_{2i}^{[qh]} - y_i^{[h]}\right) - y(x_i) = O(h^4). \tag{8.13}$$

2. *For nonlinear boundary value problems the resulting system of equation is nonlinear; for example the differential equation*

$$\begin{aligned} -y''(x) + f(x,y) &= 0 \qquad (8.14)\\ y(a) = \alpha \qquad & y(b) = \beta \end{aligned}$$

yields the nonlinear system of equations

$$By + F(y) = 0$$

with a tridiagonal matrix B and a nonlinear vector function F. This can be solved with Newton's method or with special relaxation methods. If the iterative method chosen converges, then the method is of the same order as for the linear case.

If we have to solve an eigenvalue problem, we can discretize it in the same way, leading to a matrix eigenvalue problem. For example:

$$\begin{aligned} -y''(x) + \lambda\, q(x)y(x) &= 0 \qquad (8.15)\\ y(a) = 0 \qquad & y(b) = 0 \end{aligned}$$

results in

$$\begin{aligned} y_0 &= 0,\\ \frac{-y_{i+1} + 2y_i - y_{i-1}}{h^2} + \lambda\, q_i y_i &= 0, \quad i = 1, \ldots, n, \qquad (8.16)\\ y_{n+1} &= 0. \end{aligned}$$

If all values $q_i = q(x_i) \neq 0$, then we get the standard eigenvalue problem

$$(A - \lambda I)\, y = 0 \quad \text{with} \qquad (8.17)$$

$$A = \frac{1}{h^2} \begin{pmatrix} \frac{-2}{q_1} & \frac{1}{q_1} & 0 & \cdots & 0 \\ \frac{1}{q_2} & \frac{-2}{q_2} & \frac{1}{q_2} & 0 & \\ 0 & \ddots & \ddots & \ddots & \\ \vdots & \ddots & \frac{1}{q_{n-1}} & \frac{-2}{q_{n-1}} & \frac{1}{q_{n-1}} \\ 0 & \cdots & 0 & \frac{1}{q_n} & \frac{-2}{q_n} \end{pmatrix}.$$

This is an eigenvalue problem with a tridiagonal matrix, which is normally easily solved. In Chapter 5 we solved an eigenvalue problem of this kind as an application for eigenvalues of matrices.

8.3 Shooting Methods

The idea for shooting methods is quite simple: Solve an initial value problem instead of the boundary value problem, replacing the boundary condition at the right interval end-point by an estimated initial value, which is treated as a parameter of the problem. Calculate the difference between the boundary condition at the right interval end-point and the computed solution value there, and change the estimated parameter initial value according to this difference. Then solve the initial value problem again. This looks like a nice trial and error game, but it is mathematically more complicated than one may think, because the success of the method depends on convergence and stability conditions not easily fulfilled.

The straightforward implementation of this idea results in a simple shooting method, which fails in sensitive or unstable problems. A more successful method proposed by Morrison, Riley and Zancanaro, [72], and studied more thoroughly by Keller, [60], is *multiple or parallel shooting*, which has proved to be effective even for sensitive problems, see Section 8.3.2 or for more detailed introductions [92] or [3].

8.3.1 The Simple Shooting Method

Again we consider the boundary value problem

$$\begin{array}{ll} \text{ODE} & y'' = f(x, y, y'), \\ \text{BC} & y(a) = \alpha \quad\quad y(b) = \beta. \end{array} \tag{8.18}$$

We relate this problem to an initial value problem

$$\begin{array}{ll} \text{ODE} & y'' = f(x, y, y'), \\ \text{IC} & y(a) = \alpha \quad\quad y'(a) = s, \end{array} \tag{8.19}$$

where the initial value s is now a parameter. Therefore we call the solution $y(x, s)$. By solving this initial value problem several times we will iteratively adjust the parameter s such that for $s = \bar{s}$ we have

$$y(b, \bar{s}) = \beta. \tag{8.20}$$

Then the solution $y(x, \bar{s})$ of the initial value problem (8.19) is also a solution of the boundary value problem (8.18). This method is illustrated in Figure 8.2. The initial value problem (8.19) has been solved for three values s_1, s_2 and $\bar{s}$ of s. The solution $y(x, \bar{s})$ takes on the correct value $y(b) = \beta$.

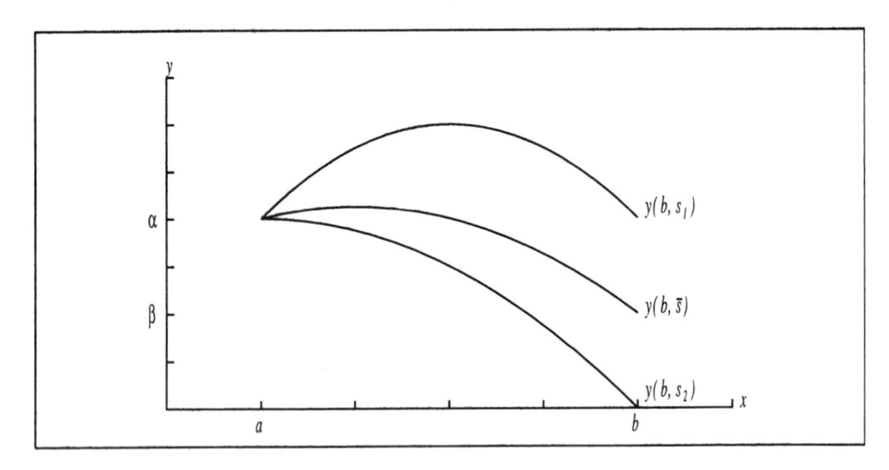

Figure 8.2. The simple shooting method

The adjustment of s to $\bar{s}$ can be represented as a problem of finding the root of a function. Determine $\bar{s}$ such that

$$F(\bar{s}) = 0, \quad \text{with} \quad F(s) := y(b, s) - \beta. \tag{8.21}$$

For each evaluation of this function F one has to solve an initial value problem from a to b.

The determination of the root $\bar{s}$ of $F(s)$ can be attempted with a *modified Newton method,* see (4.24):

$$\begin{aligned} s^{(0)} & \quad \text{given} \\ s^{(i+1)} &= s^{(i)} - \lambda \left(\Delta F(s^{(i)})\right)^{-1} F(s^{(i)}) \quad \text{with} \\ \Delta F(s^{(i)}) &:= \frac{F(s^{(i)} + \Delta s^{(i)}) - F(s^{(i)})}{\Delta s^{(i)}}. \end{aligned} \tag{8.22}$$

For each Newton step two initial value problems have to be solved, one for $s = s^{(i)}$ and one for $s = s^{(i)} + \Delta s^{(i)}$ with a small value for $\Delta s^{(i)}$ such that ΔF is a good approximation for the derivative F_s. We will not give further details of the shooting method for the general case. They can be found in [92] or in [3].

With some assumptions it can be shown that the number of roots of F is equal to the number of solutions of the boundary value problem as in the following example.

Example 8.5.

ODE $\quad y'' = y^5 - 10\,y + 1/2 \quad \text{for } x \in (0, 1),$

BC $\quad y(0) = 0 \quad y'(1) = -3.$

In contrast to (8.1) here we have a derivative value as one of the boundary conditions. This makes no difference so far as the method is concerned. The initial value problem now reads:

$$\begin{array}{ll} \text{ODE} & y'' = y^5 - 10\,y + 1/2 \quad \text{for } x \in (0,1), \\ \text{IC} & y(0) = 0 \quad y'(0) = s. \end{array}$$

By solving several initial value problems we evaluated the function

$$F(s) = y'(1,s) + 3$$

at certain points between $s = -4$ and $s = 4.5$ and found two roots $s_1 = 3.03$ and $s_2 = 4.15$, see Figure 8.3.

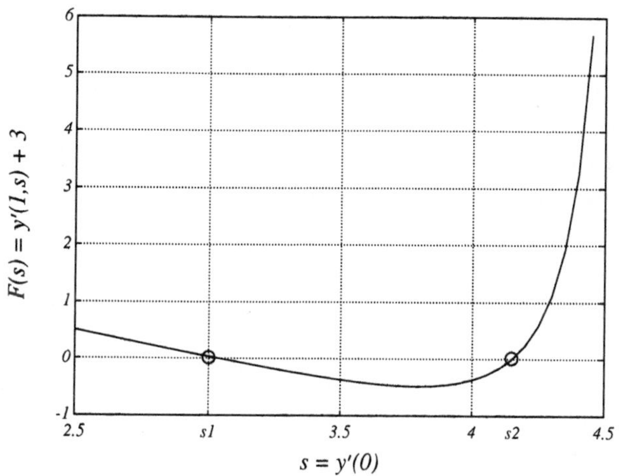

Figure 8.3. Two roots of $F(s)$

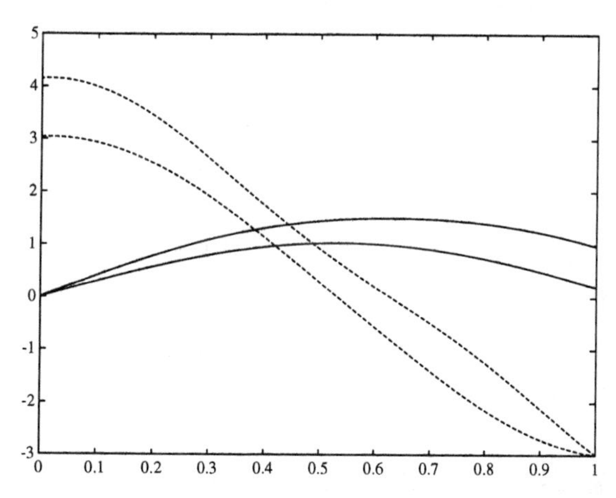

Figure 8.4. The solutions $y(x, s_1)$, $y(x, s_2)$ and their derivatives

These two roots correspond to two solutions of the boundary value problem, which can both be found with the shooting method for different initial values. Figure 8.4 shows these two solutions and their derivatives (broken curves). We will consider this problem again in Section 8.5.2. ■

The results of this example are quite satisfactory. But for many problems the simple shooting method is not stable enough. If the interval length $|b-a|$ and the Lipschitz number L, see (7.7), for the initial value problem are large, then according to (7.23) small deviations in s can yield large deviations in the solution value $y(b,s)$:

$$|y(b,s_1)-y(b,s_2)| \le |s_1-s_2| e^{L(b-a)}. \tag{8.23}$$

Then multiple shooting should be applied.

8.3.2 Multiple Shooting

Simple shooting tends to spread on long distances, therefore the hunter should prefer to shoot from shorter distances.

Therefore we divide the interval $[a,b]$ into several subintervals. Let

$$a = x_1 < x_2 < \cdots < x_{m-1} < x_m = b \tag{8.24}$$

and

$$r_k := y(x_k), \quad s_k := y'(x_k), \quad k=1,\ldots,m. \tag{8.25}$$

Now an initial value problem is solved in each subinterval. Let $y(x,x_k,r_k,s_k)$ be the solution of

$$\begin{aligned} y'' &= f(x,y,y') \;\text{ in }\; (x_k,x_{k+1}) \\ y(x_k)=r_k & \qquad y'(x_k)=s_k. \end{aligned} \tag{8.26}$$

The approximate solution for $[a,b]$ is then patched together:

$$\tilde{y}(x) := y(x,x_k,r_k,s_k) \quad \text{for} \quad x \in [x_k,x_{k+1}). \tag{8.27}$$

The initial values are combined to give a vector

$$s := (r_1,s_1,r_2,s_2,\ldots,r_{m-1},s_{m-1},r_m,s_m)^T.$$

For its $2m$ components we get $2m$ equations:

$$\begin{aligned} F_1(s) &:= y(x_2,x_1,r_1,s_1)-r_2=0, \\ F_2(s) &:= y'(x_2,x_1,r_1,s_1)-s_2=0, \\ &\;\;\cdots \\ F_{2m-3}(s) &:= y(x_m,x_{m-1},r_{m-1},s_{m-1})-r_m=0, \\ F_{2m-2}(s) &:= y'(x_m,x_{m-1},r_{m-1},s_{m-1})-s_m=0, \end{aligned} \tag{8.28}$$

$$
\begin{aligned}
F_{2m-1}(s) &:= r_1 - \alpha = 0, \\
F_{2m}(s) &:= r_m - \beta = 0.
\end{aligned}
$$

For the solution of this nonlinear system we may once more apply a modified Newton method with approximate derivative:

$$s^{(i+1)} = s^{(i)} - \lambda[\Delta F(s^{(i)})]^{-1} F(s^{(i)}). \tag{8.29}$$

Now we have to solve $m - 1$ initial value problems per Newton step, but for the subintervals which naturally are smaller. In addition two initial value problems have to be solved for each subinterval for computing the approximate ΔF to the Jacobian DF. In total $3m - 3$ 'small' initial value problems have to be solved.

(0) Select a starting value $s^{(0)} \in \mathbb{R}^{2m}$.
Select an accuracy ε.
Let $i := 0$.
(1) For $k = 1, 2, 3, \ldots, m - 1$:
Calculate $y(x_{k+1}, x_k, r_k, s_k)$ and $y'(x_{k+1}, x_k, r_k, s_k)$ by solving the initial value problem
$y'' = f(x, y, y'),\quad y(x_k) = s_k,\quad y'(x_k) = s_k$
and calculate $F(s)$ (8.28).
(2) Calculate an approximation ΔF to the Jacobian DF by solving two initial value problems per subinterval $[x_k, x_{k+1})$:
$y'' = f(x, y, y'),\ y(x_k) = r_k + \delta r_k,\ y'(x_k) = s_k,$
$y'' = f(x, y, y'),\ y(x_k) = r_k,\ y'(x_k) = s_k + \delta s_k.$
(3) Solve the $m - 1$ linear 2×2 systems
$\Delta F \Delta s = -F$.
(4) $s^{(i+1)} := s^{(i)} + \Delta s$.
(5) If $||s^{(i+1)} - s^{(i)}|| < \varepsilon$, then go to (7).
(6) $i := i + 1$.
Go to (1).
(7) End.

The algorithm is easily transferred to more general problems. For its implementation one will make use of the special form of (8.28); see [92].

8.4 Programs

There are two programs corresponding to the two methods described above, DIFFERENCE, see B.8.1, and SHOOTING, see B.8.2. They both solve boundary value problems of first order with simple shooting or finite differences and call NAG routines D02GAF and D02HAF, respectively. The differential equations must have the form

$$y_i'(x) = f_i(x, y_1(x), y_2(x), \ldots, y_n(x)), \quad i = 1, \ldots, n. \tag{8.30}$$

The solution of nonlinear boundary value problems is sensitive to parameter changes such as estimated boundary values. Because we decided to use simple NAG routines, it is important to check the results and perhaps to make several computations with different parameter values.

Both methods work with Newton's method using a finite difference approximation for the derivative F_s. The tolerance chosen serves as the estimate of the local error during the calculation as well as for the computation of the approximate derivative.

The shooting method gives the solution at equidistant points. For the finite differences method the user has to define a grid, which also may be equidistant. In addition a maximum refinement must be defined. If the grid has been refined during calculation then the solution is given on the refined grid.

The easiest way to use these programs is within the PAN system with the corresponding templates, but they can be used as simple batch programs too, with file input and output. The form of the input file is described in the program's opening comments.

The results can be seen graphically using the NAG Graphics Library. The solution and orbital curves can be drawn into a screen window or written into PostScript files. An example input file is to be seen at the end of the next section.

8.5 Applications

In this section we will first see some examples of the diversity of possible solutions. Later we will consider a worked out application. An application for the eigenvalue problem was considered in Chapter 5. Some of the examples are to be found with more details in [40].

8.5.1 No or Infinitely Many Solutions

The boundary value problem

$$\begin{array}{ll} \text{ODE} & y'' + \pi^2 y = 0, \\ \text{BC} & y(0) = 0 \quad y(1) = 1 \end{array} \tag{8.31}$$

has no solution at all, since for

$$\begin{array}{ll} \text{ODE} & y'' + \pi^2 y = 0, \\ \text{BC} & y(0) = 0 \quad y(1) = 0 \end{array} \tag{8.32}$$

infinitely many solutions exist.

8.5.2 Several Solutions

$$\begin{array}{lll} \text{ODE} & y'' = k\,y^5 - \lambda y + \mu, & \\ \text{BC} & y(0) = \alpha \quad y(1) = \beta, & \\ \text{or} & y(0) = \alpha \quad y'(1) = \gamma & \end{array} \tag{8.33}$$

can have several solutions for $k > 0$ and $\lambda > 0$, where the number of solutions grows with λ. The solution is unique if for $x \in (0,1)$: $5\,k\,y^4 > \lambda$, but that can hardly be checked before solving the ODE. For the treatment with our programs we have to transform the ODE to the system

$$\begin{aligned} y_1' &= y_2, \\ y_2' &= k\,y_1^5 - \lambda y_1 + \mu \end{aligned} \tag{8.34}$$

with the boundary conditions now having to be applied to y_1.

We already met this ODE in Example 8.5 with the second row of boundary conditions and with parameter values

$$k = 1, \quad \lambda = 10, \quad \mu = \frac{1}{2}, \quad \alpha = 0, \quad \gamma = -3.$$

Example 8.5, Page 237, shows that the uniqueness and number of solutions also depend on the boundary conditions. We will show another figure using the first row of boundary conditions in (8.33):

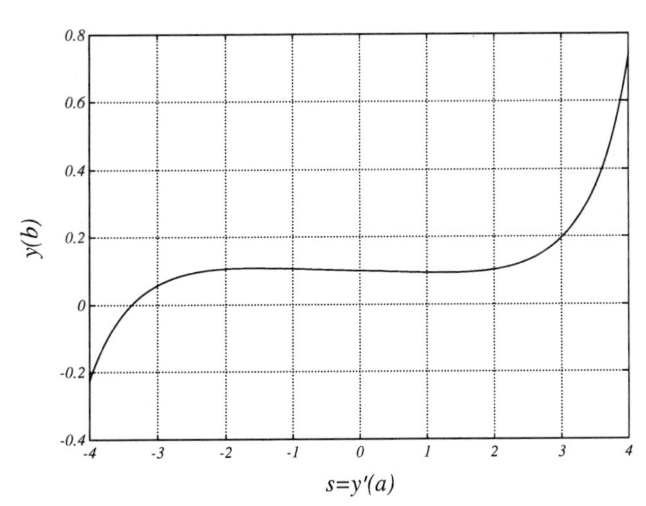

Figure 8.5. Interdependence of boundary values for Example 8.5

It is easy to see with this figure that for a small interval for the boundary value $y_1(b) \equiv y(b)$, for example $y_1(b) = 0.1$, at least three solutions exist.

We solved Example 8.5 with many different estimated values for the boundary values not given. The interesting results showed the following.

The method of finite differences generally needed more computational effort than the simple shooting method, but it always found a solution; it even found the third, which we did not find by examining $F(s)$. Both methods sometimes found different solutions for the same parameter values, as the following table will show.

Estimated BC		Shooting	Finite Differences
$y_2(0) \equiv y'(0)$	$y_1(1) \equiv y(1)$		
-5	-5	Failed	Solution 3
-1	-1	Solution 1	Solution 1
0	0	Solution 1	Solution 1
0	5	Solution 1	Solution 2
1	1	Solution 1	Solution 1
3	-3	Solution 1	Solution 3
4	-3	Solution 2	Solution 3
4.5	0	Solution 2	Solution 1
5	0	Failed	Solution 3

8.5.3 Hanging Curve of an Elastic Rope

If elastic ropes are used for mooring or towing ships, they are loaded by their own weight and by buoyancy and draft of the sea. The problem of determining the hanging curve of such a rope is a two-dimensional one, but by using the arc length s of the rope as an independent variable the problem is transformed to one of solving a system of ordinary differential equations. Let $(y(s), z(s))$ be the co-ordinates in the plane in which the rope is hanging, let $\Phi(s)$ be the angle between the tangent to the rope and the y-axis, and let $T(s)$ be the tension of the rope; then $(y(s), z(s), \Phi(s), T(s))$ must be solutions of

$$\begin{aligned} \frac{dy}{ds} &= \cos\Phi, & \frac{dz}{ds} &= \sin\Phi, \\ T\frac{d\Phi}{ds} &= \cos\Phi - a\sin\Phi|\sin\Phi|, & \frac{dT}{ds} &= \sin\Phi - b\cos\Phi|\cos\Phi| \end{aligned} \tag{8.35}$$

fulfilling the boundary conditions

$$\begin{aligned} s = 0: &\quad y(0) = z(0) = 0, \\ s = 1: &\quad y(1) = p, \quad z(1) = q, \end{aligned}$$

with parameters a, b, p and q. Let the length of the rope be normalized to 1. Then $p < 1$ is the relative horizontal distance between the end points of the rope and $q < 1$ is the relative height of the right end point, and it must be assumed that $\sqrt{p^2+q^2} < 1$. This problem looks harmless but it is again numerically sensitive to the estimate of the unknown boundary values. The quality of this estimate is more important for $\sqrt{p^2+q^2} \ll 1$ (slack rope) than for $\sqrt{p^2+q^2} \approx 1$ (taut rope). With some calculus, see [40], we obtain good estimates for the unknown boundary values. We will only report some results. In general, and in contrast to our last example, here the method of finite differences failed more often than the simple shooting method. We set $a = b = 1$ and considered two situations.

Slack Rope

The rope shall be fixed at the same height and the horizontal distance between the end points shall be three quarters of the length, which means $p = 0.75$ and $q = 0$. We do not use additional information, and set the values to be estimated to $\Phi = 0$ and $T = 1$ at both ends. For a tolerance of 10^{-6} simple shooting needs 13 Newton steps.

With the finite differences method Newton's method does not converge, even if we set the tolerance up to 0.1. We have to set these values to within one digit of the correct solution values to get convergence with this method. The grid is then refined from 11 to 73 points, which means that the computational effort is much higher than with shooting. In order to apply the program SHOOTING we have to prepare a function subroutine `shooting_1.f`:

```
subroutine fcn(x,y,f)
double precision x, y(n), f(n)
integer n
common n
f(1) = cos(y(3))
f(2) = sin(y(3))
f(3) = 1.0d0/y(4)*(cos(y(3)) - sin(y(3))*abs(sin(y(3))))
f(4) = sin(y(3)) - cos(y(3))*abs(cos(y(3)))
return
end
```

and an input file, which normally is built up by the SHOOTING template:

```
Input file for the PAN program SHOOTING:
Initial point:
0.0
Final point:
1.0
Dimension of vector of values at the initial point:
4
Vector of values at the initial point:
```

```
0 0 0 1
Dimension as above:
4
Kind of values at the initial point (0=known, 1=estimated):
0 0 1 1
Dimension as above:
4
Vector of values at the final point:
0.75 0 0 1
Dimension as above:
4
Values at the final point (0=known, 1=estimated):
0 0 1 1
Tolerance:
0.000001
Number of solution points:
11
Graphical output (no = 0, X Window = 1, PostScript = 2):
2
Output of solution curves (no = 0, yes = 1):
1
Left margin:
0.0
Right margin:
1.0
Lower margin:
-0.5
Upper margin:
0.0
Number of solution curves:
1
Dimension of vector of components:
1
Vector of components:
2
Output of an orbital curve (no = 0, yes = 1):
0
```

The results are a PostScript drawing and results like those shown in the following table:

s	y	z	Phi	T
0.000	0.000000	0.000000	-1.156636	0.846533
0.100	0.047116	-0.088087	-0.998804	0.736085

0.200	0.108424	-0.166914	-0.816286	0.619500
0.300	0.183928	-0.232195	-0.603850	0.497050
0.400	0.272346	-0.278331	-0.349243	0.372623
0.500	0.370044	-0.297456	-0.020210	0.258016
0.600	0.467431	-0.278529	0.409975	0.182037
0.700	0.550586	-0.223701	0.716867	0.167486
0.800	0.621199	-0.152972	0.834592	0.188297
0.900	0.686578	-0.077314	0.874954	0.221204
1.000	0.750000	0.000000	0.890356	0.258292

Taut Rope

A 100 m long rope shall be moored between two points such that the horizontal distance is 85 m and height difference is 50 m. That leads to parameter values $p = 0.85$ and $q = 0.5$. For this taut rope both methods yield good approximations, but again the method of finite differences needs more Newton steps and computational effort than simple shooting. The resulting rope curve is to be seen in Figure 8.6.

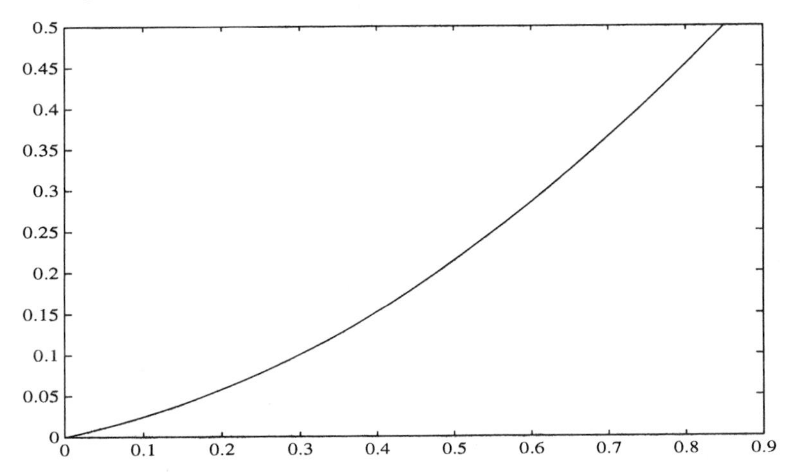

Figure 8.6. Hanging curve of an elastic rope

Input and output files of the program DIFFERENCE are similar to those for SHOOTING shown above.

8.6 Boundary Value Problems with IMSL

IMSL routines BVPFD uses finite differences and BVPMS multiple shooting for solving boundary value problems with ODEs. They do not calculate an approximate Jacobian; the user always has to program it instead.

8.6.1 Method of Finite Differences

IMSL routine BVPFD is well suited for more general problems and accordingly needs much more computational effort than D02GAF. Its counterparts with NAG are D02GBF for linear and D02RAF for nonlinear problems. These routines use the same method (routine PASVA3 of Lentini and Pereyra).

8.6.2 Multiple Shooting

IMSL has no routine which uses simple shooting like NAG routine D02HAF. IMSL routine BVPMS uses multiple shooting and is similar to NAG routine D02SAF. But D02SAF can solve much more general problems and is more complex than BVPMS.

8.7 Exercises

8.7.1 Finite Differences

1. Use Taylor series approximation to show (8.8).
2. Show similarly that the following five-point divided difference formula for $y(x)$ at the grid point $x = x_j$ for equidistant points as in (8.7)

$$S\,y_j := \frac{-y_{j-2} + 16\,y_{j-1} - 30\,y_j + 16\,y_{j+1} + y_{j+2}}{12\,h^2}$$

   is an approximation of fourth order to the second derivative:

$$y''(x_j) = S\,y_j + O(h^4).$$

3. Again for the grid (8.7), let T be defined as

$$T\,y_j := -y_{j-1} + 2\,y_j - y_{j+1}$$

   for the inner points of that grid.
   Show that T obeys a maximum principle:

$$\begin{array}{rrcll} & \text{If} \quad y_j & \geq & 0 & \text{for} \quad 0 \leq j \leq n+1 \\ & \text{and} \quad T\,y_j & \leq & 0 & \text{for} \quad 1 \leq j \leq n \\ \text{then} & \displaystyle\max_{0 \leq j \leq n+1} y_j & = & \max\{y_0,\, y_{n+1}\}. & \end{array}$$

8.7.2 Temperature Distribution in a Rod

We take up an example from [16] some forty years after the first edition of this famous book with a lot of nice examples appeared.

Find the steady temperature distribution in a homogeneous rod of length 1, in which, as a consequence of some process, heat is generated at a rate $f(y) = 1 + \exp(y)$ per unit time per unit length. If the ends of the rod are kept at given temperatures, you have to solve the following boundary value problem

$$\begin{aligned} y'' &= -f(y) \\ y(0) &= 0 \\ y(1) &= 1. \end{aligned}$$

Use PAN program DIFFERENCE to find an approximate solution and draw it.

9
Partial Differential Equations

Partial differential equations (PDEs) are one of the most important subjects of applied mathematics, with innumerable applications in science and engineering. Accordingly numerical methods for their approximate solution are of some importance. In spite of these facts this chapter is short, because the field of partial differential equations is too large to be considered in detail within a general text on numerical methods like this one.

In libraries like NAG or IMSL one normally will find only a few routines for the solution of partial differential equations. On the other hand there are a large number of software libraries, packages and programs available for the numerical solution of partial differential equations, often restricted to special classes of partial differential equations.

In this chapter we will firstly define a general partial differential equation, then look at the most important types of problems and classical examples and finally demonstrate some of the methods of solution and mention the software we use.

Our problem-solving environment PAN contains an interface to the package PLTMG which we shortly will describe as well. It is documented in detail in the electronic version of this text.

9.1 Introduction

We are looking for a function $u\colon \overline{\Omega} \subset \mathbb{R}^d \to \mathbb{R}^m$, which together with its partial derivatives solves an equation

$$F\left(x, u, \frac{\partial u}{\partial x_1}, \cdots, \frac{\partial u}{\partial x_d}, \frac{\partial^2 u}{\partial x_1^2}, \frac{\partial^2 u}{\partial x_1 x_2}, \cdots, \frac{\partial^p u}{\partial x_d^p}\right) = 0 \qquad (9.1)$$

with $x = (x_1, \cdots, x_d)^T$, under additional conditions, mostly at the boundary of the domain Ω.

If the highest derivative in (9.1) is p, then we speak of a partial differential equation of order p accordingly. A partial differential equation most often describes a physical state or process in a planar ($d = 2$) or spatial ($d = 3$) domain. The dimension of the problem increases by one to $d = 3$ or

$d = 4$ respectively, if the problem is a *time-dependent* one. Problems that are independent of time are called *stationary*.

In science, there are three main classes of physical problems which lead to the need to solve partial differential equations.

- Equilibrium problems (stationary)
- Evolutionary processes (time-dependent)
- System characteristics (eigenvalue problems)

9.1.1 Linear Partial Differential Equations of Order 2

For the remaining part of this chapter we will restrict ourselves to classical linear problems of second order in planar domains, which means $d = 2$, $m = 1$ and $p = 2$. Writing for convenience $x := x_1$, $y := x_2$ and u_{xx}, u_{xy}, u_{yy}, ..., for the partial derivatives, we obtain the general linear partial differential equation

$$Au_{xx} + 2Bu_{xy} + Cu_{yy} + Du_x + Eu_y + Fu = G \quad \forall (x,y) \in \Omega\,, \tag{9.2}$$

with functions A, B, C, D, E, F, G as coefficients, which may depend on x and y, and which should be at least piecewise continuous. For this form we arrive at the usual three types of problems:

Definition 9.1. *The linear partial differential equation of second order (9.2) with $A^2 + B^2 + C^2 \neq 0$ is called*

$$\begin{array}{lll} \textit{elliptic,} & \textit{if} \;\; AC - B^2 > 0 & \forall (x,y) \in \Omega, \\ \textit{parabolic,} & \textit{if} \;\; AC - B^2 = 0 & \forall (x,y) \in \Omega, \\ \textit{hyperbolic,} & \textit{if} \;\; AC - B^2 < 0 & \forall (x,y) \in \Omega. \end{array} \tag{9.3}$$

Additional conditions are needed for the partial differential equation being uniquely solvable. These normally are boundary conditions on $\partial\Omega$ or initial conditions on Ω. The *boundary conditions* may be of different form, for example:

$$u(x,y) = \phi(x,y) \qquad \text{(Dirichlet condition)}, \tag{9.4}$$

$$\frac{\partial u}{\partial n}(x,y) = \gamma(x,y) \qquad \text{(Neumann condition)}, \tag{9.5}$$

$$\frac{\partial u}{\partial n}(x,y) + \alpha(x,y)u(x,y) = \beta(x,y) \qquad \text{(Cauchy condition)}. \tag{9.6}$$

Here $\partial u/\partial n$ is the derivative in direction of the outer normal and ϕ, γ, α, β are given functions.

For time-dependent problems *initial conditions* are additionally imposed. They prescribe the function u or its derivative u_t on the domain Ω at initial point of time $t = 0$.

The proof of existence and uniqueness of solutions and their continuous dependence on coefficients and boundary and initial conditions are hard mathematical problems, which we will not deal with here. We will give three classical examples for the three types instead. Subsequently we will demonstrate some of the most important numerical methods applied to simple elliptic partial differential equations.

9.1.2 Poisson's and Laplace's Equation

The *Laplace operator* or *Laplacian* Δ is defined by

$$\Delta u := u_{xx} + u_{yy}. \tag{9.7}$$

One important partial differential equation with the Laplace operator is the *potential* or *Poisson's equation*:

$$-\Delta u = f \quad \text{in } \Omega. \tag{9.8}$$

It describes a gravitational or Newtonian potential u, if f is a mass density, and it describes an electric potential, if f is the electric charge.

If the right-hand side f is zero, one gets the *Laplace's equation*, which for example describes the stationary temperature distribution of a homogeneous body without sources and sinks and with fixed temperature g at the boundary:

$$\begin{aligned} -\Delta u &= 0 \quad \text{in } \Omega \\ u &= g \quad \text{on } \partial\Omega\,. \end{aligned} \tag{9.9}$$

Solutions of Laplace's equation are called *harmonic functions*. Poisson's and Laplace's equations are elliptic differential equations.

9.1.3 Time-Dependent Heat Conduction

If the temperature distribution is time dependent, the temperature $u(x, y, t)$ must fulfill the partial differential equation

$$\begin{aligned} \Delta u &= \frac{1}{\kappa}\frac{\partial u}{\partial t} \quad \text{in } \Omega, \ \forall t > 0 \\ u(x, y, 0) &= u_0(x, y) \quad \text{in } \Omega \\ u(x, y, t) &= g(x, y) \quad \text{on } \partial\Omega, \ \forall t > 0. \end{aligned} \tag{9.10}$$

It is a parabolic differential equation. κ is the ratio of the heat conductivity to the specific heat capacity. The same differential equation describes diffusion processes like diffusion of gaseous material. Then u is the concentration of this material and κ the coefficient of diffusion.

9.1.4 The Wave Equation

The hyperbolic wave equation

$$\begin{aligned} \Delta u &= \frac{1}{c^2}\frac{\partial^2 u}{\partial t^2} \quad \text{in } \Omega, \ \forall t > 0 \\ u(x,y,0) &= u_0(x,y) \quad \text{in } \Omega \\ u_t(x,y,0) &= u_1(x,y) \quad \text{in } \Omega \\ u(x,y,t) &= g(x,y) \quad \text{on } \partial\Omega\,, \ \forall t > 0 \end{aligned} \tag{9.11}$$

describes waves of sound, electromagnetic potentials or time-dependent oscillations. For acoustic waves c is the velocity of sound and u the air density.

9.2 Discretization of Elliptic Problems

We will now restrict ourselves to the consideration of numerical methods applied to the simple elliptic differential equation

$$\begin{aligned} -\Delta u &= f \quad \text{in } \Omega \\ u &= 0 \quad \text{on } \partial\Omega\,. \end{aligned} \tag{9.12}$$

As an application one may consider deflections of an perfectly elastic diaphragm, which is fixed at the boundary $\partial\Omega$ of domain Ω. We have considered this 'drum' problem several times before.

9.2.1 Difference Methods

The two-dimensional domain Ω has to be covered by a grid. Next the boundary $\partial\Omega$ has to be approximated by grid lines. Let for simplicity the grid lines be equally spaced in both directions with segment lengths h. In such a way the domain is covered by a quadratic grid, and the grid points (x_i, y_j) can be numbered like

$$\begin{aligned} x_i &:= x_0 + i \cdot h, \quad i = 0, 1, \cdots, N, \\ y_j &:= y_0 + j \cdot h, \quad j = 0, 1, \cdots, N, \\ \Omega &\subset [x_0, x_N] \times [y_0, y_N]. \end{aligned}$$

Now only inner grid points are considered for the following calculations. This situation is illustrated in Figure 9.1. Boundary points are marked by circles, inner points by crosses. The solution values of function u are approximated at all inner grid points:

$$u_{ij} \approx u(x_i, y_j).$$

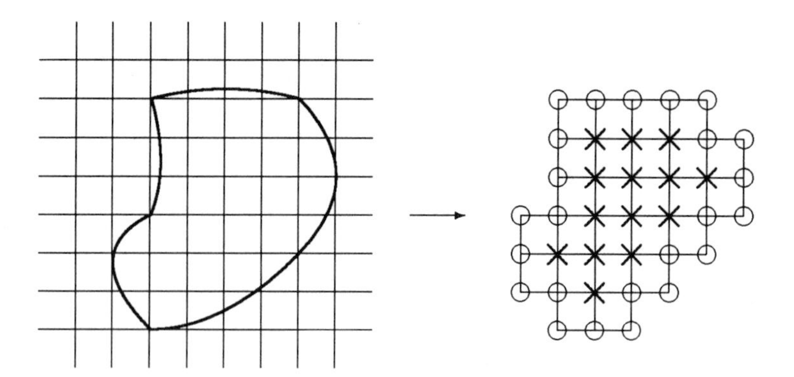

Figure 9.1. Discretization of a domain

At each of these points the partial differential equation yields a linear equation by replacing the derivatives by divided differences:

$$\begin{aligned} u_{xx}(x_i, y_j) &\approx \frac{1}{h^2}(u_{i+1,j} - 2u_{ij} + u_{i-1,j}) \\ u_{yy}(x_i, y_j) &\approx \frac{1}{h^2}(u_{i,j+1} - 2u_{ij} + u_{i,j-1}). \end{aligned} \tag{9.13}$$

(9.13) converts (9.12) into

$$-u_{i+1,j} - u_{i,j+1} + 4u_{ij} - u_{i-1,j} - u_{i,j-1} = h^2 f_{ij}, \tag{9.14}$$

with $f_{ij} := f(x_i, y_j)$ and with (i, j) taking on all values corresponding to inner grid points. If a boundary point value appears in (9.14), the corresponding term simply will be omitted, because we have $u(x_i, y_j) = 0$ for $(x_i, y_j) \in \partial\Omega$. Hence the order of the system is equal to the number of inner grid points. The matrices generated are symmetric and positive-definite. Therefore we can apply Cholesky's method for their solution. But, if the discretization is sufficiently fine (which means that h is sufficiently small) for a reasonable approximation, then we will end up with a large sparse system of linear equations with perhaps several thousand equations. Then we have to apply special methods for their solution like those described in Section 1.4. We have seen examples for such grids and for the construction of the corresponding linear system before (see Examples 1.12 and 1.13). Let us consider another simple example.

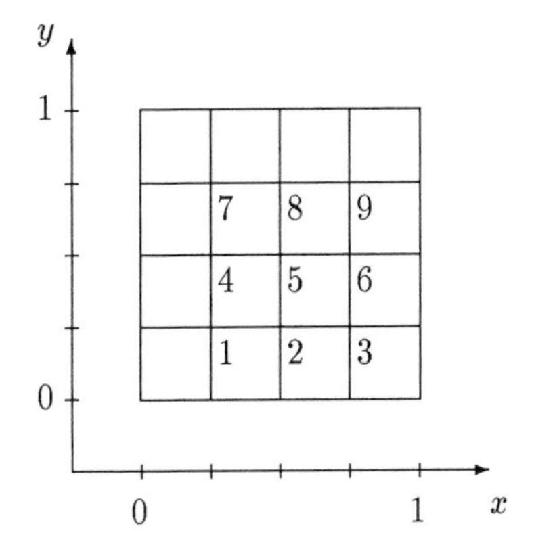

Figure 9.2. The discretized unit square

Example 9.2. Let $\Omega = (0,1) \times (0,1)$ be the unit square, and $h = 1/4$, $f \equiv 16$. Then the following system of linear equations has to be solved:

$$\begin{pmatrix} 4 & -1 & 0 & -1 & 0 & 0 & 0 & 0 & 0 \\ -1 & 4 & -1 & 0 & -1 & 0 & 0 & 0 & 0 \\ 0 & -1 & 4 & 0 & 0 & -1 & 0 & 0 & 0 \\ -1 & 0 & 0 & 4 & -1 & 0 & -1 & 0 & 0 \\ 0 & -1 & 0 & -1 & 4 & -1 & 0 & -1 & 0 \\ 0 & 0 & -1 & 0 & -1 & 4 & 0 & 0 & -1 \\ 0 & 0 & 0 & -1 & 0 & 0 & 4 & -1 & 0 \\ 0 & 0 & 0 & 0 & -1 & 0 & -1 & 4 & -1 \\ 0 & 0 & 0 & 0 & 0 & -1 & 0 & -1 & 4 \end{pmatrix} \begin{pmatrix} u_1 \\ u_2 \\ u_3 \\ u_4 \\ u_5 \\ u_6 \\ u_7 \\ u_8 \\ u_9 \end{pmatrix} = \begin{pmatrix} 1 \\ 1 \\ 1 \\ 1 \\ 1 \\ 1 \\ 1 \\ 1 \\ 1 \end{pmatrix}$$

Here we numbered the grid points (with one index only) row by row from left to right and from bottom to top, see Figure 9.2. ■

9.2.2 The Finite Element Method

This method belongs to the class of variational methods. The partial differential equation is replaced by a variational problem, which is considered in special function spaces, and approximately solved in a finite-dimensional subspace. This procedure again leads to a numerical method, for which a large sparse system of linear equations ultimately has to be solved. We will try to describe the method from a practical point of view neglecting most of the necessary mathematical assumptions.

Problem (9.12) is replaced by the variational problem

$$\min_{u \in H} \int_{\Omega} \left\{ \left(\frac{\partial u}{\partial x}\right)^2 + \left(\frac{\partial u}{\partial y}\right)^2 - 2uf \right\} \, dx \, dy. \tag{9.15}$$

The function space H for the minimization is approximated by a linear combination of a finite number of basis functions, which apart from other assumptions fulfill the zero-condition on the boundary $\partial\Omega$. Let U_n be the subspace of H, which is spanned by the functions

$$\{\phi_1, \phi_2, \cdots, \phi_n\}. \tag{9.16}$$

We are looking for an approximate solution $u_n \in U_n$:

$$u_n(x, y) = \sum_{i=1}^{n} d_i \, \phi_i(x, y). \tag{9.17}$$

u_n is determined by the coefficients d_i.

Now define:

$$s_{ij} := \int_\Omega \left\{ \frac{\partial \phi_i}{\partial x} \frac{\partial \phi_j}{\partial x} + \frac{\partial \phi_i}{\partial y} \frac{\partial \phi_j}{\partial y} \right\} dx\, dy \tag{9.18}$$

and

$$t_i := \int_\Omega \phi_i \, f \, dx\, dy. \tag{9.19}$$

Then the coefficients d_i are the solutions of the linear system

$$Sd = t \tag{9.20}$$

with

$$\begin{aligned} S &:= (s_{ij})_{i,j=1}^{n} \\ t &:= (t_1, \cdots, t_n)^T \\ d &:= (d_1, \cdots, d_n)^T. \end{aligned}$$

The finite element method is defined by special definitions of the basis functions ϕ_i. Out of the many possibilities we will describe only one.

We subdivide the domain Ω into triangles T_k of similar size and with inner angles bounded from above and below; see Figure 9.3 for an example. These triangles are also called *elements*. Two triangles must either intersect exactly at one edge or exactly at one vertex or they must be disjoint. Inner vertices are called *nodes*.

The basis functions shall be linear polynomials

$$\phi_i(x, y)\Big|_{T_k} = a_{ik}x + b_{ik}y + c_{ik} \tag{9.21}$$

in each triangle T_k. The coefficients are chosen such that each function ϕ_i is exactly equal to 1 in one node and it is equal to 0 in all other nodes. This means that we get n basis functions, if n is the number of inner nodes $P_j = (x_j, y_j)$ with

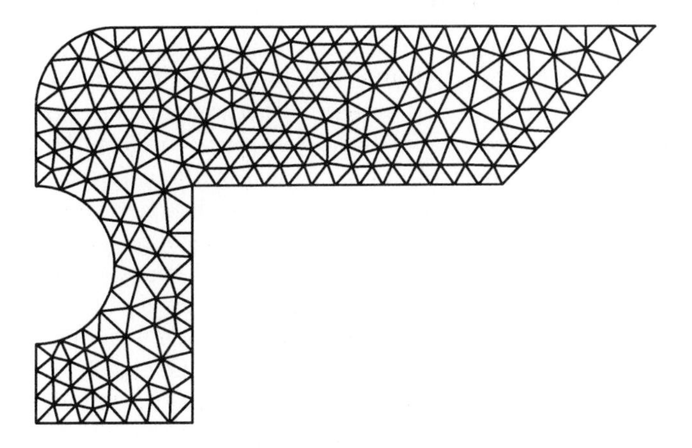

Figure 9.3. Finite element triangularization

$$\phi_i(x_j, y_j) = \begin{cases} 1 & \text{for} \quad i = j \\ 0 & \text{for} \quad i \neq j. \end{cases} \tag{9.22}$$

The method, developed along these guidelines, is not easy for the programmer, but it is efficient for the following reasons.

- Calculation of the integrals (9.18) and (9.19) is easy and fast.
- Most of the matrix elements s_{ij} are zero, because each basis function is not equal to zero only in a few triangles. Therefore we get a sparse large matrix again.
- The matrix of coefficients for the linear systems to be solved can be successively constructed from small element matrices.

On the other hand difficulties may appear with data organization as the following points must be considered:

- Regular decomposition of the domain.
- Numbering of triangles and nodes.
- Construction of tables with node coordinates and node-element correspondence, which are necessary for assembling the total system matrix from small element matrices.

The engineering formulation of the method of finite elements decomposes Ω into technical elements like plates, disks, bars for trusses etc., each of those assuming a certain type of problem. This is quite different from the mathematical formulation but it leads to the same system of linear equations.

9.2.3 Multigrid Methods

The two methods briefly described above offer different possibilities for the discretization of a partial differential equation to a linear system of equations. The solution of this system is the predominant part of the effort involved in the practical treatment of partial differential equations. Therefore the development of fast numerical methods for the solution of these large sparse systems is an important area of numerical mathematics. Multigrid methods are some of the fastest methods developed. We will give some hints to the software package PLTMG in the next section. Because PLTMG uses a multigrid method, we will shortly describe the idea of these methods. For further details we refer the reader to [49] and [69].

They presume that the solution consists of low and high frequency parts, i.e. parts with fewer and more roots, respectively. For the finite-dimensional vector approximating the solution function this is equivalent to fewer or more sign changes. Now one knows that for many problems the main part of the solution consists of low frequencies. These solution parts can be gained from a rough discretization with little effort. Then this solution is iteratively improved by high frequency parts. This is the main idea of multigrid methods.

Let us consider a simple *two-grid method*. Firstly we take an arbitrary start solution on the fine grid, e.g. $u \equiv 0$. This solution is improved on this grid with one (or more) steps of a relaxation method, see Section 1.4.2, which basically means one matrix-vector multiplication making use of the sparse structure of the system matrix. Then the residual of this solution is calculated and carried over to the coarse grid. This step is called *restriction*. On the coarse grid the residual linear system is solved. This is like an iterative refinement step, see Section 1.1.6. Then the refinement vector is transformed back to the fine grid with interpolation. There the solution is improved again by one (or more) relaxation steps. This is a cycle of steps that are iteratively repeated.

The method can be generalized to more than two grids by defining the solution method on the coarse grid recursively to be a two-grid method.

On Page 257 we will demonstrate algorithmically one step of a three-grid method for solving $Su = t$.

In Table 9.1 we compare the computing times for different methods to show how efficient the multigrid method is working on standard problems. We consider the model problem from Example 9.2, but now with many more grid lines and inner grid points. With 255 grid lines in both x and y directions we get a system of linear equations with $n = 65025$ unknowns. The computing times are seconds on a near-historic IBM 370/158.

Let $u_0 = u_0^{(\text{fine})}$ be the starting vector of approximation and $t^{(\text{fine})} := t$.

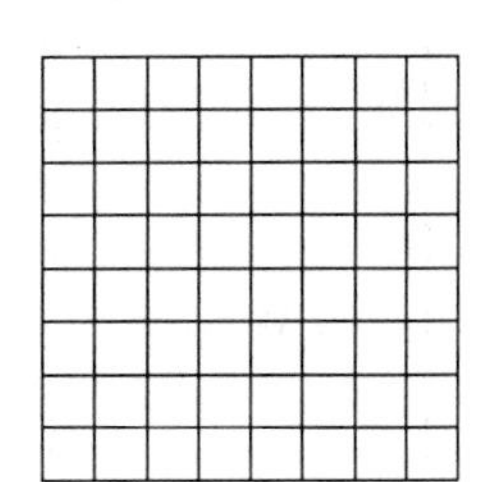

(1) **Relaxation** on the finest grid:

$$u_1^{(\text{fine})} = \text{Relax}\left(u_0^{(\text{fine})}, t^{(\text{fine})}\right)$$

and calculation of the residual vector:

$$r^{(\text{fine})} := t^{(\text{fine})} - S^{(\text{fine})} u_1^{(\text{fine})}$$

with $S^{(\text{fine})} := S$.

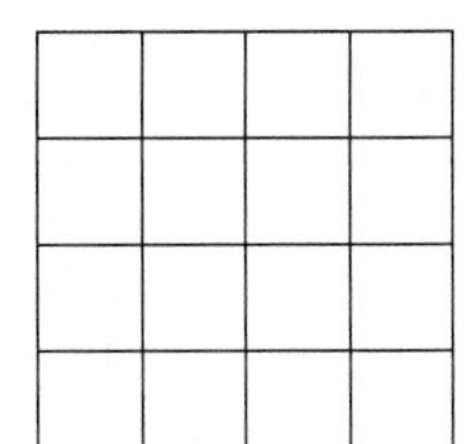

(2) **Restriction and relaxation**:

$$\begin{aligned} t^{(\text{med})} &= \text{Restrict}\left(r^{(\text{fine})}\right) \\ u_1^{(\text{med})} &= \text{Relax}\left(0, t^{(\text{med})}\right) \\ r^{(\text{med})} &= t^{(\text{med})} - S^{(\text{med})} u_1^{(\text{med})} \end{aligned}$$

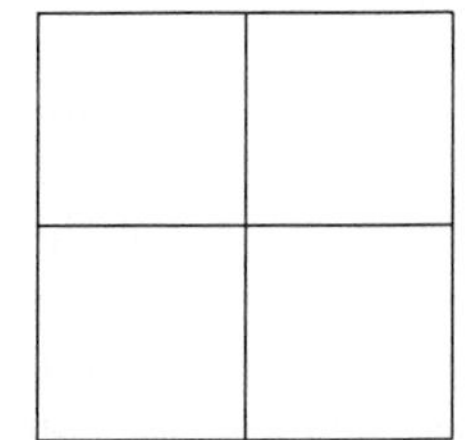

(3) **Restriction and exact solution**:

$$t^{(\text{coarse})} = \text{Restrict}\left(r^{(\text{med})}\right)$$

$$\text{Solve } S^{(\text{coarse})} u^{(\text{coarse})} = t^{(\text{coarse})}$$

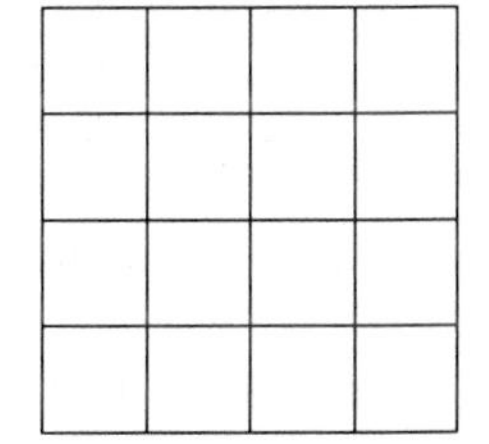

(4) **Interpolation and relaxation**:

$$\begin{aligned} u_2^{(\text{med})} &= u_1^{(\text{med})} + \text{Interp}\left(u^{(\text{coarse})}\right) \\ u_3^{(\text{med})} &= \text{Relax}\left(u_2^{(\text{med})}, t^{(\text{med})}\right) \end{aligned}$$

(5) **Interpolation and relaxation**:

$$\begin{aligned} u_2^{(\text{fine})} &= u_1^{(\text{fine})} + \text{Interp}\left(u_3^{(\text{med})}\right) \\ u_3^{(\text{fine})} &= \text{Relax}\left(u_2^{(\text{fine})}, t^{(\text{fine})}\right) \end{aligned}$$

$u_1 := u_3^{(\text{fine})}$ is the new approximate solution.

Method	Complexity	Time
Gauss–Seidel	$\sim n^2$	100 000
SOR	$\sim n^{3/2}$	1 200
Buneman	$\sim n \log n$	16
Fourier method	$\sim n \log \log n$	8
Multigrid method	$\sim n$	7

Table 9.1. Complexity and times for different solvers

SOR means *successive overrelaxation*, which is an improved Gauss–Seidel method, see [42] for example. The Buneman and Fourier methods mentioned belong to the group of *rapid elliptic solvers*, which assume a rectangular form of the domain and cannot be applied to general domains.

9.3 Software

As pointed out at the beginning of this chapter, there are a large number of programs and packages available especially designed for the solution of certain classes of partial differential equations. Most of them use the finite element method. There are reference texts containing addresses and prices with several hundred pages like the catalogue [34], which in 1976 reported about 250 program systems. We will restrict ourselves here to the two possibilities we normally use, i.e. some subroutines from the NAG Fortran Library, and the software package PLTMG, see [6].

9.3.1 NAG Routines

In Chapter D03 of the NAG Fortran Library we find 19 routines for the solution of elliptic and parabolic differential equations.

The 10 D03P routines solve linear and nonlinear parabolic problems in one space dimension using different methods:

$$\frac{\partial u}{\partial t} = \frac{1}{x^m}\frac{\partial}{\partial x}\left(x^m g(x,t,u)\frac{\partial u}{\partial x}\right) + f\left(x,t,u,\frac{\partial u}{\partial x}\right). \tag{9.23}$$

In Mark 16 of the library there are three new routines using a Keller box discretization technique for first order PDEs (with coupled differential algebraic systems) in one spatial dimension and remeshing facilities.

There are 8 routines for elliptic problems. D03EAF – as the only complete elliptic solver – solves Laplace's equation (9.9) in two dimensions.

D03EBF and D03ECF use Stone's strongly implicit procedure (SIP) to solve a system of simultaneous algebraic equations using five-point and seven-point molecules on two-dimensional and three-dimensional rectangular meshes respectively. These systems can be created with D03EEF. Routine D03FAF solves the Helmholtz equation in three dimensions. Routine D03MAF places a triangular grid over a given two-dimensional domain.

9.3.2 PLTMG: Finite Element and Multigrid Method

PLTMG is a large program for linear and nonlinear partial differential equations, which uses the method of finite elements and a multigrid method. PLTMG solves parameter-dependent problems like eigenvalue or continuation problems as well.

The general form of nonlinear boundary value problems considered is

$$-\nabla\left(a(x,y,u,\nabla u,\lambda)\right)+f(x,y,u,\nabla u,\lambda)=0 \quad \text{in } \Omega \tag{9.24}$$

with the boundary conditions

$$\begin{aligned} u &= g_1(x,y,\lambda) \quad \text{on } \partial\Omega_1 \\ a\cdot n &= g_2(x,y,\lambda) \quad \text{on } \partial\Omega_2=\partial\Omega-\partial\Omega_1 . \end{aligned} \tag{9.25}$$

Here Ω is a connected domain in $\mathbb{R}^2$, n is the outer unit normal, a is the vector $(a_1,a_2)^T$ and a_1, a_2, f, g_1 and g_2 are scalar functions. λ is a scalar continuation parameter. In addition the user may specify a functional of the solution of the form

$$\rho(u,\lambda)=\int_\Omega p_1(x,y,u,\nabla u,\lambda)\,dx\,dy+\int_{\partial\Omega} p_2(x,y,u,\nabla u,\lambda)\,ds \tag{9.26}$$

with scalar functions p_1 and p_2.

The abbreviation PLTMG means *piecewise linear triangle multigrid method*. This means that PLTMG uses linear polynomials on triangles as finite elements and solves the algebraic equations (linear or nonlinear) created with a multigrid method.

We see that PLTMG has quite general possibilities and accordingly its use is not straightforward, but it has nice possibilities of graphical results; see the example at the end of this chapter. The triangle net of Figure 9.3 was generated with PLTMG, too.

The code of PLTMG is obtainable from *Netlib*. One should begin with calling for an index by electronic mail:

```
mail netlib@research.att.com
send index for pltmg
```

The index file contains instructions for obtaining the rest of the source code. Because of the size, *ftp* should be used rather than electronic mail.

A graphical interface to PLTMG is part of our problem-solving environment PAN, see Appendix A. We hope that this interface enables an easier access for the novice, and facilitates the use of PLTMG even for the experienced user. Knowledge and experience is nevertheless important, because partial differential equations often are models for quite complicated technical systems in the real world.

9.4 PLTMG/PAN: An Example

We will show an example for the interface PLTMG/PAN. Firstly we call our graphic editor from the PAN template for the program PLTMG. With straight and curved lines we draw a two-dimensional domain similar to the car in Exercise 3.10.3, see Figure 9.4.

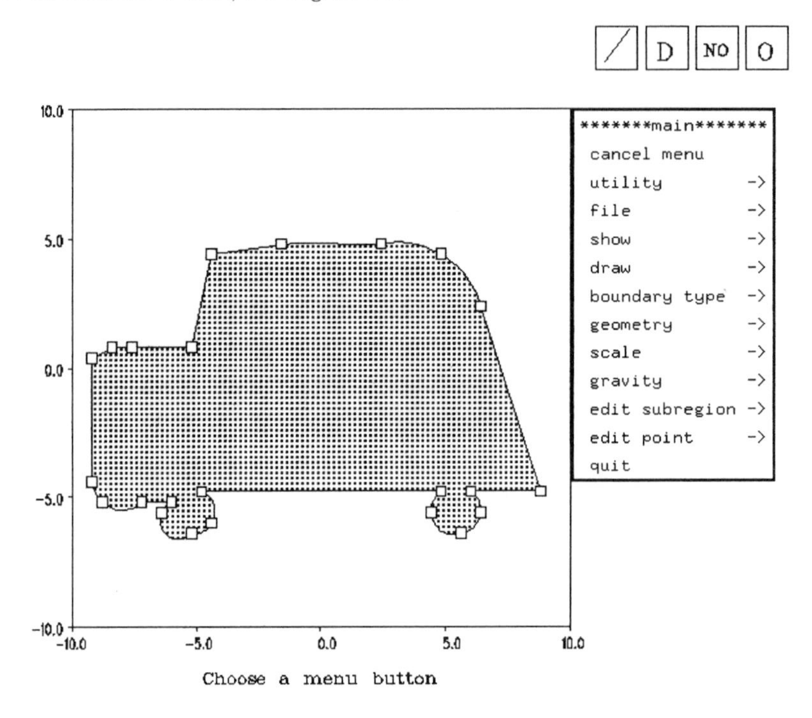

Figure 9.4. Domain design with the graph editor

With the facilities of the graph editor like zooming and creating subregions we could have been much more precise, but here the car serves merely as a domain. Now we triangulate the domain with PLTMG and obtain the upper drawing in Figure 9.5[†] (at the end of this chapter). Then we solve Poisson's

[†]These grey-shaded plots are originally nicely colored PLTMG plots.

equation (9.12) with the right-hand side $f(x,y) = 1$ and homogeneous boundary conditions. This gives the lower drawing in Figure 9.5. Then we have refined the grid adaptively getting a better solution and its error per triangle as shown in Figure 9.6. The values at the boundary should be zero. This means that the negative values at the boundary are error values. Another refinement would be necessary for getting a solution with more than two or three precise digits.

For running PLTMG we have to define three additional Fortran subroutine files called pltmg_1.f, pltmg_2.f and pltmg_3.f. The first one defines the domain Ω. Because we constructed it with the graph editor, we need not care about pltmg_1.f. The third one serves for additional user functions, which is not important for this example. The second one contains all functions defining the partial differential equations, the boundary conditions and the additional functional according to Equations (9.24)–(9.26). In this simple case the functions are u_x and u_y or constant, which gives the following file:

```
c******************** file: pltmg_2.f **********************
c
c  This file consists of the real functions a1xyuser,
c  a2xyuser, fxyuser, gxyuser, p1xyuser and p2xyuser
c  that are related to the functions a and f in the p.d.e.,
c  the functions g_i in the boundary, and the functional rho.
c
c  parameters -
c
c     x ........ coordinates of a point in element t_i.
c     y ........
c     u ........ value of the function at the point (x,y).
c     ux ....... gradient of the function at the point (x,y).
c     uy .......
c     rl ....... the continuation parameter.
c     itag ..... label assigned to element t_i that was
c                provided by the user as ITNODE(4,i).
c     itype .... specifies the expected output. If
c                = 1, function value - say f
c                = 2, the value of d(f)/d(u)
c                = 3, the value of d(f)/d(u_x)
c                = 4, the value of d(f)/d(u_y)
c                = 5, the value of d(f)/d(lambda)
c                is requested.
c******************* real function a1xyuser ****************
c
```

```
      real function a1xyuser(x,y,u,ux,uy,rl,itag,itype)
c
c     .. scalar arguments ..
      integer  itag,itype
      real     rl,u,ux,uy,x,y
c     .. executable statements
c
      go to (1,2,3,4,5),itype
    1 a1xyuser=ux
      return
    2 a1xyuser=0.0e0
      return
    3 a1xyuser=1.0e0
      return
    4 a1xyuser=0.0e0
      return
    5 a1xyuser=0.0e0
      return
      end
c
c****************** real function a2xyuser ***************
c
      real function a2xyuser(x,y,u,ux,uy,rl,itag,itype)
c
c     .. scalar arguments ..
      integer  itag,itype
      real     rl,u,ux,uy,x,y
c     .. executable statements
c
      go to (1,2,3,4,5),itype
    1 a2xyuser=uy
      return
    2 a2xyuser=0.0e0
      return
    3 a2xyuser=0.0e0
      return
    4 a2xyuser=1.0e0
      return
    5 a2xyuser=0.0e0
      return
      end
c
c************** real function fxyuser *********************
```

```
c
      real function fxyuser(x,y,u,ux,uy,rl,itag,itype)
c
c     .. scalar arguments ..
      integer  itag,itype
      real     rl,u,ux,uy,x,y
c     .. executable statements
c
        go to (1,2,3,4,5),itype
c       f = -1, because with PLTMG f is on the left-hand side
    1   fxyuser=-1.0e0
        return
    2   fxyuser=0.0e0
        return
    3   fxyuser=0.0e0
        return
    4   fxyuser=0.0e0
        return
    5   fxyuser=0.0e0
        return
        end
c
c************** real function gxyuser *********************
c
        real function gxyuser(x,y,u,rl,itag,itype)
c
c     itype .... specifies the expected output. If
c                = 1, the value g_2
c                = 2, the value of d(g_2)/d(u)
c                = 3, the value of d(g_2)/d(lambda)
c                = 4, the value of d(g_1)
c                = 5, the value of d(g_1)/d(lambda)
c                = 6, initial solution
c                is requested.
c
c     .. scalar arguments ..
      integer  itag,itype
      real     rl,u,x,y
c     .. executable statements
c
        go to(1,2,3,4,5,6),itype
    1   gxyuser=0.0e0
        return
```

```
    2   gxyuser=0.0e0
        return
    3   gxyuser=0.0e0
        return
    4   gxyuser=0.0e0
        return
    5   gxyuser=0.0e0
        return
    6   gxyuser=1.0e0
        return
        end
c
c************** real function p1xyuser *********************
c
      real function p1xyuser(x,y,u,ux,uy,rl,itag,itype)
c
c     parameters -
c
c     .. scalar arguments ..
      integer  itag,itype
      real     rl,u,ux,uy,x,y
c     .. executable statements
c
      p1xyuser=0.0e0
      return
      end
c
c************** real function p2xyuser *********************
c
      real function p2xyuser(x,y,dx,dy,u,ux,uy,rl,itag,itype)
c
c     .. scalar arguments ..
      integer  itag,itype
      real     dx,dy,rl,u,ux,uy,x,y
c     .. executable statements
c
      p2xyuser=0.0e0
      return
      end
```

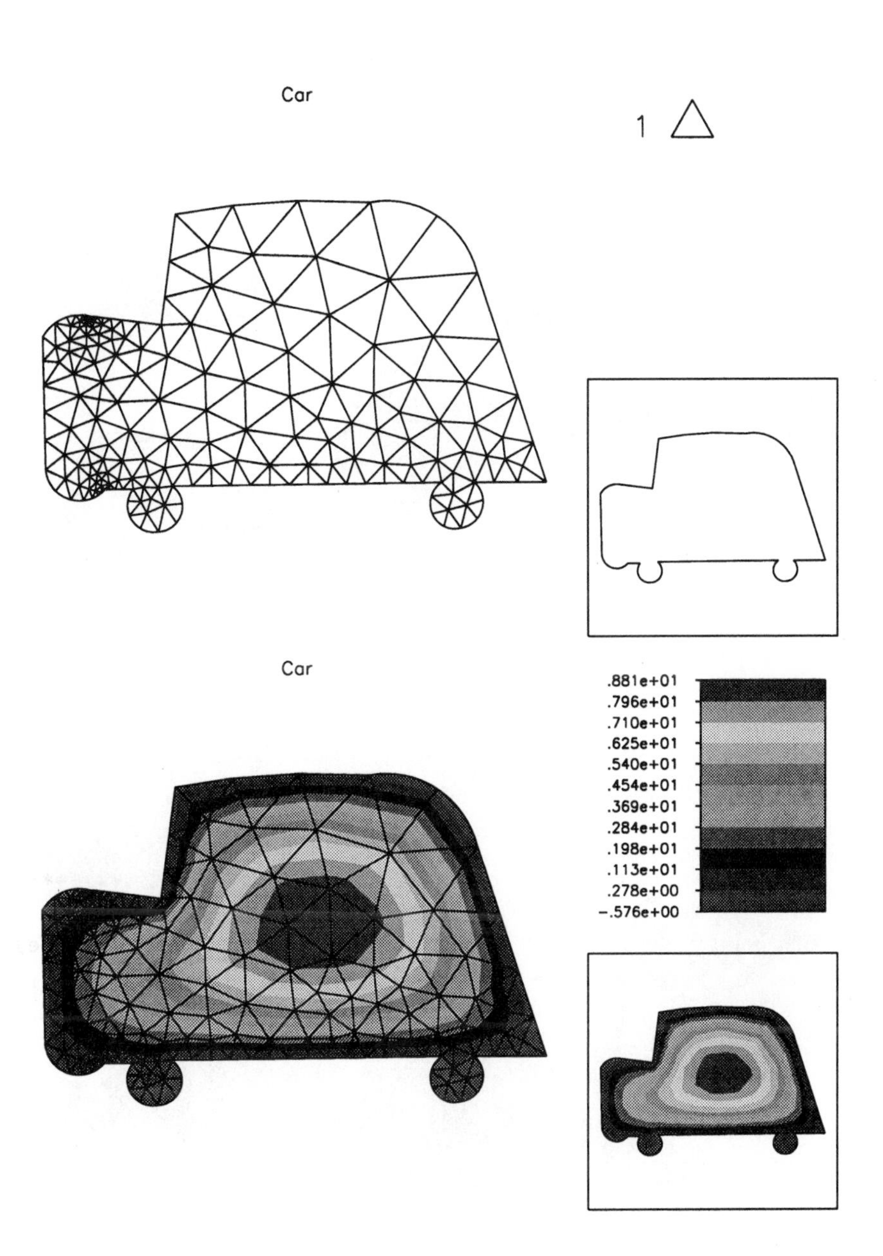

Figure 9.5. The triangulated car and PDE solution on it

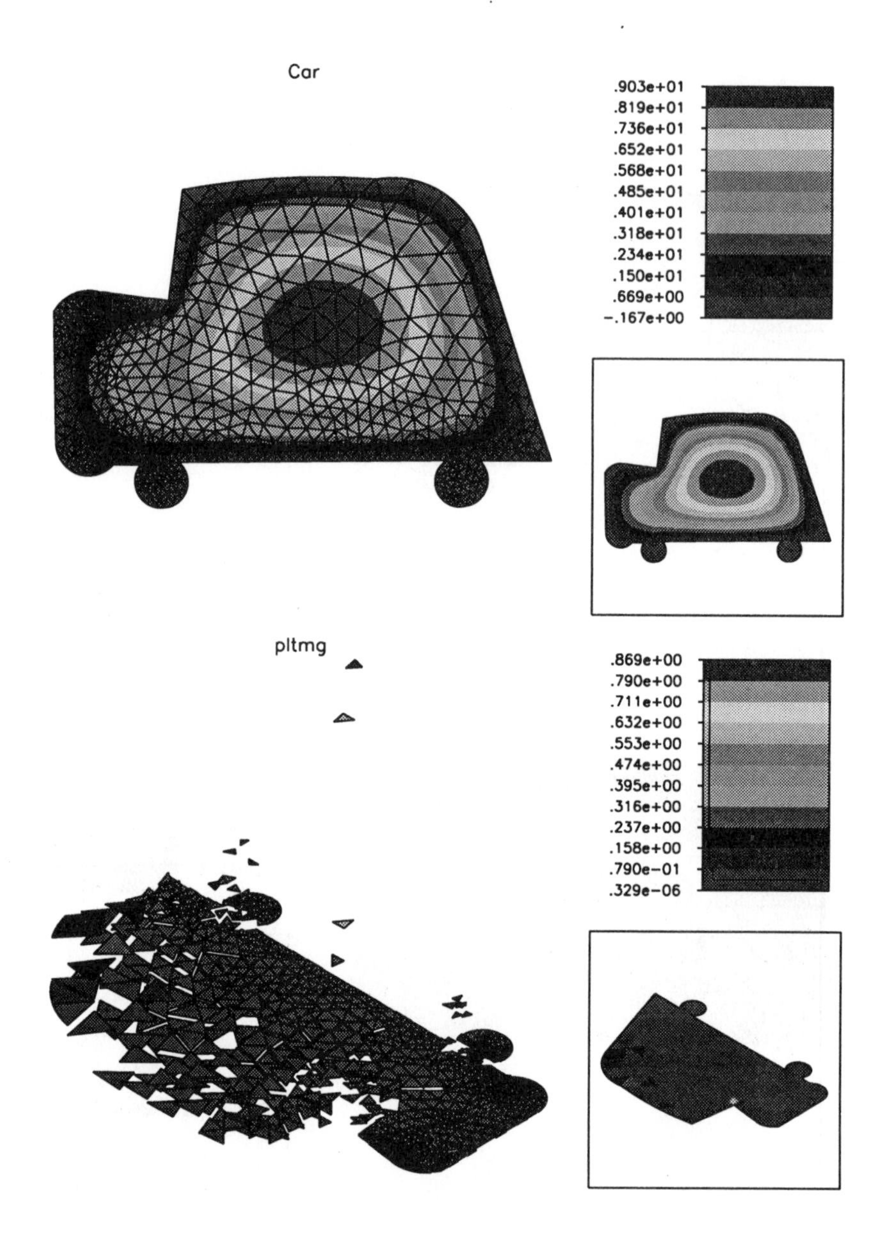

Figure 9.6. PDE solution on a refined grid and its error

Appendix A
PAN – A Problem Solving Environment

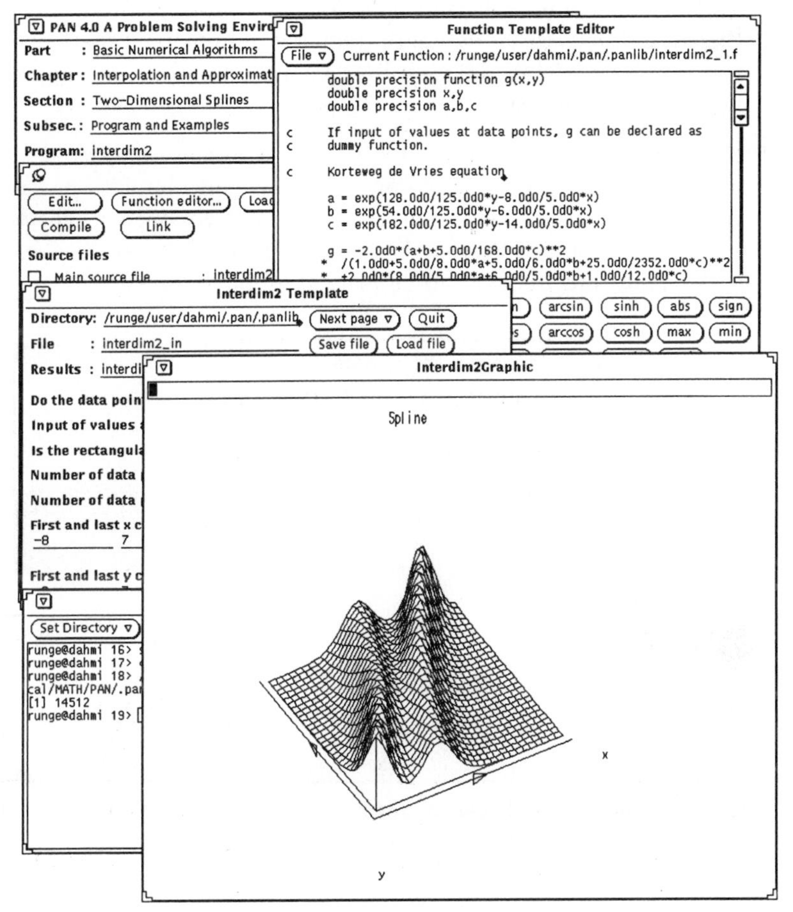

Figure A.1. The PAN system

The PAN system provides you with a graphical ('Open Look')† interface that considerably facilitates access to the example programs that have been mentioned. Together with the DVI previewer HyTEX, a program also developed by the PAN working group, it supports numerical problem solving. The system comprises the following elements:

- a management system designed for the administration of various dynamic structures and sub-windows,
- the DVI previewer HyTEX for the demonstration of Hypertext documents, i.e. a hypertext version of this textbook and a 100-page manual,
- a Fortran program collection (the example programs) representing typical solutions to the standard problems of numerical mathematics by means of the NAG Library. Here input and output are supported by a template library.

Both the management system and the previewer can be considered to be the static components. The collection of Fortran and template programs together with the tutorial represent the dynamic components. They can be extended and modified by you. Thus the user interface can be used for entirely different aims from that of numerical problem solving.

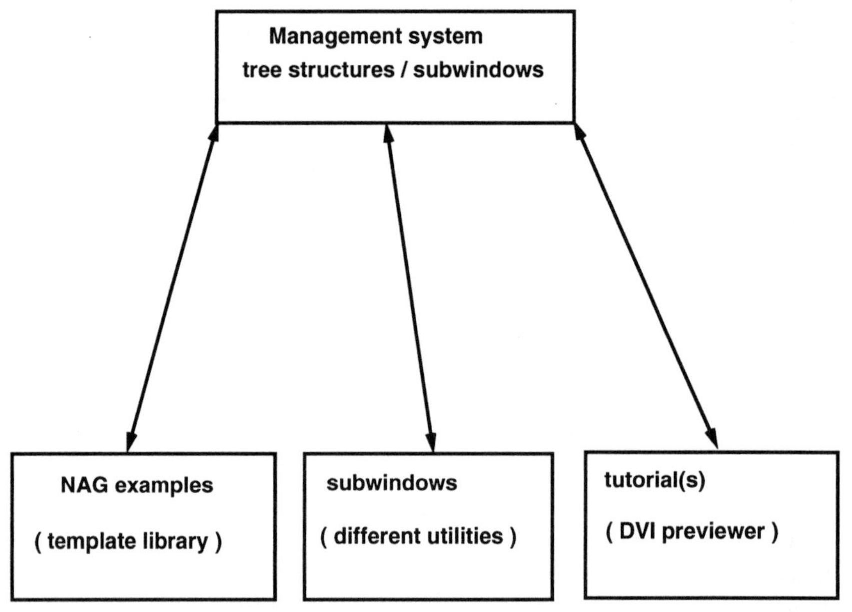

Figure A.2. PAN's components

†An OSF/Motif version and a MS Windows version are at the development stage.

A.1 Working with PAN

On entering PAN the main window becomes visible; see Figure A.3. You can choose a sector of the tutorial via the tutorial's menu; see Figure A.4. The menu's data is represented by the so-called nodes of a hypertext document.

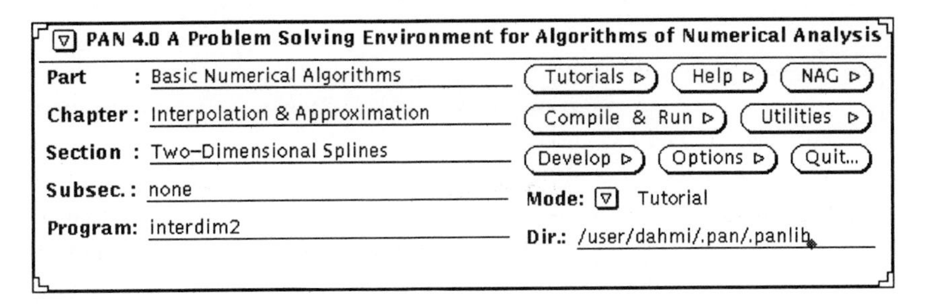

Figure A.3. PAN's main window

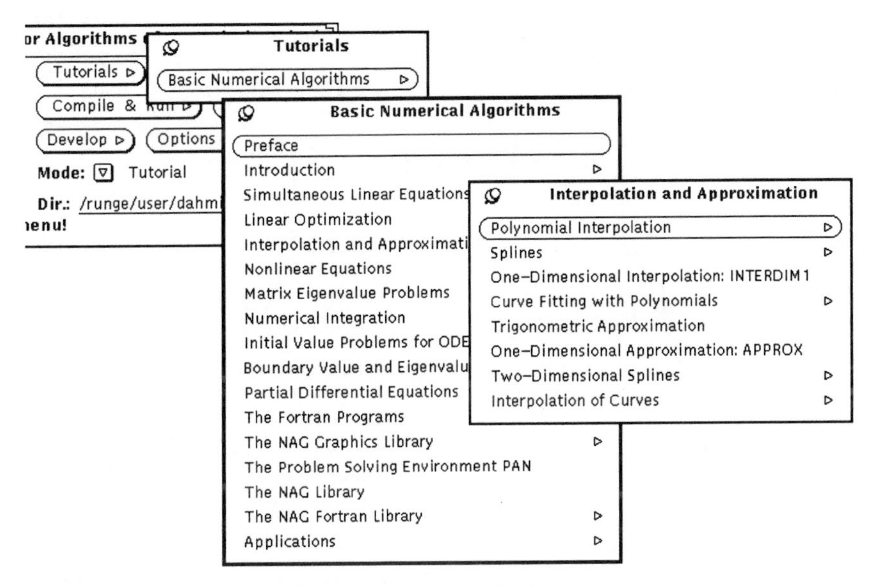

Figure A.4. 'Tutorials' menu

After the node relating to the data chosen has been selected, it will be shown in the hypertext system. An extended OSF/Motif version of HyTEX already exists; see Figure A.5. The document shown comprises information on the specific problem situation and is combined with other parts of the text by so-called references. These references are represented by bounding boxes within the text or by choice buttons in the HyTEX panel. Several browsing and navigating facilities are provided by the system – e.g. a map showing the tree structure of the document.

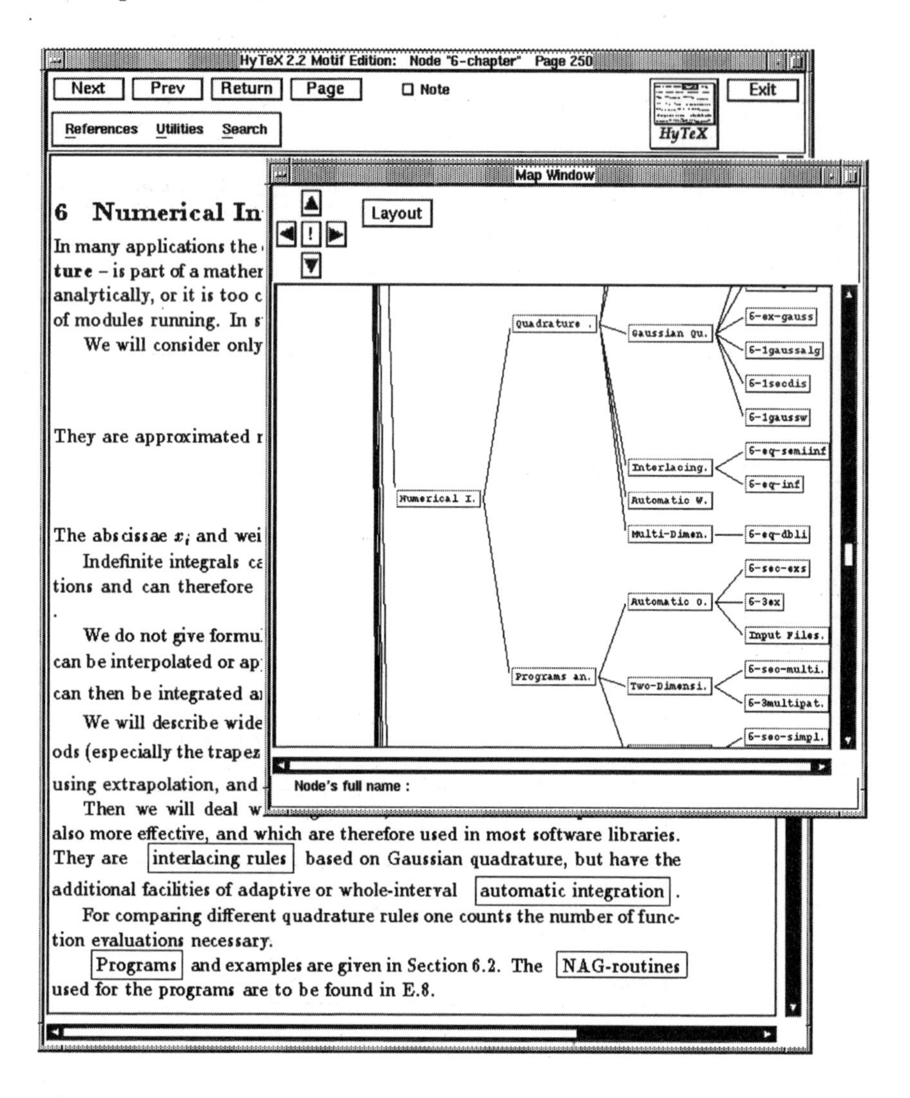

Figure A.5. HyTEX displaying the hypertext version of this book

The HyTEX previewer can be entered via external programs and it automatically reveals specific points in the document. More detailed information on writing HyTEX documents will be provided in Section A.2.

You can determine the structure of the 'Tutorials' menu according to your specific needs and requirements. Your own HyTEX documents can be integrated in the problem-solving environment by means of the 'Develop Contents (Tutorials)'.

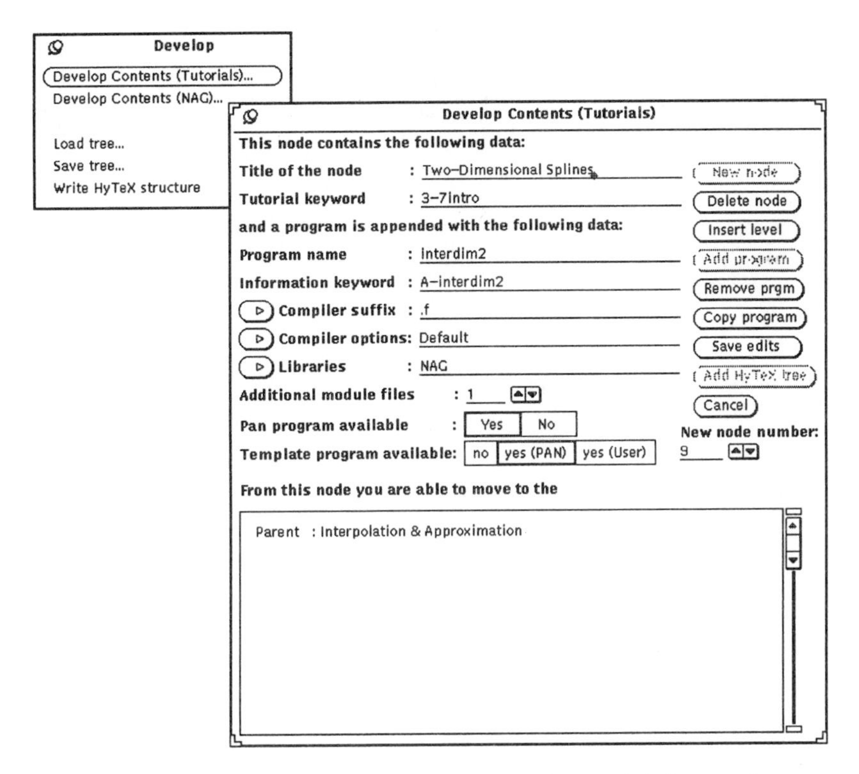

Figure A.6. The 'Develop Contents (Tutorials)' window

The Develop window enables the user to edit registered data, delete or add single nodes or subtrees or to transfer the structure of existing HyTEX documents. Via the 'Develop' menu you can save or load the different tree structures, and you are in a position to add programs to the leaves of the respective tree; see Figure A.6. However, information has to be provided on the compiler used, the options and the libraries required. In this case, pseudonyms are used for all of the options used. Thus unexperienced users can more easily develop their own programs without necessarily having supplementary knowledge about the structure of the UNIX environment. These added programs can be started via the 'Compile & Run' window (see Figure A.7), which allows you to edit, compile or link the various programs.

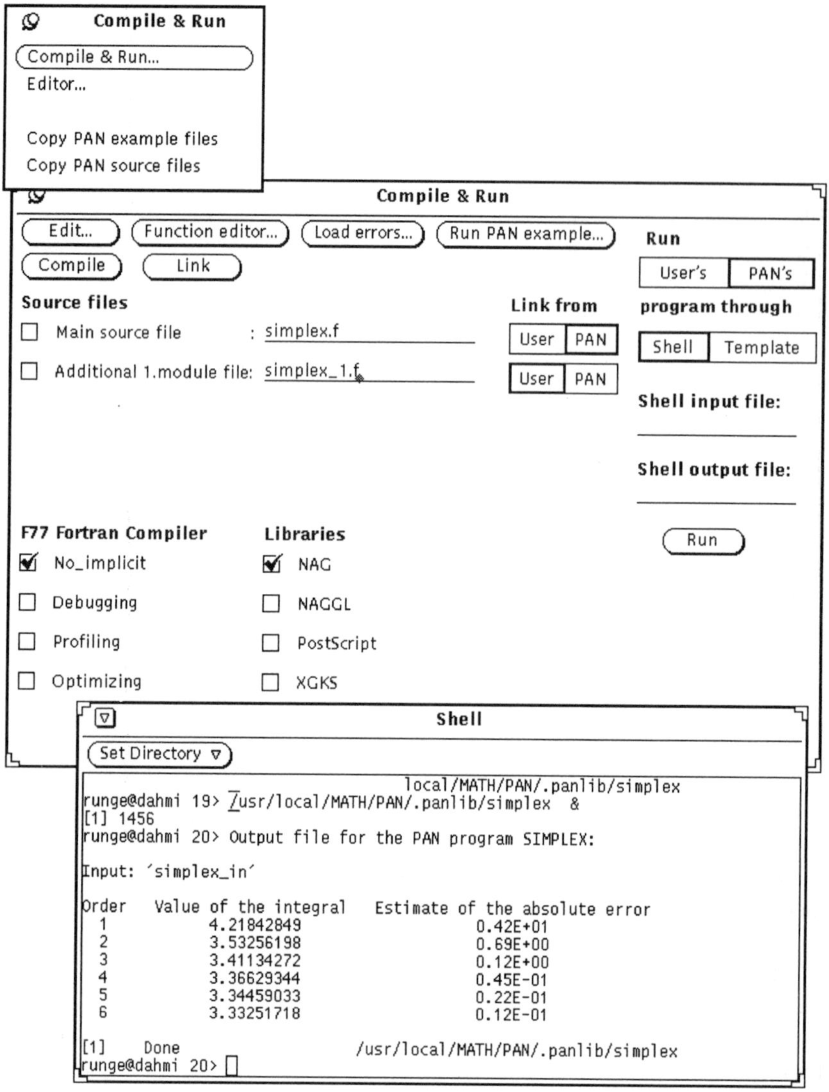

Compile & Run

Compile & Run...
Editor...

Copy PAN example files
Copy PAN source files

Compile & Run

Edit... | Function editor... | Load errors... | Run PAN example... | Compile | Link

Source files
Main source file : simplex.f
Additional 1.module file: simplex_1.f

Link from: User PAN / User PAN

Run: User's PAN's
program through: Shell Template
Shell input file:
Shell output file:
Run

F77 Fortran Compiler: No_implicit, Debugging, Profiling, Optimizing

Libraries: NAG, NAGGL, PostScript, XGKS

Shell

Set Directory

```
                                        local/MATH/PAN/.panlib/simplex
runge@dahmi 19> ~/usr/local/MATH/PAN/.panlib/simplex  &
[1] 1456
runge@dahmi 20> Output file for the PAN program SIMPLEX:

Input: 'simplex_in'

Order   Value of the integral   Estimate of the absolute error
  1           4.21842849                  0.42E+01
  2           3.53256198                  0.69E+00
  3           3.41134272                  0.12E+00
  4           3.36629344                  0.45E-01
  5           3.34459033                  0.22E-01
  6           3.33251718                  0.12E-01

[1]    Done                     /usr/local/MATH/PAN/.panlib/simplex
runge@dahmi 20>
```

Figure A.7. The 'Compile & Run' window

By so doing you can select the required program (either an example program of the Fortran programs already mentioned that are provided by the PAN system or one of your own). The program can be executed by means of a so-called template or in a shell window opened by the 'Compile & Run' window. Figure A.7 shows a program that consists of two modules. The second module ('simplex_1.f') provides a so-called function template

– an example function needed by the main module as input. The function templates can be edited using the 'function editor'; see Figure A.1.

Specific information on NAG routines can be provided through the NAG menu, which incorporates ASCII files, since not all NAG documentation is currently available in LaTeX form.

The compiler options and libraries required for the programs can be modified and extended via the 'Options' menu; see Figure A.8.

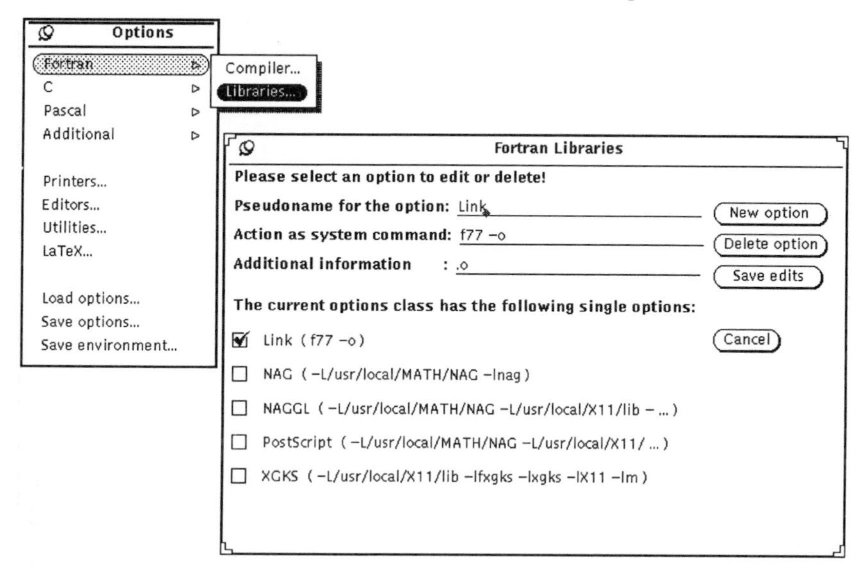

Figure A.8. The 'Options' window

Furthermore, you can choose characteristics of the user interface, such as the printer and editor used or the geometry of the subwindows, according to your needs; they will be stored in a resource file.

The creation of HyTEX documents, screendumps and a filemanager are only some of the services provided by PAN's 'Utilities'.

To help you to work with PAN, a 100-page manual in the form of a hypertext document has been created, which can be entered via the 'Help' menu; see Figure A.9. The services provided by the windows and their different items are described in detail.

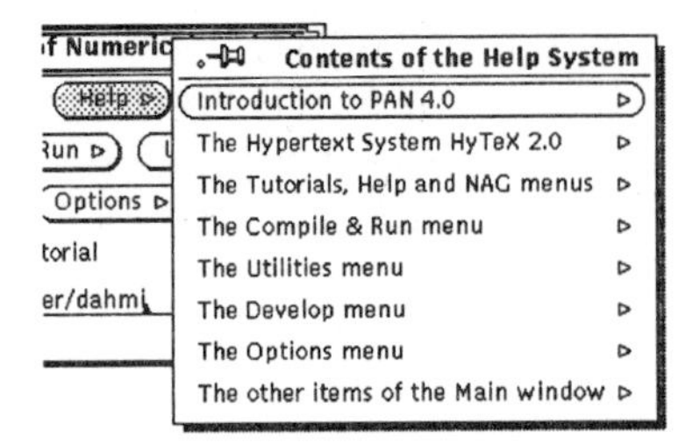

Figure A.9. The 'Help' menu

The experienced user can make use of help shortcuts. These help texts (note form or examples) explain each PAN item and are activated by placing the mouse pointer on the item and pressing the 'Help' key.

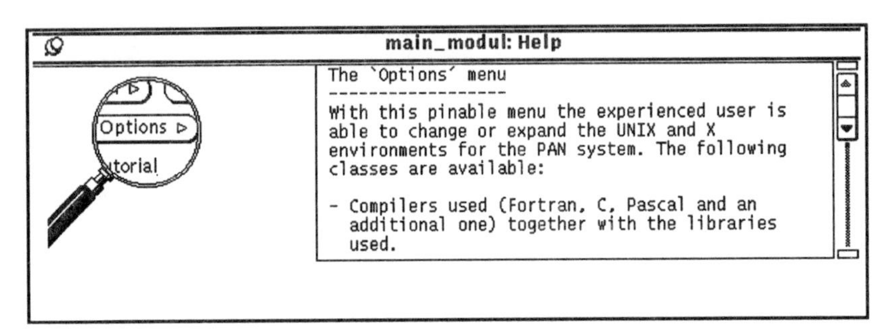

Figure A.10. A 'Help' shortcut

A demonstration (showing various steps automatically) and a guided tour (instructing the user to execute certain tasks independently) will help beginners with their first steps.

A.2 Writing a HyTEX Document

The previewer HyTEX is based upon a SunView application called **dvipage**. In order to make full use of all HyTEX services, a specific LATEX style file is required to translate documents. This style file defines the orders

```
\nodelabel{label}
\subnode(*){label}
\noderef{reference}{text}
\nodepageref{reference}{text}
```

The instructions 'nodelabel' and 'subnode' together with known LATEX structures 'chapter', 'section', 'subsection' and 'subsubsection' define the document nodes.

The position depth of the node in a tree diagram will be determined according to its respective structure. The instructions 'noderef' and 'nodepageref' will provide the defined nodes with references, which can either be integrated in the text (noderef) or in HyTEX's panel (nodepageref). 'Noderef' puts a bounding box around the phrase to be selected. 'Nodepageref' defines a choice button in the panel.

A.3 The Template Library

One of the major components of the PAN system is the template collection. For each of the example programs PAN offers a template program.

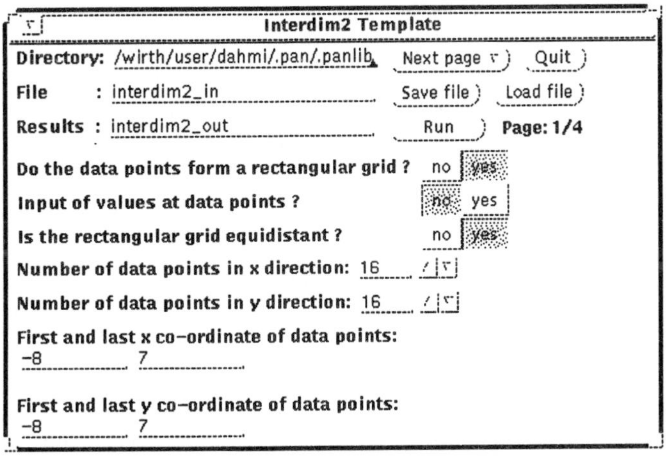

Figure A.11. A template program

These programs enable the user to load, modify and by so doing create new example input files to the example programs as well as to execute them; see Figure A.11.

With the help of the template library (a shared library provided by the PAN system) users can develop templates of their own under the XView toolkit. This will be facilitated by providing various input features (e.g. vectors) and their respective input methods. Thus, if you wish to enter a vector, you have neither to create the input items of the XView toolkit nor to implement the input function. This considerably facilitates the creation of graphical user interfaces for newly developed programs.

A.4 Software

The PAN system has been installed on Sun 4 workstations under SunOS 4 and on a LINUX PC. To install all the different elements of the PAN system the following software is needed:

- PAN management system and HyTEX
    - XView toolkit 2 or 3
    - an X11R4 or X11R5 installation
    - TEX macros, Release 2.09 or later
- Template Library and example programs
    - XView toolkit 2 or 3
    - an X11R4 or X11R5 installation
    - NAG Fortran Library, Mark 15 or later
    - NAG Graphics Library, Mark 3 or later
    - XGKS Library, Release 2.4 or later
- PLTMG
    - PLTMG 7.0

To use PAN, the NAG Libraries must be installed on your system as well as the X Libraries mentioned above. PAN includes the Fortran programs, most of which can be run on any machine as long as the NAG Fortran Library is available. Those programs which use the NAG Graphics Library can be run only if this library and the appropriate graphical interface are available as well. If the X interface is used, and if the XView toolkit is available, then the PSE PAN can be used with all its facilities including the enlarged electronic version of this text as a HyTEX-document, the template programs for comfortable input interfaces for the Fortran programs and all the management tools. For the PSE PAN there is both a demonstration and a guided tour which should facilitate the use of PAN. The total system needs disk space of about 40 MB, but by following the installation guide you need use only parts of it, the Fortran programs for example, and save a lot of the disk space necessary for the graphical parts.

PAN and HyTEX can be obtained under license for a distribution charge only; please contact the author for more information.

Norbert Köckler

FB 17 Mathematik/Informatik
University of Paderborn
D–33095 Paderborn
Germany

PAN-distribution@uni-paderborn.de

Appendix B
The Programs

This Appendix gives a short overview over the programs. It gives neither the input files nor a list of important variables. This information can be found in the headlines of the programs available with our problem-solving environment PAN and in the electronic version of this text. At the end of this Appendix we will show the names of all NAG routines used together with the calling programs in a large table.

But firstly we will give some hints on using the programs and important facts about input and output organization with or without using PAN.

B.0 Conventions

With only one exception we have used real numbers in all our mathematics, algorithms and programming. Mathematically it is often easy to generalize a method to complex numbers, and Fortran 77 contains the type `complex`. But for most applications the real case of the method is sufficient.

Type declaration of parameters of NAG routines depends on the machine and compiler used. We use double precision throughout. Within PAN we use the concise routine documents from the NAG Fortran Library for the description of the NAG routines. In Appendix C we will give an introduction to this Library.

In our programs the dimension of the arrays is in most cases defined using PARAMETER statements. It is therefore easy to change a program for smaller computers or larger applications simply by changing these parameters and re-compiling.

For all programs you will find templates in PAN. They serve as the most convenient way of providing input. If `name` is the name of the program, they produce a temporary input file `_pan_name_in`. If you want to produce an input file without using the template, you may use either this name or `name_in`. The program first looks to see if `_pan_name_in` exists, and then if `name_in` exists.

We have written the programs in standard Fortran 77 using only elementary language elements, so that even restricted compilers will normally be able to compile these programs. There are many comment lines, especially at the head of each program describing

- the purpose of the program,
- the structure of the input files, such that their use should be straightforward without much additional explanation,
- the specification part, mainly the PARAMETER definitions, and
- the NAG routine calls.

While writing these programs we tried to follow the following standards:

1. In comment lines we used ' _ ' before indices and ' ^ ' before exponents.
2. For the input files we always used file input channel 10, and we preferred unformatted input, e.g.: `read(10,*) n`.
3. We did not use `assign`, computed `goto` and arithmetic `if`-commands.

All programs use the NAG Fortran Library, some also use the NAG Graphics Library, and some others also use the XGKS Library. We have marked these additional needs by adding '(GL)' or '(GL, XGKS)' to the title of the program's section. Only PLTMG uses two more libraries, the PLTMG/PAN Library, created with the PLTMG program linked together with the PAN interface programs for PLTMG, and our special GraphEditor Library written as our graphical interface to PLTMG.

B.1 Simultaneous Linear Equations

There are five programs for the solution of a system of simultaneous linear equations

$$Ax = b.$$

The selection should depend mainly on the characteristics of the matrix of coefficients. We will get an overview of the programs from the decision tree shown in Figure B.1.

One of the characteristics of these programs is that no superfluous matrix elements are stored. Therefore it is possible to solve larger problems with SPARSE, where only single non-zero matrix elements are stored, than with BAND, where the necessary banded matrix is stored including zeros within the bandwidth, and so on. It is possible with the programs GAUSS, CHOLESKY and LSS to select as matrix the Hilbert matrix or a matrix whose elements are randomly determined. GAUSS, CHOLESKY and BAND can invert the matrix A as well as solving the linear system for several right-hand sides.

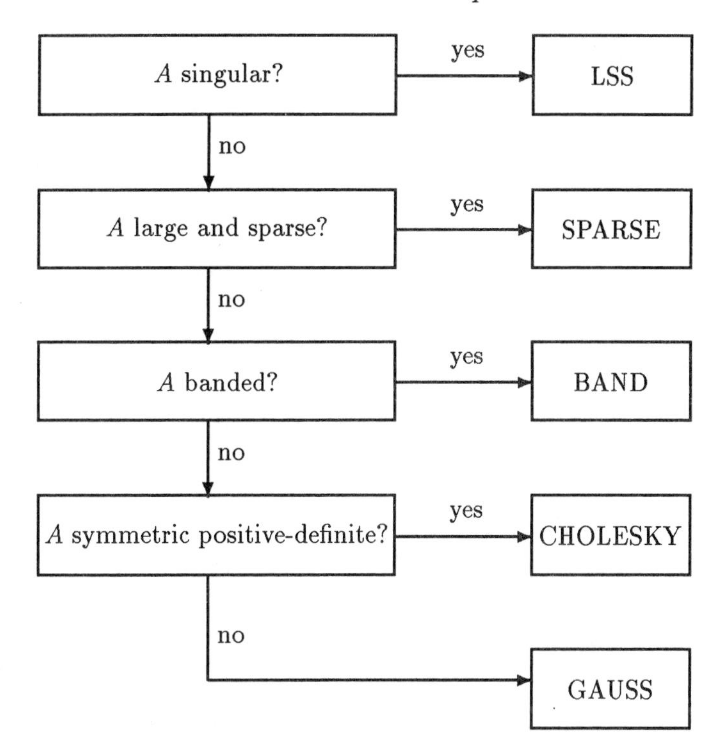

Figure B.1. Decision tree for simultaneous linear equations

B.1.1 GAUSS

GAUSS solves a system of simultaneous linear equations $Ax = b$ with a non-singular quadratic matrix A and one or several right-hand sides b, using Gaussian elimination.

NAG routines used: F04AEF, G05CAF, G05CCF.

B.1.2 CHOLESKY

CHOLESKY solves a system of simultaneous linear equations $Ax = b$ with a symmetric positive-definite matrix A and one or several right-hand sides b, using Cholesky factorization.

NAG routines used: F03AEF, F04AFF, G05CAF, G05CCF, X02AJF.

B.1.3 BAND

BAND solves a system of simultaneous linear equations $Ax = b$ with a band matrix A with $m1$ sub- and $m2$ super-diagonals. In the case of a symmetric positive-definite matrix A may have m main diagonals or a variable row bandwidth $nrow(i), i = 1, ..., n$. Several right-hand sides b are possible.

NAG routines used: F01LBF, F01MCF, F04ACF, F04LDF, F04MCF.

B.1.4 SPARSE

SPARSE solves a system of simultaneous linear equations $Ax = b$ with a sparse quadratic matrix A.

NAG routines used: F01BRF, F01MAF, F04AXF, F04MAF, X02AJF.

B.1.5 LSS

LSS solves homogeneous and non-homogeneous systems of linear equations $Ax = b$ for arbitrary matrices A using the method of least squares.

For homogeneous systems the rank is determined as well as the dimension of the solution space. Then basic vectors of this space are given as results.

For non-homogeneous systems the solution of $Ax = b$ is replaced by one of

$$\min \|Ax - b\|_2, \quad \text{see Section 1.5.}$$

If it is not unique, $\|x\|_2$ is also minimized.

NAG routines used: F04JAF, F04JDF, G05CAF, G05CCF.

B.2 Linear Optimization

This chapter contains only one program solving linear optimization problems with the well known simplex algorithm. It should not be used for large sparse problems.

B.2.1 LOPT

LOPT solves linear optimization problems of the form:
Minimize the linear function

$$z(x) = b_1 x_1 + \cdots + b_n x_n$$

subject to the bounds

$$l_i \leq x_i \leq u_i, \quad i = 1, \ldots, n$$

and linear constraints

$$l_i \leq (A\,x)_i \leq u_i, \quad i = n+1, \ldots, n+m$$

where $m = 0$ is possible and A is a rectangular matrix with m rows, n columns.

LOPT can be used to search only a feasible point. In that case the linear objective function z can be omitted.

The results can be given in short or long form. For the latter case the output form of NAG routine E04MBF is used.

NAG routine used: E04MBF.

B.3 Interpolation and Approximation

The programs for this chapter cover:

- INTERDIM1: One-dimensional interpolation.
- APPROX: One-dimensional approximation.
- INTERDIM2: Two-dimensional interpolation or approximation.
- INTERPAR: m-dimensional curve interpolation.

Unfortunately at the moment one-dimensional approximation is only possible with functions, not with tables of data, but this will be changed during the next few months, and will be available with the next version of PAN.

B.3.1 INTERDIM1 (GL, XGKS)

INTERDIM1 interpolates the table (x_i, y_i), $i = 0, \ldots, n$, alternatively with polynomials, splines or rational functions. In the case of spline or polynomial interpolation it is also possible to evaluate derivatives and integrals of the interpolating function.

The graphical system GKS is used for the output and input of evaluation points and the possible input of values.

NAG routines used: E01AEF, E01BAF, E01RAF, E01RBF, E02AHF, E02AJF, E02AKF, E02BBF, E02BCF, E02BDF, J06APF, J06XAF, J06XBF, J06XFF.

B.3.2 APPROX (GL, XGKS)

APPROX calculates either the Fourier or Chebyshev coefficients of a real function f, and evaluates the series. Therefore let x_i, $i = 0, \ldots, n$, be equidistant points in the first case and Chebyshev points in the second case in an interval $[x_{\min}, x_{\max}]$. Then the table $(x_i, f(x_i))$ is approximated.

The graphical system GKS is used for the output and for the input of evaluation points.

NAG routines used: C06FAF, E02ADF, E02AFF, E02AKF, J06APF, J06XAF, J06XBF, J06XFF, X01AAF.

B.3.3 INTERDIM2

INTERDIM2 solves two different problems depending on the form of input data. Either it determines a smooth interpolating spline surface S from a table $(x_i, y_j, f_{i,j})$, $i = 1, \ldots, n$, $j = 1, \ldots, l$, where the x_i and y_i are monotonically increasing and define a rectangular grid in the x-y-plane, or it determines a smooth approximating spline surface S from a table (x_i, y_i, f_i), $i = 1, \ldots, m$, where (x_i, y_i) are arbitrary points in $\mathbb{R}^2$. In the second case the Euclidean norm of the residual vector

$$\sqrt{\sum_{i=1}^{m}(S(x_i, y_i) - f_i)^2}$$

becomes minimal, and if this least squares solution is not unique, then also the coefficients of the B-spline representation become minimal relative to the Euclidean norm. In addition, NAG routine E02DAF allows weights for this least squares solution, all of which we set to $w_i = 1$.

NAG routines used: E01DAF, E02DAF, E02DEF, E02ZAF, J06AHF, J06HDF, J06WAF, J06WCF, J06WDF, J06WZF, J06XGF.

B.3.4 INTERPAR

INTERPAR interpolates the points $x_k = (x_{1,k}, \ldots, x_{m,k})$, $k = 0, \ldots, n$, with a smooth curve, where the x_k are elements of $\mathbb{R}^m$ with $m \geq 2$. Therefore a parametrization $x_1 = x_1(t), \ldots, x_m = x_m(t)$ is used with

$$\begin{aligned} t_0 &= 0 \\ t_k &= t_{k-1} + \sqrt{(x_{1,k} - x_{1,k-1})^2 + \cdots + (x_{m,k} - x_{m,k-1})^2}, \quad k = 1, \ldots, n. \end{aligned}$$

Then the m tables $(t_k, x_{1,k}), \ldots, (t_k, x_{m,k})$, $k = 0, \ldots, n$, are interpolated by cubic splines, giving a curve $x(t) = (x_1(t), \ldots, x_m(t))^T$.

The graphical system GKS is used for the output and for the possible input of the data points, if $m = 2$.

NAG routines used: E01BAF, E02BBF, J06APF, J06CCF, J06XAF, J06XBF, J06XFF.

B.4 Nonlinear Equations

The programs for this chapter cover:

- DIRECT: One-dimensional root finding, which means: a single root of a real function f is looked for.
- NLIN: Multi-dimensional root finding, which means: a system of non-linear equations is solved, f is a vector-valued function.

B.4.1 DIRECT

DIRECT determines a single root of a continuous real function $f : \mathbb{R} \to \mathbb{R}$, using different iterative methods with the following termination criteria:

$$|x_i - x_{i-1}| \leq \varepsilon \text{ or } |f(x_i)| < \eta.$$

x_i is the approximation of the root s after the ith iteration step. The user can either give a good approximate value for the root or an interval with a sign change of f or a quite arbitrary interval with a step length for a find algorithm.

NAG routines used: C05ADF, C05AGF, C05AJF, X02AJF.

B.4.2 NLIN

NLIN solves a of a system of nonlinear equations, i.e. it determines a single zero of a vector function $f = (f_1, \ldots, f_n)^T$, where f_i are functions depending on n variables $(x_1, \ldots, x_n)$, i.e. $f : \mathbb{R}^n \to \mathbb{R}^n$.

NLIN has a simple and a parameter controlled version. It uses the Jacobian for its iteration process. This can be given via subroutine FM by the user or calculated approximately by NLIN.

NAG routines used: C05NBF, C05NCF, C05PBF, C05PCF, C05ZAF, X02AJF.

B.5 Matrix Eigenvalue Problems

There are five programs in this chapter. They are probably best described by the decision tree on Page 165. Here we only repeat their purpose:

- STDEV: Standard eigenvalue problem without special assumptions.
- SYMEV: Standard eigenvalue problem with symmetric matrix.
- GEV: Generalized eigenvalue problem.
- SPARSEEV: Standard eigenvalue problem with sparse matrix.
- SVD: Singular value decomposition of an arbitrary matrix.

B.5.1 STDEV

STDEV solves the eigenvalue problem $Ax = \lambda x$ with a quadratic matrix $A \in \mathbb{R}^{n,n}$, a complex vector $x \in \mathbb{C}^n$ and a complex scalar $\lambda \in \mathbb{C}$.

NAG routines used: F01AKF, F01APF, F01ATF, F01AUF, F02AQF, G05CAF, G05CCF, X02AJF, X02BHF.

B.5.2 SYMEV

SYMEV solves the eigenvalue problem $Ax = \lambda x$ with a symmetric matrix $A \in \mathbb{R}^{n,n}$, a vector $x \in \mathbb{R}^n$ and a scalar $\lambda \in \mathbb{R}$.

NAG routines used: F01AJF, F02AMF, G05CAF, G05CCF, X02AJF, X02AKF.

B.5.3 GEV

GEV solves the eigenvalue problem $Ax = \lambda Bx$ with quadratic matrices $A, B \in \mathbb{R}^{n,n}$, a complex vector $x \in \mathbb{C}^n$ and a complex scalar $\lambda \in \mathbb{C}$.

NAG routines used: F01AEF, F01AFF, F01AJF, F02AMF, F02BJF, G05CAF, G05CCF, X02AJF, X02AKF.

B.5.4 SPARSEEV

SPARSEEV calculates m eigenvalues and the corresponding eigenvectors of a symmetric sparse matrix A, distinguishing the following cases of eigenvalues to be determined:

- the one of largest absolute value,
- the one of smallest absolute value,
- the ones furthest from a user-defined value σ,
- the ones nearest to a user-defined value σ.

NAG routines used: F02FJF, F04MBF, X02AJF.

B.5.5 SVD

SVD calculates the singular value decomposition

$$A = U\,S\,V^T$$

of an arbitrary real matrix A with m rows and n columns. If $k = \min(m, n)$, then the k first columns of U contain the left-hand singular vectors and the first k rows of V^T the right-hand singular vectors of A. S is a diagonal matrix with k rows and columns.

NAG routines used: F02WEF, G05CAF, G05CCF.

B.6 Numerical Integration

This chapter contains the following four programs:

- ADAPT: one-dimensional adaptive integration. ADAPT allows infinite intervals.
- PATT: one-dimensional automatic whole-interval integration.
- MULTIPATT: two-dimensional integration by repeated application of the one-dimensional automatic whole-interval integration.
- SIMPLEX: multi-dimensional integration over a simplex.

B.6.1 ADAPT

ADAPT performs an adaptive integration of a function $f : \mathbb{R} \to \mathbb{R}$, taken over a finite, semi-infinite or infinite interval $[a, b]$. In the first case the 10-point formula of Gauss and the 21-point formula of Kronrod are used. In the other two cases the corresponding 7- and 15-point formulae are used.

NAG routines used: D01AJF, D01AMF.

B.6.2 PATT

PATT computes the integral of a function $f : \mathbb{R} \to \mathbb{R}$, taken over a finite interval $[a, b]$ using Patterson's method.

NAG routine used: D01ARF.

B.6.3 MULTIPATT

MULTIPATT evaluates the double integral of a function $f : \mathbb{R}^2 \to \mathbb{R}$, using repeated application of Patterson's method.

NAG routine used: D01DAF.

B.6.4 SIMPLEX

SIMPLEX determines a sequence of approximations to the integral of a function $f : \mathbb{R}^n \to \mathbb{R}$, taken over a simplex $S \in \mathbb{R}^n$ with $n + 1$ vertices.

NAG routine used: D01PAF.

B.7 Initial Value Problems

The programs of this chapter solve the initial value problem for a system of ordinary differential equations

$$\begin{aligned} \frac{dy}{dx} &= f(x, y) \\ \text{with initial condition} & \\ y(x_0) &= y_0 \end{aligned} \tag{B.1}$$

with a real independent variable x and a vector $y = (y_1, .., y_n)$ of real functions $y_i = y_i(x) : \mathbb{R} \to \mathbb{R}$.

B.7.1 RUKU (GL)

RUKU solves problem (B.1) using a Runge–Kutta method.

NAG routines used: D02PAF, J06AAF, J06AHF, J06CCF, J06WAF, J06WBF, J06WCF, J06WDF, J06WZF, J06XGF, J06YRF, X02AJF.

B.7.2 ADAMS (GL)

ADAMS solves problem (B.1) using a method of Adams type. This method is recommended for high accuracy demands, large systems or long intervals.

NAG routines used: D02QGF, D02QWF, J06AAF, J06AHF, J06CCF, J06WAF, J06WBF, J06WCF, J06WDF, J06WZF, J06XGF, J06YRF, X02AJF.

B.7.3 BDF (GL)

BDF solves problem (B.1) using a backward differentiation formulae (BDF). This method is recommended for weakly stiff systems.

NAG routines used: D02QDF, D02QQF, J06AAF, J06AHF, J06CCF, J06WAF, J06WBF, J06WCF,J06WDF J06WZF, J06XGF, J06YRF, X02AJF.

B.7.4 STIFFTEST

STIFFTEST solves problem (B.1) like the other three programs in this chapter. But STIFFTEST especially estimates the global discretization error of the Runge–Kutta method used and it estimates the stiffness of the system.

A stiffness value is computed at the end of each integration step. This value is in the range $[0, 1]$, and, in general, the larger the value the stiffer the system of differential equations. Normally, for a stiff system the stiffness value increases as the integration proceeds. If the value exceeds 0.75 in the second half of the integration interval, then the system should be integrated using a BDF method. Even if the stiffness value is greater than 0.5, the use of a BDF method may save computation time. If the final stiffness value is much smaller than the values obtained during the integration then the problem may not be stiff. Further checks should be made, possibly by varying the tolerance. Note that, in any case, the returned stiffness value depends on the tolerance and usually decreases if the tolerance is increased. If the value is small enough, the computed solution is correct, because a Runge–Kutta method is used.

NAG routines used: D02BDF, X02AJF.

B.8 Boundary Value Problems: BVP

The two programs of this chapter solve the boundary value problem for a system of ordinary differential equations

$$\frac{dy}{dx} = f(x, y) \tag{B.2}$$

with real variable $x \in [a, b]$, and vector $y = (y_1, .., y_n) \in \mathbb{R}^n$ of functions $y_i : \mathbb{R} \to \mathbb{R}$ using two different methods. In both cases we have to define $2n$ possible boundary values; n of them must be given as known, the other n must be given as estimated.

The two methods used are quite different, so in delicate problems only one may be successful.

B.8.1 DIFFERENCE (GL)

DIFFERENCE solves (B.2) using a finite difference method.

NAG routines used: D02GAF, J06AAF, J06AHF, J06CCF, J06WAF, J06WBF, J06WCF, J06WDF, J06WZF, J06XGF, J06YRF.

B.8.2 SHOOTING (GL)

SHOOTING solves (B.2) using a shooting method, where a Runge–Kutta–Merson method is used for the initial value problems.

NAG routines used: D02HAF, J06AAF, J06AHF, J06CCF, J06WAF, J06WBF, J06WCF, J06WDF, J06WZF, J06XGF, J06YRF.

B.9 Partial Differential Equations

B.9.1 PLTMG (GL, GraphEditor, XGKS)

PLTMG is a package for solving boundary value problems of the form:

$$-\nabla\left(a(x,y,u,\nabla u,\lambda)\right)+f(x,y,u,\nabla u,\lambda)=0 \quad \text{in } \Omega$$

with the boundary conditions

$$\begin{aligned} u &= g_1(x,y,\lambda) \quad \text{on } \partial\Omega_1 \\ a\cdot n &= g_2(x,y,\lambda) \quad \text{on } \partial\Omega_2=\partial\Omega-\partial\Omega_1. \end{aligned}$$

Here Ω is connected domain in $\mathbb{R}^2$, n is the outer unit normal, a is the vector $(a_1,a_2)^T$ and a_1, a_2, f, g_1 and g_2 are scalar functions. λ is a scalar continuation parameter. In addition, the user may specify a functional of the solution of the form

$$\rho(u,\lambda)=\int_\Omega p_1(x,y,u,\nabla u,\lambda)\,dx\,dy+\int_{\partial\Omega} p_2(x,y,u,\nabla u,\lambda)\,ds$$

with scalar functions p_1 and p_2.

B.10 Program Overview

Program	NAG routines called
GAUSS	F04AEF, G05CAF, G05CCF
CHOLESKY	F03AEF, F04AFF, G05CAF, G05CCF, X02AJF
BAND	F01LBF, F01MCF, F04ACF, F04LDF, F04MCF
SPARSE	F01BRF, F01MAF, F04AXF, F04MAF, X02AJF
LSS	F04JAF, F04JDF, G05CAF, G05CCF
LOPT	E04MBF
INTERDIM1	E01AEF, E01BAF, E01RAF, E01RBF, E02AHF, E02AJF, E02AKF, E02BBF, E02BCF, E02BDF, J06APF, J06XAF, J06XBF, J06XFF
APPROX	C06FAF, E02ADF, E02AFF, E02AKF, J06APF, J06XAF, J06XBF, J06XFF, X01AAF
INTERDIM2	E01DAF, E02DAF, E02DEF, E02ZAF, J06AHF, J06HDF, J06WAF, J06WCF, J06WDF, J06WZF, J06XGF
INTERPAR	E01BAF, E02BBF, J06APF, J06CCF, J06XAF, J06XBF, J06XFF
DIRECT	C05ADF, C05AGF, C05AJF, X02AJF
NLIN	C05NBF, C05NCF, C05PBF, C05PCF, C05ZAF, X02AJF
STDEV	F01AKF, F01APF, F01ATF, F01AUF, F02AQF, G05CAF, G05CCF, X02AJF, X02BHF
GEV	F01AEF, F01AFF, F01AJF, F02AMF, F02BJF, G05CAF, G05CCF, X02AJF, X02AKF
SYMEV	F01AJF, F02AMF, G05CAF, G05CCF, X02AJF, X02AKF
SPARSEEV	F02FJF, F04MBF, X02AJF
SVD	F02WEF, G05CAF, G05CCF

Program	NAG routines called
ADAPT	D01AJF, D01AMF
PATT	D01ARF
MULTIPATT	D01DAF
SIMPLEX	D01PAF
RUKU	D02PAF, J06AAF, J06AHF, J06CCF, J06WAF, J06WBF, J06WCF, J06WDF, J06WZF, J06XGF, J06YRF, X02AJF
ADAMS	D02QGF, D02QWF, J06AAF, J06AHF, J06CCF, J06WAF, J06WBF, J06WCF, J06WDF, J06WZF, J06XGF, J06YRF, X02AJF
BDF	D02QDF, D02QQF, J06AAF, J06AHF, J06CCF, J06WAF, J06WBF, J06WCF, J06WDF, J06WZF, J06XGF, J06YRF, X02AJF
STIFFTEST	D02BDF, X02AJF
SHOOTING	D02HAF, J06AAF, J06AHF, J06CCF, J06WAF, J06WBF, J06WCF, J06WDF, J06WZF, J06XGF, J06YRF
DIFFERENCE	D02GAF, J06AAF, J06AHF, J06CCF, J06WAF, J06WBF, J06WCF, J06WDF, J06WZF, J06XGF, J06YRF

Appendix C
The NAG Fortran Library

All NAG routines used in our programs are described in PAN in machine-readable concise form organized through a cascade menu. Here we will give an introduction to the NAG Fortran Library; this is based on the Essential Introduction in the Fortran Library Manual, Mark 16, and the Introduction in the Fortran Library Concise Reference, Mark 14.

C.1 Introduction

The NAG Fortran Library is a comprehensive collection of Fortran 77 **routines** for the solution of numerical and statistical problems. The word 'routine' is used to denote 'subroutine' or 'function'.

The Library is divided into chapters, each devoted to a branch of numerical analysis or statistics; for details see Section C.2.1. The **NAG Fortran Library Manual** is the principal documentation for the Library. It has the same chapter structure as the Library. The chapters occur in alphanumeric order. General introductory documents and indexes are placed at the beginning of the Manual. A list of all chapters can be found in Section C.2.3. Each chapter consists of the following documents:

- **Chapter Introduction**, e.g. Introduction – D01;
- **Chapter Contents**, e.g. Contents – D01;
- **routine documents**, one for each documented routine.

A routine document has the same name as the routine which it describes. It is divided into nine sections, which are described in Section C.2.2. The concise routine documents contained in the 'On-Line Information Supplement' (see Section C.1.1) as well as in PAN give details of

- the purpose of the routine;
- its parameter list;
- a summary of the specification of each parameter; and
- a summary of the meaning of each error-exit,

with the parameter specifications consisting of

- a number indicating the position of the parameter in the argument-list;
- the parameter class (e.g. input, output, workspace etc.);
- the type of the parameter and whether it is a scalar or an array;
- the parameter name; and
- a description.

C.1.1 Supplementary Documentation

In addition to the full Manual, NAG provides the following alternative forms of documentation:

- the **Introductory Guide**
- the **On-line Information Supplement**

These forms of documentation follow the same structure as the Manual.

The **Introductory Guide** contains all the general introductory documents, indexes, chapter introductions and chapter contents, from the full Manual.

The **On-line Information Supplement** is a machine-based 'Help' system, which describes the subject areas covered by the Library, advises on the choice of routines, and gives essential programming details for each documented routine. It contains a machine-readable version of Sections 1, 2, 5 and 6 of each routine document.

C.1.2 Marks of the Library

Periodically a new **Mark** of the NAG Fortran Library is released: new routines are added, corrections or improvements are made to existing routines; occasionally routines are withdrawn if they have been superseded by improved routines.

At each Mark, the documentation of the Library is updated. Marks are numbered; Mark 16 was released in Winter 1993/94.

C.1.3 Notation

To illustrate the notation used in the routine documents, let us consider a typical NAG Fortran Library routine such as F04AEF, which is the first NAG routine to be mentioned in this text.

Each parameter is classified as to whether it is input, output or otherwise. The following is a list of all the classifications with their meanings.

Input parameter: you must – or may need to – assign values to these parameters on or before entry to the routine, and these values are unchanged on exit from the routine (e.g. the parameter N in F04AEF).

Output parameter: you need not assign values to these parameters on or before entry to the routine. The routine may assign values to them (e.g. the parameter C in F04AEF).

Input/output parameter: you must – or may need to – assign values to these parameters on or before entry to the routine, and the routine may then change these values (e.g. the parameter IFAIL in F04AEF).

Workspace: array parameters which are used as workspace by the routine. You must supply arrays of the correct type and dimension, but you need not be concerned with their contents (e.g. the parameter WKSPCE in F04AEF).

User workspace: array parameters which are passed by the Library routine to an external procedure parameter. They are not used by the routine, but you may use them to pass information between your calling program and the external procedure.

Dummy: a simple variable which is not used by the routine. A variable or constant of the correct type must be supplied, but its value need not be set. (A dummy parameter is usually a parameter which was required by an earlier version of the routine and is retained in the parameter-list for compatibility.)

External procedure: a subroutine or function which must be supplied. Usually it must be supplied as part of your calling program, in which case its specification includes full details of its parameter-list and specifications of its parameters (all enclosed in a box). Its parameters are classified in the same way as those of the Library routine, but because you must write the procedure rather than call it, the significance of the classification is different.

Input: values may be supplied on entry, which your procedure must not change.

Output: you may or must assign values to these parameters before exit from your procedure.

Input/Output: values may be supplied on entry, and you may or must assign values to them before exit from your procedure.

Occasionally the procedure can be supplied from the NAG Fortran Library, and then you only need to know its name.

The Manual uses bold italics to distinguish terms which have different interpretations in different implementations, such as

real for ***real*** or ***double precision***, and
complex for ***complex*** or ***complex*16*** (or equivalent)

Another important bold italicized term is ***machine precision***, which denotes the relative precision to which floating-point numbers are stored in the computer, e.g. in an implementation with approximately 16 decimal digits of precision, ***machine precision*** has a value of approximately 10^{-16}. The precise value is returned by X02AJF.

The parameter name is followed by a specification for the parameter. An array is indicated by the parameter name having brackets and appropriate dimensions (see the parameters A(IA,N) and WKSPCE(N) in F04AEF). For arguments of type CHARACTER the associated length is appended to the argument name.

A constraint or constraints in the specification of an Input parameter is usually indicated by an expression such as N$\geq$ 0 in F04AEF. If the routine is called with an invalid value for the parameter (e.g. N $<$ 0), the routine will usually take an error exit, returning a nonzero value of IFAIL.

The phrase 'suggested value:' introduces a suggestion for a reasonable initial setting for an Input parameter (e.g. accuracy or maximum number of iterations) in case you are unsure what value to use; you should be prepared to use a different setting if the suggested value turns out to be unsuitable for your problem.

In a few instances the name of a procedure which the user must supply is fixed and does not appear in the argument list. For these cases the procedure is specified after the argument list and before the error message summaries.

C.1.4 Conventions, Routine Names and Precision

The NAG Fortran Library is developed in both single precision and double precision versions. On most systems only one precision of the Library is available; the precision chosen is that which is considered most suitable in general for numerical computation (double precision on most systems).

On some systems both precisions are provided: in this case, the double precision routines have names ending in F (as in the documentation), and the single precision routines have names ending in E. Thus in DEC VAX/VMS implementations:

D01AJF is a routine in the double precision implementation;
D01AJE is the corresponding routine in the single precision implementation.

The F06 Chapter (Linear Algebra Support Routines) contains all the Basic Linear Algebra Subprograms (BLAS) with NAG-style names as well as with the actual BLAS names, e.g. F06AAF (SROTG/DROTG). The names in brackets are the equivalent single and double precision BLAS names respectively. The F07 Chapter (Linear Equations (LAPACK)) and the F08 Chapter (Least-squares and Eigenvalue Problems (LAPACK)) contain routines derived from the LAPACK project. Like the BLAS, these routines have NAG-style names as well as LAPACK names, e.g. F07ADF

(SGETRF/DGETRF). Details regarding these alternate names can be found in the relevant Chapter Introductions.

You must know which implementation, which precision and which Mark of the Library you are using or intend to use. To find out which implementation, precision and Mark of the Library is available at your site, you can run a program which calls the Library routine A00AAF (or A00AAE in some single precision implementations). This routine has no parameters; it simply outputs text to the Library advisory message unit. An example of the output is:

```
*** Start of NAG Library implementation details ***
Implementation title: SUN 4
           Precision: double
        Product Code: FLSOL16D
                Mark: 16A
*** End of NAG Library implementation details ***
```

(The implementation code can be ignored, except possibly when communicating with NAG; see Section C.4.2.)

In addition to those Library routines which are documented and are intended to be called by users, the Library also contains many auxiliary routines. In general, you need not be concerned with them at all, although you may be made aware of their existence if, for example, you examine a memory map of an executable program which calls NAG routines. The only exception is that when calling some Library routines, you may be required or allowed to supply the name of an auxiliary routine from the Library as an external procedure parameter. The routine descriptions give the necessary details. In such cases, you only need to supply the name of the routine; you never need to know details of its parameter-list.

NAG auxiliary routines have names which are similar to the name of the documented routine(s) to which they are related, but with last letter Z, Y, and so on, e.g. D01BAZ is an auxiliary routine called by D01BAF.

C.1.5 Input/Output in the Library

Most NAG Fortran Library routines perform no output to an external file, except possibly to output an error message. All error messages are written to a logical error message unit. This unit number (which is set by default to 6 in most implementations) can be changed by calling the Library routine X04AAF.

All output from the Library is formatted.

Some NAG Fortran Library routines may optionally output their final results, or intermediate results to monitor the course of computation. All output other than error messages is written to a logical advisory message unit. This unit number (which is also set by default to 6 in most implementations) can be changed by calling the Library routine X04ABF. Although it is logically distinct from the error message unit, in practice the two unit numbers may be the same.

There are only a few Library routines which perform input from an external file: the unit number is always a parameter to the routine, and all input is formatted.

You must ensure that the relevant Fortran unit numbers are associated with the desired external files, either by an OPEN statement in your calling program, or by operating system commands.

C.1.6 Error Handling and the Parameter IFAIL

NAG Fortran Library routines may detect various kinds of error, failure or warning conditions. Such conditions are handled in a systematic way by the Library. They fall roughly into three classes:

(i) an invalid value of a parameter on entry to a routine;

(ii) a numerical failure during computation (e.g. approximate singularity of a matrix, failure of an iterative process to converge);

(iii) a warning that although the computation has been completed, the results cannot be guaranteed to be completely reliable.

All three classes are handled in the same way by the Library, and are all referred to here simply as errors.

The error-handling mechanism uses the parameter IFAIL , which occurs in the calling sequence of most Library routines (almost always it is the last parameter). IFAIL serves two purposes:

(i) it allows users to specify what action a Library routine should take if it detects an error;

(ii) it reports the outcome of a call to a Library routine, either success or failure (IFAIL $\neq 0$, with different values indicating different reasons for the failure, as explained in Section 6 of the routine document).

For this purpose IFAIL must be assigned a value before calling the routine; since IFAIL is reset by the routine, it must be passed as a variable, not as an integer constant (see C.2.2, 6).

C.1.7 Array Parameters

Most array parameters have dimensions which depend on the size of the problem. In Fortran terminology they have adjustable dimensions: the dimensions occurring in their declarations are integer variables which are also parameters of the Library routine.

For example, a Library routine might have the specification:

```
SUBROUTINE <name> (M, N, A, B, LDB)
INTEGER       M, N, A(N), B(LDB,N), LDB
```

For a one-dimensional array parameter, such as A in this example, you must ensure that the dimension of the array, as declared in your calling (sub)program, is at least as large as the value you supply for N. It may be larger; but the routine uses only the first N elements.

For a two-dimensional array parameter, such as B in the example, you must supply the first dimension of the array B, as declared in your calling (sub)program, through the parameter LDB, even though the number of rows actually used by the routine is determined by the parameter M. You must ensure that the first dimension of the array is at least as large as the value you supply for M. You must also ensure that the second dimension of the array, as declared in your calling (sub)program, is at least as large as the value you supply for N. It may be larger, but the routine only uses the first N columns.

Many NAG routines are designed so that they can be called with a parameter like N in the above example set to 0 (in which case they would usually exit immediately without doing anything). If so, the declarations in the Library routine would use the assumed size array dimension, and would be given as:

```
INTEGER        M, N, A(*), B(LDB,*), LDB
```

However, the original declaration of an array in your calling program must always have constant dimensions, greater than or equal to 1.

C.1.8 Programming Advice

The NAG Fortran Library and its documentation are designed on the assumptions that users know how to write a calling program in Fortran.

When programming a call to a routine, read the routine document carefully, especially the description of the **Parameters**. This states clearly which parameters must have values assigned to them on entry to the routine, and which return useful values on exit. The most common types of programming errors in using the Library are connected to incorrect use of parameters.

If a call to a Library routine results in an unexpected error message from the system (or possibly from within the Library), **check** the following.

- Has the routine been called with the correct number of parameters?
- Do the parameters all have the correct type?
- Have all array parameters been dimensioned correctly?

- Is your program in the same precision as the NAG Library routines to which your program is being linked?

C.2 The Construction of the NAG Fortran Library

C.2.1 Systematics

The name of a routine consists (with very few exceptions) of six letters, which form three parts, for example for F04AEF:

$$\underbrace{\mathrm{F\,0\,4}}_{(1)}\quad\underbrace{\mathrm{A\,E}}_{(2)}\quad\underbrace{\mathrm{F}}_{(3)}$$

These three parts have the following meaning:

1. The first three letters define the **Chapter** that the routine belongs to, here 'Chapter F04'. The chapters are listed in Section C.2.3.
2. These two digits uniquely define the routine within the chapter.
3. The last letter is mostly 'F' (for double precision versions) or 'E' (for single precision versions), but is sometimes different for service or system routines not described in the manual.

C.2.2 The Routine Document Sections

1. **Purpose**
   One or two sentences define the purpose of the routine precisely.
2. **Specification**
   The head lines of the routine are given with all its parameters and their specifications.
3. **Description**
   Now the purpose is described in more detail. The algorithms are mentioned, the most important formulae are given, and references are given for users interested in the mathematical or algorithmic background.
4. **References**
   References to background literature are given.
5. **Parameters**
   All parameters are described in separate sections. They are classified as described above. Relations between parameters are given as well as constraints.
6. **Error Indicators**
   The parameter IFAIL is always the error indicator. Usually it has one of the values 0, 1 or −1 with the following meaning:

IFAIL = 0	Hard Fail Option Execution is interrupted, and an error message is given containing the value of IFAIL. The channel for error messages can be defined by routine X04AAF.
IFAIL = −1	Soft Fail Option Execution is interrupted, an error message is given containing the value of IFAIL, and control is given back to the calling program.
IFAIL = 1	Soft Fail Option As with IFAIL=−1, but without error message.

After every error interruption the parameter IFAIL contains an error flag value, or else it contains the value IFAIL=0. With the 'Soft Fail Option' it is important to check the value of IFAIL before the program takes up further actions. We always have used the 'Soft Fail Option' in our programs.

7. **Accuracy**
   Whenever possible, results about the expected precision of results are given here. They most often depend on the machine precision as well as on routine parameters.
8. **Further Comments**
9. **Example**
   For each routine you will find a small example program with input data and results which clarify the usage of the routine.

C.2.3 The Library Chapters

Chapter A02	Complex Arithmetic
Chapter C02	Zeros of Polynomials
Chapter C05	Roots of One or More Transcendental Equations
Chapter C06	Summation of Series
Chapter D01	Quadrature
Chapter D02	Ordinary Differential Equations
Chapter D03	Partial Differential Equations
Chapter D04	Numerical Differentiation
Chapter D05	Integral Equations
Chapter E01	Interpolation
Chapter E02	Curve and Surface Fitting
Chapter E04	Minimizing or Maximizing a Function
Chapter F01	Matrix Operations, Including Inversion
Chapter F02	Eigenvalues and Eigenvectors
Chapter F03	Determinants
Chapter F04	Simultaneous Linear Equations
Chapter F05	Orthogonalization

13: IBB – INTEGER. *Input*
On entry: the first dimension of the array BB as declared in the (sub)program from which F04AEF is called.
Constraint: IBB $\geq$ max(1,N).

14: IFAIL – INTEGER. *Input/Output*
On entry: IFAIL must be set to 0, -1 or 1. For users not familiar with this parameter (described in Chapter P01 in the NAG Fortran Library Manual or this HELP system) the recommended value is 0.
On exit: IFAIL = 0 unless the routine detects an error (see Section D).

D. Error Indicators and Warnings

Errors detected by the routine:

If on entry IFAIL = 0 or -1, explanatory error messages are output on the current error message unit (as defined by X04AAF).

IFAIL = 1
The matrix A is singular, possibly due to rounding errors.

IFAIL = 2
Iterative refinement fails to improve the solution, i.e. the matrix A is too ill-conditioned.

IFAIL = 3
On entry, N $<$ 0,
or M $<$ 0,
or IA $<$ max(1,N),
or IB $<$ max(1,N),
or IC $<$ max(1,N),
or IAA $<$ max(1,N),
or IBB $<$ max(1,N).

C.4 Concluding Remarks

C.4.1 Summary for New Users

If you are unfamiliar with the NAG Library and are thinking of using a routine from it, please follow these instructions.

1. Read the whole of the **Essential Introduction**.
2. Consult the **Contents Summary, KWIC Index** or **GAMS Index** to choose an appropriate chapter or routine.
3. Read the relevant **Chapter Introduction**.
4. Choose a routine, and read the **routine document**. If the routine does not after all meet your needs, return to steps 2 or 3.

5. Read the **Users' Note** for your implementation.
6. Consult local documentation, which should be provided by your local support staff, about access to the NAG Library on your computing system.

C.4.2 Contact between Users and NAG

For further advice or communication about the NAG Fortran Library, you should first turn to the staff of your local computer installation. This covers such matters as:

- obtaining a copy of the Users' Note for your implementation;
- obtaining information about local access to the Library;
- seeking advice about using the Library;
- reporting suspected errors in routines or documents;
- making suggestions for new routines or features;
- purchasing NAG documentation.

Your installation may have advisory and/or information services to handle such enquiries. In addition NAG asks each installation mounting the Library to nominate a NAG site representative, who may be approached directly in the absence of an advisory service. Site representatives receive information from NAG about confirmed errors, the imminence of updates, and so on, and will forward users' enquiries to the appropriate person in the NAG organization if they cannot be dealt with locally. If you are unable to make contact with your local site representative please contact NAG directly. Full details of the NAG Response Centers are given in the Users' Note.

Appendix D
The NAG Graphics Library

The NAG Graphics Library is a collection of over 100 carefully designed routines which provides the Fortran 77 programmer with a convenient and versatile means of producing a graphical representation of numerical and statistical results. It is available for a wide spectrum of machines from supercomputers to personal computers.

Logically, the Graphics Library is composed of two types of routines: high level and low level. The high level routines are written in terms of a small set of primitive, or low level routines, known collectively as the NAG Graphical Interface. There is a version of the NAG Graphical Interface for each plotting package or protocol supported (e.g. GKS, Adobe PostScript). The Interface translates drawing requests into a form which can be interpreted by the underlying plotting software. A number of interfaces for screen and hardcopy are supported by NAG as part of the standard service.

Although the NAG Graphics Library has been designed for use primarily by Fortran 77 programmers, it is possible to access the full functionality of the Library from C. An optional set of C header files may be purchased for use with the Library.

We use the NAG Graphics Library on a SUN SPARCstation under the X Window System for display graphics as well as for PostScript paper plots. Our programs INTERDIM1, INTERDIM2, INTERPAR, APPROX, RUKU, ADAMS, BDF, DIFFERENCE and SHOOTING include calls to the NAG Graphics Library. Some of them, which allow graphical input, also include calls directly to the underlying XGKS Library. To write programs with such a mixture you should be a sophisticated user of both the NAG Graphics Library and GKS. The advantage of a library like NAG Graphics Library over a kernel system like GKS is the bundling of several elementary routine calls into one. Most of the figures in this text have been generated using the NAG Graphics Library.

First we will describe the structure of the library with a simple example program, then go through the chapters of the library giving a short introduction to each of them.

D.1 An Example Program Using the NAG Graphics Library

Your program must contain the following sequence of actions.

1. Initialize the graphical output device. For our example we assume that you have to call subroutine XXXXXX to perform the initialization of your chosen graphical device. Then you have to initialize the NAG Graphics Library by calling subroutine J06WAF.
2. Define a plotting surface, a data region or window, and a graphical viewport by calls to J06WBF and J06WCF.
3. Draw the actual picture by certain calls to J06 routines.
4. Complete your picture
   - *either* by closing the graphical interface by calling J06WZF,
   - *or* by clearing the screen (or taking fresh paper) and selecting a new frame for another picture by calling J06WDF. Then you start again with 2.

For further details we refer the reader to the documentation which is supplied with the NAG Graphics Library. We will look at the NAG Graphics Library parts of a simple drawing program, where you can easily see what is meant by the different steps.

```
C      Initialize your graphical output device
       CALL XXXXXX
C      Initialize the NAG Graphics Library
       CALL J06WAF
C      -----------------------------------
C      Define a data region,
C      specifying a margin for annotation
       MARGIN = 1
       XMIN =   0.0D0
       XMAX =  10.0D0
       YMIN =   1.0D0
       YMAX =   5.0D0
       CALL J06WBF (XMIN,XMAX,YMIN,YMAX,MARGIN)
C      Set the current viewport in normalized
C      device co-ordinates
       CALL J06WCF (0.0D0,1.0D0,0.0D0,1.0D0)
C      -----------------------------------
C      Select high character and marker quality
       CALL J06XFF (2)
C      Draw a scaled border to fit the data region
       CALL J06AEF
C      Draw a smooth curve through a set of six
```

```
C       data points (xx,yy) with Bessel's method
        METHOD=1
        N = 6
        IFAIL = 0
        CALL J06CAF (XX,YY,N,METHOD,IFAIL)
C       Mark the points
C       Symbols only:
        LINE = 0
C       Define a plotting symbol to mark the data points
        MARKER = 2
        IFAIL = 0
        CALL J06BAF (XX,YY,N,LINE,MARKER,IFAIL)
C       Draw a title within the plot margin
        CALL J06AHF ('Bessel curve')
C       ------------------------------------
C       Terminate NAG Graphics Library
        CALL J06WZF
```

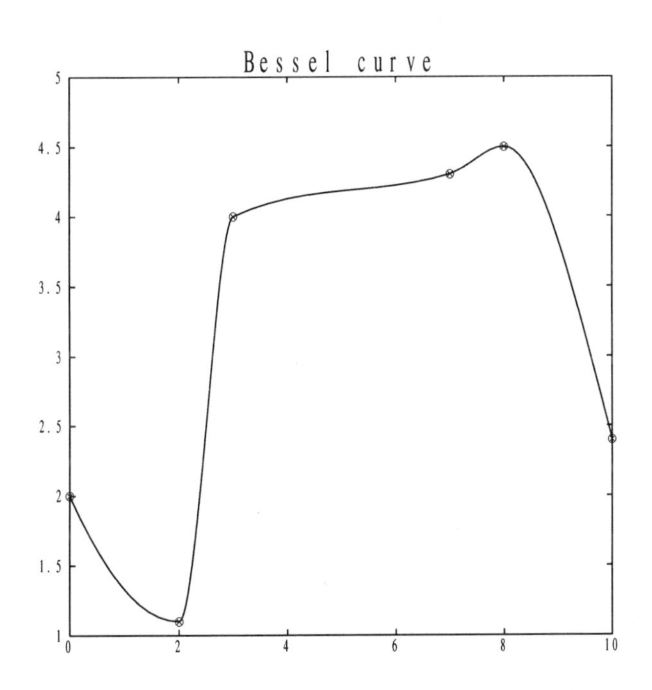

Figure D.1. Simple plot generated by NAG Graphics Library

If we define the data points as in the following table we get Figure D.1.

0.0	2.0	3.0	7.0	8.0	10.0
2.0	1.1	4.0	4.3	4.5	2.4

D.2 The Contents of the NAG Graphics Library

The routines of the NAG Graphics Library are grouped in chapters which will be described briefly in the following sections.

D.2.1 J06A – Axes, Grids, Borders and Titles

Chapter J06A provides facilities for drawing axes, grids and borders for linear or logarithmic graphs, and centred titles. Examples can be seen in the program in Section D.1.

D.2.2 J06B – Point Plotting and Straight Line Drawing

Chapter J06B deals with the graphical representation of discrete points specified by two-dimensional data values (x, y), and the drawing of straight lines to connect them. An example can be seen in the program in Section D.1.

D.2.3 J06C – Curve Drawing

Chapter J06C deals with the drawing of curves through a set of data points in a two-dimensional plane, distinguishing the following possibilities.

- *Single-valued curves.* Data points can be interpolated using cubic spline interpolation, possibly taking conditions like monotonicity into account.
- *Parametric curves.* The same interpolation methods are used for observational data points (x_i, y_i), $i = 1, 2, \ldots, n$, from two independent variables x and y, which are regarded as single-valued functions depending on the same parameter t. Different parametrizations and the option of closing the curve are supported.

An example can be seen in the program in Section D.1.

D.2.4 J06D – ODE Graphics

Chapter J06D deals with the drawing of single-valued curves to represent the solution of a system of ordinary differential equations. A mixed use of the ODE solving routines of the NAG Fortran Library and routine J06DAF is necessary.

D.2.5 J06E – General Function Drawing

Chapter J06E deals with the drawing of functions of one independent variable, $f(x)$, and curves with parametric equations $x = u(t)$ and $y = v(t)$. The single-valued functions f, u, and v are assumed to be continuous over the range of the plot, but their properties are otherwise considered to be unknown. Routine J06EAF is used to draw most of the functions in the figures of Chapter 4, for example.

D.2.6 J06F – Special Function Drawing

Chapter J06F deals with the drawing of cubic spline and Chebyshev polynomial functions. As such there is a partial overlap between routines in this chapter and our programs INTERDIM1 and APPROX. But INTERDIM1 and APPROX use graphical input, which is not supported by the NAG Graphics Library; furthermore, the NAG Graphics Library does not include Fourier approximation with trigonometric functions, which is available in APPROX.

D.2.7 J06G – Contouring

Chapter J06G deals with the two-dimensional representation of a single-valued function of two variables $z(x, y)$. Contouring means that lines of equal function value are traced in the (x, y) plane. There are routines for contouring a specified function z, or a specified set of data points (x_i, y_i, z_i), $i = 1, 2, \ldots, n$, where the values z_i are observations of some underlying function $z(x, y)$ from which a contour plot is to be drawn.

The data points may be given on the nodes of a regular or irregular rectangular grid or they may be scattered.

D.2.8 J06H – Surface Viewing

Chapter J06H deals with the three-dimensional graphical representation of a single-valued function of two variables $z(x, y)$. The function $z(x, y)$ or a set of data values can be represented graphically in the form of a surface view. When the data are derived from a set of three-dimensional data values (x_k, y_k, z_k) a three-dimensional histogram or block diagram may be more appropriate. Routine J06HDF is used for drawing the waves on Page 114, the trough function on Page 142, and the eigenfunctions of the diaphragm on Page 172.

D.2.9 J06J – Data Presentation

Chapter J06J deals with the presentation of one-, two- and three-dimensional data.

One-dimensional data specified by data values x_i, $i = 1, 2, \ldots, n$, may be represented graphically by pie charts, marker diagrams or bar charts.

Two-dimensional data specified by data values (x_i, y_i), $i = 1, 2, \ldots, n$, can be classified as data of one or two independent variables. Data of one independent variable can be represented by various types of line and bar charts. Data of two independent variables can be represented graphically by block diagrams.

Three-dimensional data specified by (x_i, y_i, z_i), $i = 1, 2, \ldots, n$, may be represented graphically by scatter or frequency distribution diagrams.

D.2.10 J06K – Vector Field Plotting

Chapter J06K deals with the display of two- and three-dimensional vector field data. The vectors are represented by arrows whose size and shape indicate their magnitude and orientation.

D.2.11 J06S – Statistical Graphics

Chapter J06S deals with the graphical representation of statistical properties of a set of data. It is of special interest to those users who solve statistical problems with the NAG Fortran Library.

D.2.12 J06T–J06Z – Elementary Routines

There are many elementary routines in the remaining chapters, which are relevant for designing the plot in detail. Some of them may be found in our example programs.

Chapter J06T contains a number of utilities concerned with the drawing of text and markers, and the definition of margins around data regions. Chapter J06V contains input/output utility routines. Chapter J06W contains routines for the control of plotting; see our example programs. The routines in Chapter J06X allow the user to set attributes. The routines in Chapter J06Y are elementary routines, so-called basic plotting primitives, which are similar to typical routines of a kernel system like GKS. In Chapter J06Z we find graphical interface utilities mainly for drawing of strings.

Bibliography

1. Abaffy J. and Spedicato E. (1989). *ABS Projection Algorithms. Mathematical Techniques for Linear and Nonlinear Equations.* Horwood, Chichester.
2. Anderson E., Bai Z., Bischof C., Demmel J., Dongarra J. J., Du Croz J., Greenbaum A., Hammarling S., McKenney A., Ostrouchov S., and Sorensen D. (1992). *LAPACK User's Guide.* SIAM, Philadelphia.
3. Ascher U. M., Mattheij R. M. M. and Russell R. D. (1988). *Numerical Solution of Boundary Value Problems for Ordinary Differential Equations.* Prentice-Hall, Englewood Cliffs.
4. Atkinson K. E. (1989). *An Introduction to Numerical Analysis.* Wiley, New York.
5. Axelsson O. (1985). A Survey of Preconditioned Iterative Methods for Linear Systems of Algebraic Equations. *BIT* **25**, 166–187.
6. Bank R. E. (1994). *PLTMG: A Software Package for Solving Elliptic Partial Differential Equations. Users' Guide 7.0.* SIAM, Philadelphia.
7. Barsky B. A. (1988). *Computer Graphics and Geometric Modeling Using Beta-splines.* Springer, Berlin.
8. Bartels R. H., Beatty J. C. and Barsky B. A. (1987). *An Introduction to Splines for use in Computer Graphics and Geometric Modeling.* Kaufmann, Los Altos.
9. Brent R. P. (1971). An Algorithm with Guaranteed Convergence for Finding a Zero of a Function. *Comp. J.* **14**, 422–425.
10. Brigham E. O. (1988). *The Fast Fourier Transform and its Applications.* Prentice-Hall, Englewood Cliffs.
11. Broyden C. G. (1965). A Class of Methods for Solving Nonlinear Simultaneous Equations. *Math. Comp. J.* **19**, 577–593.
12. Broyden C. G. (1969). A New Method of Solving Nonlinear Simultaneous Equations. *Comp. J.* **12**, 94–99.
13. Bus J. C. P. and Dekker T. J. (1975). Two Efficient Algorithms with Guaranteed Convergence for Finding a Zero of a Function. *ACM Trans. Math. Software* **1**, 330–345.

14. Char B. W., Geddes K. O., Gonnet G. H., Leong B. L., Monagan M. B. and Watt S. M. (1992). *Maple V Language Reference Manual.* Springer, New York.
15. Clenshaw C. W. (1960). Curve Fitting with a Digital Computer. *Comp. J.* **2**, 170–173.
16. Collatz L. (1966). *The Numerical Treatment of Differential Equations.* Springer, Berlin.
17. Coppins R. and Wu, N. (1981). *Linear Programming and Extensions.* McGraw-Hill New York.
18. Cox M. G. (1975). An Algorithm for Spline Interpolation. *J. Inst. Math. Appl.* **15**, 95–108.
19. Cullen M. R. (1985). *Linear Models in Biology.* Horwood, Chichester.
20. Dahlquist G. and Bjørck Å. (1974). *Numerical Methods.* Prentice-Hall, Englewood Cliffs.
21. De Boor C. (1978). *A Practical Guide to Splines.* Springer, New York.
22. De Doncker E. (1979). New Euler MacLaurin Expansions and Their Application to Quadrature over the s-Dimensional Simplex. *Math. Comp.* **33**, 1003–1018.
23. Dew P. M. and James K. R. (1986). *Introduction to Numerical Computation in Pascal.* Springer, New York.
24. Dewey B. R. (1988). *Computer Graphics for Engineers.* Harper and Row, New York.
25. Dierckx P. *FITPACK User Guide.* Part 1: *Curve Fitting Routines.* Report TW 89, 1987. Part 2: *Surface Fitting Routines.* Report TW 122, 1989. Katholieke Universiteit Leuven, Department of Computer Science Celestijnenlaan 200A, B-3030 Leuven (Belgium).
26. Dongarra J. J., Du Croz J., Duff I. S. and Hammarling S. (1979). *LINPACK User's Guide.* SIAM, Philadelphia.
27. Doubleday W. G. (1975). Harvesting in Matrix Population Models. *Biometrics* **31**, 189–200.
28. Duff I. S. (1977). MA28 – a Set of Fortran Subroutines for Sparse Unsymmetric Linear Equations. A.E.R.E. Report R.8730, H.M.S.O. London.
29. Duff I. S., Erisman A. M. and Reid J. K. (1986). *Direct Methods for Sparse Matrices.* Clarendon Press, Oxford.
30. Fletcher R. (1981). *Practical Methods of Optimization. Vol. 1: Unconstrained Optimization. Vol. 2: Constrained Optimization.* Wiley, Chichester.
31. Forsythe G. E. (1957). Generation and Use of Orthogonal Polynomials for Data Fitting with a Digital Computer. *J. SIAM* **5**, 74–88.

32. Fox L. and Mayers D. F. (1987). *Numerical Solution of Ordinary Differential Equations.* Chapman and Hall, London.
33. Francis J. (1961/62). The QR Transformation. A Unitary Analogue to the LR Transformation. *Comp. J.* **4**, 265–271 and 332–345.
34. Fredriksson B. and Mackerle J. (1976). *Structural Mechanics Finite Element Computer Programs. Surveys and Availability.* Department of Mechanical Engineering, Linköping University.
35. Fuchssteiner B. (1987). Solitons in Interaction. *Progr. Theoret. Phys.* **78**, 1022–1050.
36. Fuchssteiner B., Wiwianka W., Gottheil K., Kemper A., Kluge O., Morisse K., Naundorf H., Oevel G. and Schulze Th. (1993). *Multi Processing Algebra Data Tool, Benutzerhandbuch, MuPAD Version 1.1.* Birkhäuser, Basel.
MuPAD 1.2 User Manual. Birkhäuser, Basel. To appear soon.
37. Gaffney P. (1988). When Things go Wrong ... in: Griffiths D. F. and Watson G. A. (eds.): *Numerical Analysis 1987.* Longman, Harlow.
38. Gentleman W. M. (1969). An Error Analysis of Goertzel's (Watt's) Method for Computing Fourier Coefficients. *Computer J.* **12**, 160–165.
39. Gill Ph. E., Murray W. and Wright M. H. (1981). *Practical Optimization.* Academic Press, London.
40. Gladwell I. and Sayers D. K., (eds.) (1980). *Computational Techniques for Ordinary Differential Equations.* Academic Press, London.
41. Golub G. H. and Kahan W. (1965). Calculating the Singular Values and Pseudoinverse of a Matrix. *SIAM J. Num. Anal.* **2**, 205–224.
42. Golub G. H. and Ortega J. M. (1991). *Scientific Computing and Differential Equations.* Academic Press, San Diego, London.
43. Golub G. H. and Reinsch C. (1971). Singular Value Decomposition and Least Squares Solution. Contribution I/10 in [99].
44. Golub G. H. and Van Loan C. F. (1983). *Matrix Computations.* North Oxford Academic, Oxford and John Hopkins Press, Baltimore.
45. Grandine T. A. (1990). *The Numerical Methods Programming Projects Book.* Oxford University Press, Oxford.
46. Graves-Morris P. R. and Hopkins T. R. (1981). Reliable Rational Interpolation. *Numer. Math.* **36**, 111–128.
47. Greenough C. and Robinson K. Finite Element Library. Rutherford Appleton Laboratory. The Numerical Algorithm Group Ltd. Wilkinson House, Jordan Hill Road, Oxford OX2 8DR, U.K.
48. Grundmann A. and Möller H. M. (1978). Invariant Integration Formulas for the n-Simplex by Combinatorial Methods. *SIAM J. Num. Anal.* **15**, 282–290.

49. Hackbusch W. (1985). *Multigrid Methods and Applications.* Springer Berlin.

50. Hairer E., Nørsett S. P. and Wanner G. (1987). *Solving Ordinary Differential Equations I: Nonstiff Problems.* Springer, Berlin.

51. Hairer E. and Wanner G. (1991). *Solving Ordinary Differential Equations II: Stiff and Differential-Algebraic Problems.* Springer, Berlin.

52. Hämmerlin G. and Hoffmann K.-H. (1991). *Numerical Mathematics.* Springer, Berlin.

53. Herchenröder Th., Köckler N. and Wendt D. (1992). PAN – A Problem Solving Environment for Numerical Analysis. In: Hodnett Frank (ed.). *Proc. Sixth European Conference on Mathematics in Industry.* Teubner, Stuttgart, 183–186.

54. Heuser H. (1989). *Gewöhnliche Differentialgleichungen.* Teubner, Stuttgart.

55. Hill D. R. (1988). *Experiments in Computational Matrix Algebra.* Random House, New York.

56. Hopkins T. and Phillips Ch. (1988). *Numerical Methods in Practice Using the NAG Library.* Addison-Wesley, Wokingham.

57. Hoppensteadt F. C. and Peskin C. S. (1992). *Mathematics in Medicine and the Life Science.* Springer, New York.

58. *IMSL User's Manual.* Customer Relations, 14141 Southwest Freeway, Suite 3000, Sugar Land, Texas 77478–3498, USA.

59. Kågstrøm B. (1986). RGSVD – An Algorithm for Computing the Kronecker Structure and Reducing Subspaces of Singular Matrix Pencils $A - \lambda B$. *SIAM J. Sci. Stat. Comp.* **7**, 185–211.

60. Keller H. B. (1984). *Numerical Methods for Two-Point Boundary Value Problems.* SIAM, Philadelphia.

61. Köckler N. (1990). *Numerische Algorithmen in Softwaresystemen.* Teubner, Stuttgart.

62. Köckler N. and Simon M. (1991). Parallel Singular Value Decomposition with Cyclic Storing. *Parallel Comput.* **19**, 39–47.

63. Kronecker L. (1881). *Monatsber. Königl. Preuss. Akad. Wiss. Berlin* **535**.

64. Kronrod A. S. (1966). *Nodes and Weights of Quadrature Formulas.* Consultants Bureau, New York.

65. Lamport L. (1992). LaTeX. Addison-Wesley, Reading.

66. Lawson C. L. and Hanson R. J. (1974). *Solving Least Squares Problems.* Prentice-Hall, Englewood Cliffs.

67. Lyness J. N. (1983). When not to use an Automatic Quadrature Routine. *SIAM Rev.* **25**, 63–87.

68. Magid A. R. (1985). *Applied Matrix Models.* Wiley, New York.

69. McCormick S. F. (ed.) (1987). *Multigrid Methods.* SIAM, Philadelphia.

70. Moler C. B. *MATLAB – User's Guide.* The Math Works Inc., Cochituate Place, 24 Prime Park Way, Natick, Mass. 01760, USA.

71. Moler C. B. and Stewart G. W. (1973). An Algorithm for Generalized Matrix Eigenvalue Problems. *SIAM J. Numer. Anal.* **10**, 241–256.

72. Morrison D. D., Riley J. D. and Zancanaro J. F. (1962). Multiple Shooting Method for Two-Point Boundary Value Problems. *Comm. ACM* **12**, 613–614.

73. Munksgaard N. (1980). Solving Sparse Symmetric Sets of Linear Equations by Preconditioned Conjugate Gradients. *ACM Trans. Math. Software* **6**, 206–219.

74. *NAG Fortran Library Manual.* The Numerical Algorithm Group Ltd. Wilkinson House, Jordan Hill Road, Oxford OX2 8DR, U.K.

75. *NAG Graphics Library Manual.* The Numerical Algorithm Group Ltd. Wilkinson House, Jordan Hill Road, Oxford OX2 8DR, U.K.

76. Patterson T. N. L. (1968). The Optimum Addition of Points to Quadrature Formulae. *Math. Comp.* **22**, 847–856.

77. Phillips J. (1986). *The NAG Library. A Beginners' Guide.* Clarendon Press, Oxford.

78. Piessens R., de Doncker-Kapenga E., Überhuber C. W. and Kahaner D. K. (1983). *QUADPACK. A Subroutine Package for Automatic Integration.* Springer, Berlin.

79. Prenter P. M. (1975). *Splines and Variational Methods.* Wiley, New York.

80. Press W. H. (1986). *Numerical Recipes.* Cambridge University Press, Cambridge.

81. Rabinowitz Ph., (ed.) (1988). *Numerical Methods for Nonlinear Algebraic Equations.* Gordon and Breach, New York.

82. Rao C. R. and Mitra S. K. (1971). *Generalized Inverse of Matrices and its Applications.* Wiley, New York.

83. Rice J. R. and Boisvert R. F. (1985). *Solving Elliptic Problems Using ELLPACK.* Springer, New York.

84. Runge C. (1903). Über die Zerlegung empirisch gegebener periodischer Funktionen in Sinuswellen. *Z. Math. Phys.* **48**, 443–456.

85. Runge C. (1905). Über die Zerlegung einer empirischen Funktion in Sinuswellen. *Z. Math. Phys.* **52**, 117–123.

86. Rutishauser H. (1969). Computational Aspects of F. L. Bauer's Simultaneous Iteration Method. *Num. Math.* **13**, 4–13.

87. Sewell G. (1985). *Analysis of a Finite Element Method: PDE/PROTRAN.* Springer, New York.

88. Shoup T. E. (1979). *A Practical Guide to Computer Methods for Engineers.* Prentice-Hall, Englewood Cliffs.

89. Smith B. T., Boyle J. M., Dongarra J. J., Garbow B. S., Ikebe Y., Klema V. C. and Moler C. B. (1988). *Matrix Eigensystem Routines – EISPACK Guide.* Springer, Berlin.

90. Spiegel Murray R. (1981). *Applied Differential Equations.* Prentice-Hall, Englewood Cliffs.

91. Stiefel E. (1960). Note on Jordan Elimination, Linear Programming and Tschebyscheff Approximation. *Num. Math.* **2**, 1–17.

92. Stoer J. and Bulirsch R. (1993). *Introduction to Numerical Analysis.* Springer, New York.

93. Swift A. and Lindfield G. R. (1978). Comparison of a Continuation Method with Brent's Method for the Numerical Solution of a Single Nonlinear Equation. *Comp. J.* **21**, 359–362.

94. Tischel G. (1980). *Angewandte Mathematik.* Diesterweg, Frankfurt.

95. Twizell E. H. (1988). *Numerical Methods, with Applications in the Biomedical Sciences.* Ellis Horwood, Chichester.

96. Townend M. S. and Pountney D. C. (1989). *Computer-aided engineering mathematics.* Ellis Horwood, Chichester.

97. Wilkinson J. H. (1963). *Rounding Errors in Algebraic Processes.* Wiley, New York.

98. Wilkinson J. H. (1978). *The Algebraic Eigenvalue Problem.* Clarendon Press, Oxford.

99. Wilkinson J. H. and Reinsch C. (1971). *Linear Algebra. Handbook for Automatic Computation II.* Springer, Berlin.

100. Wolfram S. (1991). *Mathematica – a system for doing mathematics by computer.* Addison-Wesley, Redwood City.

Index

The author has tried to provide as complete an index as possible. Therefore the index is rather long, and it contains items such as 'blue whale population', which you might not expect at a first glance. But perhaps you remember later on that there was an example about blue whales and you want to read it again.
A page number is given in italics (e.g., '*345*') when it represents the main source of information about that item.

A

B

C

D

J

K

L

O

Q

R

T

U

V

W

X

Z